Frontiers of Combining Systems

APPLIED LOGIC SERIES

VOLUME 3

Managing Editor

Dov M. Gabbay, *Department of Computing, Imperial College, London, U.K.*

Co-Editor

Jon Barwise, *Department of Philosophy, Indiana University, Bloomington, IN, U.S.A.*

Editorial Assistant

Jane Spurr, *Department of Computing, Imperial College, London, U.K.*

SCOPE OF THE SERIES

Logic is applied in an increasingly wide variety of disciplines, from the traditional subjects of philosophy and mathematics to the more recent disciplines of cognitive science, computer science, artificial intelligence, and linguistics, leading to new vigor in this ancient subject. Kluwer, through its Applied Logic Series, seeks to provide a home for outstanding books and research monographs in applied logic, and in doing so demonstrates the underlying unity and applicability of logic.

Frontiers of Combining Systems

First International Workshop, Munich, March 1996

edited by

FRANS BAADER

LuFG Theoretical Computer Science,
Technical University of Aachen

and

KLAUS U. SCHULZ

CIS,
University of Munich

Springer Science+Business Media, LLC

A C.I.P. Catalogue record for this book is available from the Library of Congress

ISBN 978-94-010-6643-3   ISBN 978-94-009-0349-4 (eBook)
DOI 10.1007/978-94-009-0349-4

Logo design by L. Rivlin

EDITORIAL PREFACE

The editors of the Applied Logic Series are pleased to present the third volume in the series, this one on Frontiers of Combining Logics and Systems. With the proliferation of important logical systems in recent years, an important problem has emerged: how to combine simpler logical systems into more complex, and now to understand the properties of the new in terms of the properties of its parts. This book is the first devoted entirely to this problem.

The Editors

TABLE OF CONTENTS

vi

REFEREES

The program committee gratefully acknowledges the help of the following referees. We would also like to thank the anonymous referee of the publishing house.

H. Baumeister	R. Caferra	E. Contejean
D. Fehrer	T. Frühwirth	U. Furbach
W. Gehrke	M. Henz	M. Hofmann
K. Homann	X. Huang	J. Hudelmaier
U. Hustadt	F. Jacquemard	M. Kerber
C. Kirchner	A. Kutzner	R. Letz
J. J. Lu	D. Lugiez	C. Lynch
J. Marcinkowski	K. Marriott	R. Matthes
E. Monfroy	J. Montelius	J. Nieren
T. Noll	A. Nonnengart	S. E. Panitz
J. van de Pol	C. Prehofer	C. Ringeissen
M. Rusinowitch	V. Saraswat	M. Schütz
J. Schumann	W. Schreiner	H. Simonis
W. Snyder	V. Sofronie	M. Staudt
B. Steffen	K. Stokkermans	F. Stolzenburg
G. Stumme	M. Vittek	E. Waller
A. Weber	S. Wilke	T. Wilke

PROGRAM COMMITTEE

Franz Baader, RWTH Aachen
Peter Baumgartner, Universität Koblenz
Patrick Blackburn, Universität des Saarlandes
Alexander Bockmayr, MPI Saarbrücken
Alexandre Boudet, LRI Orsay
Jacques Calmet, Universität Karlsruhe
Alain Colmerauer, LIM Marseille
Dov Gabbay, Imperial College London
Hélène Kirchner, CRIN & INRIA Lorraine
Hans Jürgen Ohlbach, MPI Saarbrücken
Jochen Pfalzgraf, RISC Linz
Maarten de Rijke, University of Warwick
William Rounds, University of Michigan
Manfred Schmidt-Schauß, Universität Frankfurt
Klaus Ulrich Schulz, Universität München

LOCAL ORGANISATION

Klaus Ulrich Schulz & Stephan Kepser, Universität München

PREFACE

In many areas of Logic, Computer Science, and Artificial Intelligence, there is a need for specialized formalisms and inference mechanisms to solve domain-specific tasks. For this reason, various methods and system have been developed that allow for an efficient and adequate treatment of such restricted problems. In most realistic applications, however, one is faced with a complex combination of different problems, which means that a system tailored to solving a single problem can only be applied if it is possible to combine it both with other specialized systems and with general purpose systems. There are already relevant contributions that successfully solve particular instances of this combination and integration problem in different areas. More recently, the problem of developing general frameworks for combining formalisms, inference methods, and systems has been addressed. Depending on the nature of the given subsystems, and on the intended behaviour of the combined system, combination of systems can be an extremely ambitious goal, which explains why there are still many open problems concerning the possible—and necessary—forms of interaction and coordination between subsystems. Consequently, combination of formal systems and algorithms, their logical and algebraic background, as well as the general architecture of complex and interacting systems is a very important and active research area.

The first international workshop "Frontiers of Combining Systems" (FroCoS'96) intended to create a common forum for these research activities, and to stimulate an interdisciplinary discussion that focuses on different aspects of the combination problem. The Call for Papers for FroCoS'96 mentioned the following as possible, but not exclusive, topics of interest for the workshop:

- combination of logics (e.g., modal logics, logics in AI,...),

- combination of constraint solving techniques (for unification and matching problems, general symbolic constraints, numerical constraints,...) and combination of decision procedures,

- integration of equational and other theories into deductive systems (e.g. theory resolution, constraint resolution, constraint paramodulation,...),

- combination of term rewriting systems,

- integration of data structures (e.g., sets, multisets, lists) into Constraint Logic Programming formalisms and deduction processes,

- hybrid systems in computational linguistics, knowledge representation, natural language semantics, and human computer interaction,

- logic modeling of multi-agent systems.

In order to achieve our goal of stimulated interdisciplinary discussion and collaboration on combination problems, the workshop was intended to

- provide for a series of invited talks that illustrate motivations, perspectives, results, methods, and limitations of contributions to the combination problem coming from different areas, and

- support the distribution of technical papers that offer new solutions to interesting instances of the combination problem, investigate general aspects of the combination problem, or describe interesting applications of combined systems.

The present volume contains the proceedings of FroCoS'96. We are glad that two of the invited speakers, Bruno Buchberger and Dov M. Gabbay, agreed to contribute articles to this collection. The accepted technical papers cover most—if not all—of the relevant areas that have been addressed in the Call for Papers, but there is a clear emphasis on modal and intuitionistic logic, automated deduction, and constraint logic programming. A comparison of the different contributions also shows that the problem of combining systems has been investigated on various conceptual levels. At one end, combination of logical systems is studied with an emphasis on formal properties using tools from mathematics and logics. On the other end of the spectrum, the combination of software tools necessitates discussing physical connections and appropriate communication languages. Between these two extremes lies the combination of algorithms such as decision procedures for inference problems in combined logical systems.

The workshop took place on March 26–29, 1996, in Munich. It attracted 39 submissions, with authors from Australia, Austria, Belgium, Brasil, Bulgaria, Egypt, France, Germany, Italy, Japan, Portugal, Russia, Spain, Sweden, the United Kingdom, Taiwan, and the USA. With a few exceptions, each paper was reviewed by at least three experts. On the basis of this evaluation, the program committee selected 18 papers for presentation at the workshop and for inclusion in the proceedings. Invited talks were given by Bruno Buchberger, Alain Colmerauer, Dov M. Gabbay, Uwe Glässer, and Mark Stickel.

We should like to thank the Münchener Universitätsgesellschaft, the Deutsche Forschungsgemeinschaft (DFG), and the Centrum für Informations- und Sprachverarbeitung (CIS, SIL) for their kind and generous support. Thanks also to the program committee, to the additional referees, and in particular to Stephan Kepser and to the other people involved in the local organization of the workshop.

Munich, March 1996

Franz Baader and Klaus U. Schulz (Co-chairs of FroCoS'96)

DOV M. GABBAY

AN OVERVIEW OF FIBRED SEMANTICS AND THE COMBINATION OF LOGICS

Abstract. This paper presents an overview of the authors methodology of Fibred Semantics and The Combinations of Logics presented in a series of papers under the same title. We explain the ideas behind fibring and illustrate them in several case studies. We include fibring modal and intuitionistic logics, fibring with fuzzy logics as well as self fibring of predicate logics.

1. Introduction

The purpose of this paper is to present an intuitive explanation of the author's fibred semantics methodology for combining logics and systems, and to give a brief overview of the main results, ideas and open problems in the area.

The problem of combining logics and systems is central to modern logic, both pure and applied.

The need to combine logics arises both from applications and from within logic itself as a discipline. As logic is being used more and more to formalise field problems in philosophy, language, artificial intelligence, logic programming and computer science, the kind of logics required become more and more complex. An increasing number of features from the application area need to be formally represented. These features are highly mutually interactive and a formal study of their combined nature becomes necessary. A methodology for combining systems has now become essential for most logic applications.

A simple example will illustrate the complexity of the problem we are facing. There exist well known and well studied logical systems and theories of belief and knowledge. Such systems use notation such as $B\varphi$ (φ is believed) and $K\varphi$ (φ is known). The formal theory of such notions has been studied in the past 30 years. Similarly there are well known systems

1

F. Baader and K.U. Schulz (eds.), Frontiers of Combining Systems, 1–55.
© 1996 *Kluwer Academic Publishers.*

of temporal logics, with operators such as $G\varphi$ (φ will always be true) and $H\varphi$ (φ has always been true). Such systems were developed in connection with the study of language and some problems in the philosophy of time. With the rise of artificial intelligence as a consumer of logic, various theories and models of reactive agents were put forward. Such theories involve a cycle of temporally evolving and changing repositories of knowledge and beliefs, as well as theories of actions for agents making use of their changing knowledge and beliefs.

Clearly any logical system modelling reactive agents should be a combined system of logics of knowledge, belief, time and modal logics (of actions).

It is therefore clear that a good understanding and practical methodology for combining logics can provide the consumer of logic (e.g. a robotics designer) with tools which make him more competent in designing his own system. Such a methodology can help decompose the problem of designing a complex system into developing components (logics) and combining them.

Furthermore, such a methodology might allow the designer to use existing components (which may be very well understood and/or efficiently implemented) and put them together into a desired useful combined system, at a relatively acceptable intellectual and practical cost.

In fact, a good combining methodology allows the study of the complex application area in terms of some of its pure components, yielding both theoretical and practical benefits.

A good methodology of combining systems should study *transfer theorems*. Such theorems prove that if the components satisfy property $\mathcal{P}$, then we are assured that the combined system satisfies (a possibly slightly different) property $\mathcal{P}'$. Moreover, transfer theorems can also help validate existing complex systems by verifying properties of their components.

It turns out that our fibring methodology does not only help realise many of the above expectations, but also seems to bring out unexpected theoretical properties of pure logic itself, showing new connections between existing themes of logic.

The following is a list of the main areas where the fibring methodology makes a serious contribution.

- Straightforward combination of logics such as multimodal logics or modal intuitionistic logics.
- Combining a metalevel language with its object level language. For example the problem of bringing a consequence relation into the object language as a conditional.
- Allowing for a single language to be syntactically ambivalent, and regarding the various ambivalent 'aspects' as a combined system. For example allowing classical logic formulas to apply to themselves (e.g. $\varphi(\varphi)$).

- Modelling the temporal behaviour of a system by fibring it with a temporal logic.
- Making a system fuzzy by fibring it with a fuzzy logic.

The power of the concept of fibring is also emphasised by some other recent research activities that arose independently from our own work. In (Pfalzgraf, 1991; Pfalzgraf & Stokkermans, 1994) the introduction of *logical fibrings* has been strongly motivated by the classical notion of fibre bundles and sheaves. Concrete applications aim at modelling logical control of cooperating robots (agents) scenarios. In (Baader & Schulz, 1995a; Baader & Schulz, 1995; Baader & Schulz, 1995b; Kepser & Schulz, 1996), another variant of fibring is used for combining solution domains and constraint solvers for symbolic constraints.

Our plan for this paper is first to give quick details of the above areas in subsections 1.1–1.5 and then explain the idea of the fibring methodology in Section 2. Further sections present sample case studies and the final section concludes with a discussion.

1.1. COMBINING PURE LOGICAL SYSTEMS

Problems of combining two logics arise both in pure logical theory and in practical applications. We give several typical examples

1. *Modal Intuitionistic Logics*
   The logic of modality □ was originally philosophically motivated. The same is true for intuitionistic logic. Intuitionistic logic arose as the logic for constructive mathematics. It was natural therefore, for philosophers and pure logicians, to try and study the nature of modal necessity □ and possibility ◊ from a constructive point of view. Such studies gave rise to systems involving both intuitionistic implication ⇒ and the modal operators □ and ◊. In the past 30 years many modal intuitionistic systems were put forward, by philosophically motivated considerations. These systems were formulated directly for the mixed language and were not presented as combined systems. Various methods were used in the presentation, some semantical, some theoretical, some straightforward and some rather roundabout and highly individual. Among the early philosophically motivated systems by famous logicians were Fitch (Fitch, 1969), Bull (Bull, 1965) and Ono (Ono, 1977; Ono, 1987).
   Later, as modal logic has had serious applications in computer science, a new wave of interest in modal intuitionistic logic arose, by virtue of its applicability. The modality could stand for possible algorithmic options and the intuitionistic proof theory for the constructive nature of the

execution. Sample papers in this category are Wijesekera (Wijesekera, 1990) and A. K. Simpson (Simpson, 1994).

After a while logicians got fascinated by modal intuitionistic logics for formal technical reasons. Many papers were written without any special philosophical or field applications motivation. Among them are (Amati & Pirri, to appear; Font, 1986; Fischer-Servi, 1977; Fischer-Servi, 1984; Fischer-Servi, 1980; Došen, 1985; Suzuki, 1990; Božić & Došen, 1984; Ewald, 1986; Ono, 1987; Suzuki, 1988).

Among the technical problems involved in the modal intuitionistic logic area are:

- proof theory for the system. Hilbert formulation, Gentzen formulation, tableaux formulation, etc.;

- Kripke type semantics for the system;

- decision procedures;

- finite model property;

- comparison with other systems of modal intuitionistic logics, in terms of expressive power and relative interpretability.

We shall see that our fibring methodology can yield many of the above systems and results automatically and in bulk. Our main paper in this area is (Gabbay, 1993), which includes a detailed discussion of the main systems in the literature. In this survey in Section 3, we quote the main theorems of our paper.

2. *syntactical Modality*

Here we mean that a modal operator may arise naturally in another logic by a syntactic definition. For example in Girard linear logic, which is independently and very deeply motivated, some modalities (the exclamation !) arise naturally by definition. Studying their properties is of course the study of a combined system. Another example is the modality $\Box A = \mathrm{def}(A \Rightarrow A) \Rightarrow A$ in Anderson and Belnap's *Entailment and Relevant Logics* (Anderson & Belnap, 1975). The above does turn out to be some sort of modal-relevant system. See also the paper of Meyer and Mares (Meyer & Mares, 1993).

Modalities can also be obtained as by-products of translations and interpretations. We mention two well-known examples.

(a) the strict implication of modal **S4** for boxed formulas is intuitionistic implication, thus giving rise to a combined modal intuitionistic logic.

(b) the semantics translation of modal logic into classical predicate logic gives rise to a modal intuitionistic logic when classical predicate logic is weakened into intuitionistic predicate logic.

3. *Multimodal Logics*

These arise from applications where several modal operators interact. The application may be a philosophical study of some notions and its paradoxes, such as systems of obligation and permission, see (Jones & Pörn, 1985; Jones & Pörn, 1986) or a direct formalisation of some practical application area, e.g. logical omniscience, see (Fagin *et al.*, 1990).

A large very useful area of multimodal logic is dynamic logic. In fact, some applied logicians believe that multimodal logic is *the logic* for computer science applications.

The bulk of the literature on combined systems is mainly concerned with the presentation of individual combined systems. There are a few papers on methodological questions and transfer theorems. Besides my own papers on fibring methodology, there are some theoretical studies of multiple modalities as well as some algebraic studies of transfer properties. See (Fischer-Servi, 1977; Fitting, 1969; Fine & Schurz, to appear; Kracht & Wolter, 1991; Goranko & Passy, 1992). Again, I am happy to say, that the fibred semantics methodology gives bulk results in multimodal logics. See (Gabbay, 1993).

1.2. COMBINING META-LEVEL WITH THE OBJECT LEVEL

This is a new application of the fibred methodology.

Consider a consequence relation of the form $\Delta\!\!\mid\sim B$ between wffs of a logic. Assume it is a non-monotonic consequence relation on classical language satisfying Reflexivity ($\Delta\!\!\mid\sim A$ if $A \in \Delta$); Restricted monotonicity ($\Delta\!\!\mid\sim A$ and $\Delta\!\!\mid\sim B$ imply $\Delta, A\!\!\mid\sim B$) and Cut ($\Delta\!\!\mid\sim A$ and $\Delta, A\!\!\mid\sim B$ imply $\Delta\!\!\mid\sim B$). See (Gabbay, 1985; Kraus *et al.*, 1990). We would like to bring $\mid\sim$ into the language as a connective $A > B$ and find a conservative extension $\mid\sim^*$ of $\mid\sim$ such that the following holds

$$- \ \varnothing\!\!\mid\sim^* A > B \text{ iff } A\!\!\mid\sim B.$$

The above means that $>$ represents $\mid\sim$ in the language itself.

The fibred semantics methodology allows us to do this in a methodological way.

Delgrande has already observed (Delgrande, 1988) the formal similarities between non-monotonic consequence relations $\mid\sim$ and the conditional $>$. Of course $>$ is an object level connective and $\mid\sim$ is a metalevel consequence. Our fibred methodology turns $\mid\sim$ into a conditional. In our paper (Gabbay, 1995d) we indeed get the traditional semantics for the conditional out of the fibred semantics of the construction that brings $\mid\sim$ into the object level.

1.3. SELF-FIBRING OF PREDICATE LOGICS

This is another surprising application of the fibred methodology. It allows us to give solid and meaningful semantics to expressions of the form $A(t, \varphi(f(\alpha))$ where $A(x, y)$ is a formula, $\alpha, \varphi(x)$ are formulas and $f(x)$ a function symbol. As you can see, we have substituted formulas within other formulas and other function symbols.

Such expressions and languages are widely used in logic, philosophy and linguistics and have extensive and important applications. We mention a few typical examples:

- Logics of non-denoting singular terms
- In the logical analysis of language, we try and analyse statements like
   'John said that Mary believed that he did not love her'
   Here we use the predicates, *say* (t, φ) and *believe* (t, φ).
- In logic programming there is a widespread use of metapredicates such as
   Demo(A, B) and
   Y if *Not*(X)

   Ordinary clauses for the transitive closure of a relation R are written as
   $$(cl(Z))(X, Y) \text{ if } Z(X, Y)$$
   $$(cl(Z))(X, Y) \text{ if } Z(X, V) \wedge ((cl(Z))(V, X))$$
   or expressions like

   $$\forall x. \text{ bel(John, friend (John, } x)) \rightarrow \text{ exists } (y, \text{ loves } (x, y))),$$

- Tarski truth predicate
   $$\mathbf{T}(\varphi) \leftrightarrow \varphi$$

   and the diagonalisation function symbol

   $$d(\varphi) = \sim \varphi(\varphi)$$

We make sense of the above expressions using fibred semantics.

1.4. TEMPORALISING A SYSTEM

This example is very practical. Given a system $\mathcal{S}$, we can describe it in some logic $\mathbf{L}$. We now let the system vary in time. We can get snapshots $\mathcal{S}_t$ of it at different times t. What would be a suitable logic to describe the evolving and changing system?

The answer is the combination of $\mathbf{L}$ with a suitable temporal logic of our choice. The fibred semantics methodology will tell us what is the semantics of the new combined logic and the availability of transfer theorems will tell us what properties to expect of the combined system, given what we know of the components. This is done in, e.g. (Finger & Gabbay, to appear) and (Finger & Gabbay, 1992).

1.5. MAKING YOUR LOGIC FUZZY

This is the most satisfactory application of fibred semantics. Fuzzy logic has been controversial among pure logicians for a long time. It lacked respectability. People had the impression that we can take any system and make it fuzzy in an ad hoc way, by taking any $\{0,1\}$ function we can find and turning it into a fuzzy $[0,1]$ function. There is no methodology and no logic.

Take modal logic for example. Its models have the form (S, R, a, h) where S is the set of possible worlds, $a \in S$, R is a crisp $\{0,1\}$ relation and h is the assignment, giving for each $t \in S$ and atomic q a crisp $\{0,1\}$ value $h(t,q) \in \{0,1\}$.

How can we make modal logic fuzzy?

We can turn h into a $[0,1]$ function. We can turn R into a $[0,1]$ function. We can even make fuzzy which world is the actual world. This seems arbitrary. See (Fitting, 1991; Fitting, 1992; Fitting, 1994) for examples of where it is done. Can we make sense of it?

The answer is yes. The very same fibring methodology used to combine intuitionistic and modal logics and used to combine two modalities, can be used here to combine modal logic with Lukasiewicz infinite valued logic.

Depending on how we fibre we can get the above (or all) different options of making modal logic fuzzy.

2. The Idea of Fibring

The basic problem of combining systems can be formally understood as follows:

COMBINING SYSTEMS PROBLEM

We are given two systems $\mathcal{S}_1$ and $\mathcal{S}_2$ in languages $\mathbf{L}_1$ and $\mathbf{L}_2$. We combine the languages to form $\mathbf{L}$. We ask the following:

1. How can we define a system $\mathcal{S}$ for the combined language, which is a conservative extension of each $\mathcal{S}_1$ and $\mathcal{S}_2$? This is not an easy problem, because $\mathcal{S}_1$ and $\mathcal{S}_2$ may be presented to us in two completely different and incompatible ways. For example even though $\mathcal{S}_1$ and $\mathcal{S}_2$ may be two modal logics with modalities $\square_1$ and $\square_2$, the system for $\square_1$ may

be defined via some algebraic semantics while the system for $\square_2$ may be a tableaux system.

2. How many options for $\mathcal{S}$ are there and how do they relate to each other?

3. If we want $\mathcal{S}_1$ and $\mathcal{S}_2$ to interact, how do we do that? Can we develop a methodology for interaction and have a set of prearranged and well understood 'interactive axioms'?

4. Suppose we know that both $\mathcal{S}_1$ and $\mathcal{S}_2$ satisfy some property $\mathcal{P}$. Can we prove transfer theorems which ensure that the combined $\mathcal{S}$ satisfies property $\mathcal{P}'$ (a variation of $\mathcal{P}$)?

5. Suppose we are given a system $\mathcal{S}$ directly defined in a language $\mathbf{L}$. Assume $\mathbf{L}$ is a syntactic combination of $\mathbf{L}_1$ and $\mathbf{L}_2$. Can we decompose $\mathcal{S}$ into $\mathcal{S}_1$ and $\mathcal{S}_2$, where $\mathcal{S}_i$ are some projections onto the sublanguages $\mathbf{L}_i$, and then can we reconstruct $\mathcal{S}$ back as some combination of $\mathcal{S}_1$ and $\mathcal{S}_2$ with possibly additional interaction axioms? If $\mathcal{S}_i$ happens to be well understood systems, we would have certainly 'simplified' $\mathcal{S}$.

The fibring methodology we are about to describe allows one to combine systems through their semantics and is a very successful framework for answering the above questions.

2.1. APPRECIATION OF THE DIFFICULTIES INVOLVED IN COMBINING SYSTEMS

The problem of combining systems can be pretty difficult in practice, not only because the two systems to be combined may be presented in two completely different ways, but also because that even when they are presented in the same way, it is not clear how to combine them. The next examples will illustrate.

Example 2.1 (Two modal logics) *Let $\square_1$ and $\square_2$ be two modalities. We define two systems as follows:*

Let $\mathcal{S}_1$ be the system obtained by adding to classical propositional logic the modality $\square_1$ with the following axioms and rules:

1. $\square_1 A$, *where A is substitution instances of truth functional tautology;*

2. $\square_1(A \rightarrow B) \rightarrow (\square_1 A \rightarrow \square_1 B)$

3. $\square_1(\square_1(A \rightarrow B) \rightarrow (\square_1 A \rightarrow \square_1 B))$

4. $\square_1 A \rightarrow \square_1 \square_1 A$

5. $\square_1(\square_1 A \rightarrow \square_1 \square_1 A)$

6. $\square_1 A \rightarrow A$

7. *Modus ponens.*

$$\vdash A; \vdash A \rightarrow B \ imply \ \vdash B.$$

Note that we do not require necessitation. So although $\square_1 A \rightarrow A$ is a theorem $\square_1(\square_1 A \rightarrow A)$ is not provable.

Let S_2 be the system obtained by adding $\Box_2$ to classical propositional logic. The theorems of S_2 are defined semantically as follows.

Let E^2 be Euclidean plane with the usual topology. An assignment h is a function giving to each atom q, a set of points $h(q) \subseteq E^2$.

h can be extended to all wffs of S_2 as follows

$$
\begin{aligned}
h(\neg A) &= E^2 - h(a) \\
h(A \wedge B) &= h(A) \cap h(B) \\
h(A \vee B) &= h(A) \cup h(B) \\
h(A \to B) &= (E^2 - h(A)) \cup h(B) \\
h(\Box_2 A) &= \textit{Topological interior of } h(A).
\end{aligned}
$$

Let $\vDash_2 A$ mean that for all h, $h(A) = E^2$.

Problem: *Combine S_1 and S_2 into a system S which is the smallest logical system for the combined language which is a conservative extension of both S_1 and S_2.*

The two systems are presented in totally different ways. How are we going to combine them?

We shall fibre (combine) them in the next subsection by observing that the $\Box_2$ modality is really **S4** modality, complete for the class of Kripke models (S, R, a, h) where R is reflexive and transitive and that the $\Box_1$ modality is complete for all Kripke models (S, R, a, h) such that R is transitive and aRa holds. Without this extra knowledge we cannot combine them. With this extra knowledge we can apply our methodology and combine them, as we shall see in the next subsection.

Even when the two systems are presented in the same way, it is not clear what to do and how to combine them. In the next example, we try to combine classical implication with intuitionistic implication. The two systems will be presented in the same way, with classical implication having one more axiom than intuitionistic implication. We cannot be more 'compatible' than that, can we?

We shall combine them in a straightforward way by taking the union of the axioms, taking each axiom for its own language. The next example will show that we still get into trouble!

See F. del Cerro and A. Herzig (del Cerro & Herzig, 1996) on how to do it correctly.

Example 2.2 (Fibring counterexample) *Let $\mathbf{L}_1$ be the language with $\wedge_1$ and $\to_1$ and $\mathbf{L}_2$ with $\wedge_2$ and $\to_2$. $\wedge_i$ is intended to be conjunction and $\to_i$ is intended to be implication. Let $\vdash_1$ and $\vdash_2$ be the respective consequence relations for $\mathbf{L}_1$ and $\mathbf{L}_2$, with $\vdash_1$ intending to define the $\wedge, \to$ fragment of intuitionistic logic and $\vdash_2$ intended to define the $\wedge, \to$ fragment of classical*

*logic. Both $\mathrel{\vdash}_i$ can be defined as the smallest consequence relations satisfying
Reflexivity, Monotonicity and Cut and closed under the additional rules
below. $\mathrel{\vdash}_2$, which supposed to define classical implication can be assumed to
be closed also under Peirce's rule $(A \twoheadrightarrow_2 B) \twoheadrightarrow_2 A \mathrel{\vdash}_2 A$.*

- $A \wedge_i B \mathrel{\vdash}_i C$ iff $A \mathrel{\vdash}_i B \twoheadrightarrow_i C$
- $A \wedge_i B \vdash_i A$
- $A \wedge_i B \vdash_i B$
- $$\frac{A \mathrel{\vdash}_i B;\ A \mathrel{\vdash}_i C, B \wedge_i C \mathrel{\vdash}_i D}{A \mathrel{\vdash}_i D}.$$

*Let us now combine the languages and let $\mathrel{\vdash}$ be the smallest consequence
relation in the combined language closed under both rules, for the respective
languages. We show that $\twoheadrightarrow_1$ and $\twoheadrightarrow_2$ become equal.*

Since $A \twoheadrightarrow_i B \mathrel{\vdash}_i A \twoheadrightarrow_i B$ we get $(A \twoheadrightarrow_i B) \wedge_i A \mathrel{\vdash}_i B$.
We also have:

1. $(A \twoheadrightarrow_1 B) \wedge_2 A \mathrel{\vdash} A$
2. $(A \twoheadrightarrow_1 B) \wedge_2 A \mathrel{\vdash} A \twoheadrightarrow_1 B$
3. $(A \twoheadrightarrow_1 B) \wedge_1 A \mathrel{\vdash} B$

Hence from (1)–(3)
$$(A \twoheadrightarrow_1 B) \wedge_2 A \mathrel{\vdash} B$$

and hence
$$(A \twoheadrightarrow_1 B) \mathrel{\vdash} (A \twoheadrightarrow_2 B).$$

*The above is impossible since $\twoheadrightarrow_1$ can be classical implication and $\twoheadrightarrow_2$ in-
tuitionistic implication, and we know from **S4** Kripke model semantics that
they do not collapse when combined..*

2.2. THE BASIC IDEA OF FIBRING

This subsection explains how fibring works. We do it in two stages. First
we take a well known logic, modal logic, and show how it works in that
case, and then we give a general schematic definition.

Let $\mathbf{L}_1$ and $\mathbf{L}_2$ be two modal languages with $\square_1$ and $\square_2$. We assume that
$\mathbf{L}_1$ and $\mathbf{L}_2$ share (are built up from) the same set Q of atomic propositions.
Consider two logics $\mathrel{\vdash}_1$ in $\mathbf{L}_1$ and $\mathrel{\vdash}_2$ in $\mathbf{L}_2$. These are our *systems* $\mathcal{S}_1$ of
$\mathbf{L}_1$ and $\mathcal{S}_2$ of $\mathbf{L}_2$. To be specific let $\mathcal{S}_1$ be modal logic $\mathbf{K}_1$ for $\square_1$ and let
$\mathcal{S}_2$ be modal logic **S4** for $\square_2$, as described in Example 2.1. We have to
say how these logics are *presented* to us. To simplify matters let us assume
that the logics are presented to us via classes $\mathcal{K}_1$ and $\mathcal{K}_2$ respectively, of
Kripke models for which the logics are sound and complete. Indeed for
our fibring methodology to work there is no need to assume the classes

are frame classes or normal or anything special. Let $\mathcal{K}_1$ be the class of all Kripke models of the form (S, R, a, h), where R is transitive and aRa holds. Let $\mathcal{K}_2$ be the class of all models where R is reflexive and transitive. To distinguish between the classes, we write $\mathbf{m}^1 = (S^1, R^1, a^1, h^1)$ for a model in $\mathcal{K}_1$ and $\mathbf{m}^2 = (S^2, R^2, a^2, h^2)$ for a model in $\mathcal{K}_2$.

To explain in principle how fibring works, consider a mixed wff of the form $\alpha = \Diamond_1 \Box_2 q$.

We say this wff α is in the language $\mathbf{L}_{(1,2)}$, namely we have outer connectives of $\mathbf{L}_1$ and inside there are connectives of $\mathbf{L}_2$. We shall define later the language $\mathbf{L}_{(x_1, \ldots, x_n)}$ in general, where $(x_1, \ldots, x_n)$ is an alternating sequence of numbers in $\{1, 2\}$.

We now motivate the definition of fibred models for formulas of $\mathbf{L}_{(1,2)}$, by looking at our example α.

Let us consider α as a formula of $\mathcal{S}_1$ (since its outer connective is $\Diamond_1$). From the point of view of $\mathbf{L}_1$ this formula has the form $\Diamond_1 p$, where $p = \Box_2 q$ is *atomic*, since $\mathcal{S}_1$ does not recognise $\Box_2$.

To give a model for $\Diamond_1 p$ we take any $\mathcal{S}_1$ model. Take for example $\mathbf{m}^1 = (S^1, R^1, a^1, h^1)$ and check whether $ra^1 \vDash_1 \Diamond_1 p$.

For this to hold we need to check whether for some $t \in S^1$ such that $a^1 R^1 t$, we have $t \vDash_1 p$, i.e. whether $t \vDash_1 \Box_2 q$. Since $\Box_2$ is not in the language of $\mathcal{S}_1$, we do not know how to evaluate it! We observe, however, that we need no more than a truth value for $t \vDash_1 \Box_2 q$. We need an answer, yes or no.

The basic idea of fibring is to associate, with each $t \in S^1$, a model $\mathbf{m}_t^2 = (S_t^2, R_t^2, a_t^2, h_t^2)$ of $\mathcal{S}_2$, and get our answer by evaluating at the associated model i.e. we have

(*) $t \vDash_1 \Box_2 q$ iff $a_t^2 \vDash_2 \Box_2 q$.

Of course $\Box_2 q$ can be evaluated at $\mathbf{m}_t^2$ because it is in the right language.

Let $\mathbf{F}^1$ be the fibring function, i.e. let $\mathbf{f}^1(t) = \mathbf{m}_t^2$.

We can now say that our fibred semantics for the combined language $\mathbf{L}_{(1,2)}$, has models of the form $(S^1, R^1, a^1, h^1, \mathbf{F}^1)$, where $\mathbf{F}^1$ is as above and we use (*) above in our evaluation. Figure 1 below describes the model schematically:

In principle fibred models for formulas with nestings of modalities (e.g. $\Diamond_1 \Box_2 \Diamond_1 \Diamond_1 q$) can be defined inductively by iterating the process of fibring.

We shall see later that these models can be (just for the case of modal logic) greatly simplified, using various technical devices, which will lead us to the notion of *SFM*-models.

Before we do that let us discuss the general case. The above idea of fibring can be schematically generalised, and the rest of this section is going to give an intuitive description of general fibring.

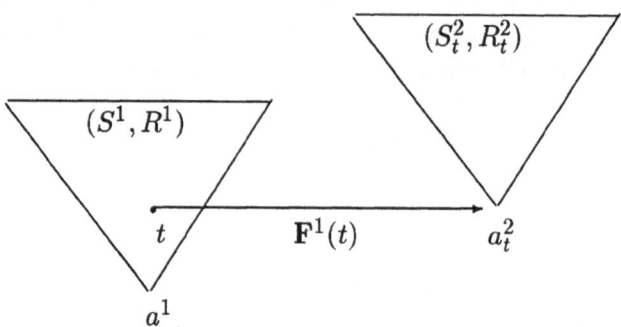

Figure 1.

2.3. GENERAL FIBRING

Since the fibring methodology is applicable in diverse and unconnected areas of logic, the best approach is to give a very general definition of fibring and then show how the fibred semantics can be simplified for each case (area) it is used. The general definition will highlight the basic general assumptions needed to execute the fibring methodology.

Let $\mathbf{L}_1$ and $\mathbf{L}_2$ be two languages, and let $\mathcal{S}_1$ and $\mathcal{S}_2$ be two systems (usually logics, but they could be any systems, for example theorem provers or programming environments). To be specific we advise the reader to think of three typical examples:

- combining two modal logics $\mathcal{S}_1$ and $\mathcal{S}_2$
- $\mathcal{S}_1$ a logic programming language
  $\mathcal{S}_2$ a C-language
  For example, we want to allow C-code to be put into PROLOG code[1]
- Combining two partially ordered systems, with equality $=$ and ordering $\leq$.

We assume the systems satisfy the following assumptions:

Basic assumption for general fibring

1. SYNTACTICAL ASSUMPTION
   The expressions E^i of the language $\mathbf{L}_i$ is built up using $\mathbf{L}_i$-constructors (connectives) from a set of atomic units Q_i. It is sometimes convenient to assume that $Q_1 = Q_2 = Q$, i.e. the languages share the atoms. We

---

[1]Our fibring methodology has been widely applied in the logic context. I believe it can also be applied for general systems as well. The details for programming systems have not yet been worked out. I am asking the reader to think of the PROLOG-C example because it can help understand the logic case studies.

schematically write $E^i = E^i(q_1^i, \ldots, q_n^i)$ to indicate that E^i is built up from the atoms $Q_1^i, \ldots, q_n^i \in Q_i$. Let $\mathbb{C}^i$ be the set of constructors of S_i. We assume $\mathbb{C}^1 \cap \mathbb{C}^2 = \varnothing$.

2. SEMANTICAL ASSUMPTION

The system S_i is 'characterised' (in case of logics S_i this means completeness) by a class $\mathcal{K}_i$ of models $\{\mathbf{m}_1^i, \mathbf{m}_2^i, \ldots\}$. Each model $\mathbf{m}_n^i$ is built up from a set S_n^i of basic semantic components $t \in S_n^i$. In many cases the model $\mathbf{m}_n^i$ comes with a distinguished point $a_n^i \in S_n^i$ (the actual world in Kripke models). This distinguished point plays a role in the semantic evaluation in the model.

For technical reasons it is convenient to assume that for $m \neq n$, $S_n^i \cap S_m^i = \varnothing$. Let $S^i = \bigcup_n S_n^i$. We also assume that $S^1 \cap S^2 = \varnothing$. This can always be achieved by renaming.[2]

3. THE EVALUATION FUNCTION

An evaluation function val^i is available which can give values ($\top$ or $\bot$ or solutions or any other output) to the following:

(a) $val^i(\mathbf{m}_n^i, E^i)$.

This is the value of a general expression E^i of S_i, evaluated at a model $\mathbf{m}_n^i$. In many cases (e.g. modal logic) $val^i(\mathbf{m}_n^i, E^i)$ is defined to be $val^i(a_n^i, E^i)$ where a_n^i is the distinguished point of $\mathbf{m}_n^i$.

(b) $val^i(t, E^i), t \in S_n^i$.

The value of a general expression at a unit $t \in S_n^i$.

(c) The value $val^i(\mathbf{m}_n^i, E^i(q_1, \ldots, q_k))$ and $val^i(t, E^i)$ is functionally reducible through some possibly algorithmic, inductive process to the values $\{val^i(t, q_j)\}$ where $t \in S^i$ and q_j are the atoms appearing in E. These latter values are arbitrary within an acceptable family of assignments. Let us write this as $\{val^i(t, q)\} \in Acceptable^i$.

Note that there may be some metapredicates involved in the formulation of *Acceptable*, relating $val^i(t, q)$ with $val^i(s, q)$. The most common (and in fact most general) is to impose an ordering $\prec$ on S^i and some relation on val (also denoted by $\prec$) and require something like

$- t \prec s$ implies $val^i(t, q) \prec val^i(s, q)$.

4. DEFINITION OF THE FIBRED LANGUAGE

We now define the fibred languages $\mathbf{L}_{(x_1, \ldots, x_n)}$, as follows:

$-$ Let $\mathbf{L}_{(1)}$ be $\mathbf{L}_1$ and $\mathbf{L}_{(2)}$ be $\mathbf{L}_2$.

---

[2]Note that the nature of this assumption is not clear for the case of programming languages. See previous footnote.

- Let $\bar{y} = (1, y_1, \ldots, y_k)$. Let $\mathbf{L}_{(2)*\bar{y}}$ be defined as the family of all expressions of the form $\alpha \in \mathbf{L}_{(y_1,\ldots,y_k)}$ or $\alpha = E^2(q_1/A_1, \ldots, q_n/A_n)$ where $E^2(q_1, \ldots, q_n) \in \mathbf{L}_2$ and $A_1, \ldots, A_n$ are in $\mathbf{L}_{\bar{y}}$, and q_j/A_j indicate the substitution of A_j for q_j in E^2.

  In other words, $\mathbf{L}_{(2,1,y_1,\ldots,y_k)}$ is the set of all expressions with outer constructor from $\mathbf{L}_2$ and with no more than $k+2$ nested alternation of constructors from $\mathbf{L}_1$ and $\mathbf{L}_2$.

- We similarly define $\mathbf{L}_{(1,2,y_1,\ldots,y_k)}$.

- Let $\mathbf{L}_\infty = \bigcup_{\bar{y}} \mathbf{L}_{\bar{y}}$.

5. THE FIBRING FUNCTION $\mathbf{F}$

We are now ready for our last assumption for fibring. We require that the values $val^2(\mathbf{m}_n^2, E^2)$, for models $\mathbf{m}_n^2$ of S_2 and E^2 of $\mathbf{L}_2$ are acceptable values for the function $val^1(t, q)$, for $t \in S^1$ and q atomic. In symbols:

- For all $\mathbf{m}_n^2$ and E^2, $\{val^2(\mathbf{m}_n^2, E^2)\} \in Acceptable^1$.

Let us see what this means. Take for example $\mathbf{L}_1$ as intuitionistic logic and take a Kripke model for intuitionistic logic, $(S, \leq, a, h)$. the assignments satisfy persistence:

$$t \leq s \text{ and } h(t, q) = 1 \text{ imply } h(s, q) = 1.$$

So if we fibre models $\mathbf{m}_t$ to t and $\mathbf{m}_s$ to s then for any formula α of $\mathbf{L}_2$ (the other language, which can be for example, modal logic) we must have $\mathbf{m}_t \vDash_2 \alpha$ implies $\mathbf{m}_s \vDash_2 \alpha$. This has to be ensured, so we cannot fibre arbitrarily in the intuitionistic case. In general the nature of the allowed fibring function $\mathbf{F}$ has to be worked out for each application with possibly some correctness theorems involved.

Consider the case of PROLOG and C-languages.

If we encounter a C-code α in the middle of a PROLOG program, then $val^C(\alpha)$ will probably be a numerical value. This value must be usable to the *Prolog* program, which only recognised predicates. So α may be an equation or constraints verified in C, etc.

Under the above assumptions we can define the family of fibred models for $\mathbf{L}_{(1,2)}$. These have the form $(\mathbf{m}_n^1, \mathbf{F}_n^{(1,2)})$, where $\mathbf{F}_n^{(1,2)}$ is a fibring function on S_n^1, assigning for each $t \in S_n^1$ a model $\mathbf{m}_t^2 \in \mathcal{K}_2$. The basic evaluation clause for $t \in S^1$ is

- $val^{(1,2)}(t, E^2) = val^2(\mathbf{m}_t^2, E^2)$

6. SIMPLIFYING THE FIBRING FUNCTION.

The fibring function can be simplified, for the following reasons:

(a) Since $S_n^1, n = 1, 2, \ldots$ are all disjoint, the domain of $\mathbf{F}_n^{(1,2)}$ can be $S^1 = \bigcup_n S_n^1$.

(b) Since for each t, $\mathbf{F}_n^{(1,2)}(t)$ is a model $\mathbf{m}_t^2$, this model must have an index n_t, i.e. $\mathbf{m}_t^2 = \mathbf{m}_{n_t}^2$.

It is sufficient to regard $\mathbf{F}_n^{(1,2)}$ as giving either the numerical value n_t or some $a \in S_{n_t}^2$. Since the model $S_1^2, S_2^2, \ldots$ are pairwise disjoint, any $a_t \in S^2$ will characterise a model $\mathbf{m}_{n_t}^2$ uniquely. In fact we mentioned in item 2 above that the models $\mathbf{m}_n^i$ can come with a distinguished element $a_n^i \in S_n^i$. We can let $\mathbf{F}^{(1,2)}(t)$ be $a_{n_t}^2 \in S^2$.

(c) Also because we can always duplicate names of models, we can always assume that

(*) $t \neq s, t, s, \in S^1$ imply $n_t \neq n_s$.

Thus we can view the fibring function $\mathbf{F}^{(1,2)}$ as a one-to-one function from S^1 into S^2 satisfying (*) above.

(d) Since the models $\mathbf{m}_n^1$, $n = 1, 2, \ldots$ can themselves be duplicated, we can assume that if the $t \in S_k^1$ and $s \in S_m^1$ and $k < m$ then the fibred models $\mathbf{F}^{(1,2)}(t) = a_{n_t}^2$ and $\mathbf{F}^{(1,2)}(s) = a_{n_s}^2$ satisfy $n_t < n_s$.

(e) Furthermore a function $\mathbf{F}^{(1,2)}$ can be similarly defined from S^2 to S^1 for the fibred language $\mathbf{L}_{(2,1)}$. The two functions can be joined as one function

$$\mathbf{F} : S^1 \cup S^2 \mapsto S^1 \cup S^2$$

satisfying the following

 - If $t \in S^1$ then $\mathbf{F}(t) = a_{n_t}^2 \in S^2$.
 - If $t \in S^2$ then $\mathbf{F}(t) = a_{n_t}^1 \in S^1$.
 - $\mathbf{F}$ is one-to-one and satisfies the condition in (d).

We shall see that for each particular fibring application, astonishing simplifications are possible, showing connections between fibring and other known concepts in logic.

7. GENERAL DEFINITION OF FIBRING.
We can now define the fibred model for the mixed language $\mathbf{L}_\infty = \bigcup_{\bar{x}} \mathbf{L}_{\bar{x}}$ as follows:

(a) A fibred model for $\mathbf{L}$ has the form $\mathbf{n} = (\mathbf{m}_n^i, \mathbf{F})$ where $\mathbf{F}$ is the simplified fibring function as in (6e) above. Note that although $\mathbf{F}$ was defined in (6e) for the case of $\mathbf{L}_{(1,2)}$ and $\mathbf{L}_{(2,1)}$, the form it was given in (6e) allows it to be used for any $\mathbf{L}_{\bar{x}}$.

(b) We now define $val(\mathbf{n}, E)$ as follows.
We can assume that $E \in \mathbf{L}_{(x_1, \ldots, x_n)}$. E has the form $E =$

$E^i(q_j/A_j)$ where $A_j \in \mathbf{L}_{(x_1,\ldots,x_n)}$. We know that the $val^i(\mathbf{m}_n^i, E^i)$ depends directly on

$$V = \{val^i(t, q_j) \mid t \in S^1, i = 1, 2, \ldots\}.$$

Let $f_{E^i}(V)$ symbolise this dependency. Therefore if we replace $val^i(t, q_j)$ by the new $V' = \{val(t, A_j)\}$, which we can assume we know how to get, then the same dependency will give us

$$val(\mathbf{n}, E) = f_{E^1}(V').$$

8. The notion of dovetailing.

We saw in item 6 above that the function $\mathbf{F}$ can be viewed as a function giving for each $t \in S^1 \cup S^2$, an element $\mathbf{F}(t) \in S^1 \cup S^2$ such that if $t \in S^i$ then $\mathbf{F}(t) \in S^j, i \neq j$.

Consider the case where the language $\mathbf{L}_1$ and $\mathbf{L}_2$ share the same set of atoms Q.

We can compare the values $val^i(t, q)$ and $val^j(\mathbf{F}(t), q)$ for atom q. These two values need *not* be identical. If we require from our fibring function $\mathbf{F}$ that for each $t \in S^i, \mathbf{F}(t) \in S^j$ and each $q \in Q$ we have

† $val^i(t, q) = val^j(\mathbf{F}(t), q)$

Then this fibring case is referred to as *dovetailing*. We shall see later that dovetailing allows for serious simplifications in the combined system.

2.4. CASE STUDY: MODAL LOGIC FIBRING AND DOVETAILING

We take this opportunity to show how the simplifications can be done in the case of fibring modal logics.

In the modal logic case each model involved has the form $\mathbf{m}_n^i = (S_n^i, R_n^i, a_n^i, h_n^i)$.

Since we assumed that all S_n^i are pairwise disjoint, we can put both semantics together in one big set of possible worlds W.

$$W = \bigcup_{(i,n)} S_n^i.$$

The relations R_n^i can be all unified under a single relation $R \subseteq W^2$,

$$R = \bigcup_{(i,n)} R_n^i.$$

This can be done without loss of generality. For any two elements $t, s \in W, tRs$ can hold only if they are both in the same S_n^i and only if $tR_n^i s$

holds. Thus $R_n^i = R \upharpoonright (S_n^i)^2$. Can we retrieve S_n^i out of W? The answer is yes if we record the actual worlds a_n^i. We can assume, from modal logic considerations, that the model (S_n^i, R_n^i, a_n^i) has the property that each $t \in S_n^i$ is accessible from a_n^i via a chain of elements i.e.

$$\forall t \exists k \exists t_1, \ldots, t_k (a_n^i R_n^i t_1 \wedge \ldots t_{k-1} R_n^i t_k \wedge t_k R_n^i t).$$

With this assumption S_n^i is retrievable from a_n^i using R_n^i. So if we let $W_i = \{a_n^i \mid n = 1, 2, \ldots\}$ and $W_a = W_1 \cup W_2$ (a for 'actual') be the subsets of W for all actual worlds, then knowing W_a, we can get back our models. For each $x \in W_a$ let $S^x = \{t \mid \exists k \exists t_1, \ldots, t_k (xRt_1 \wedge \ldots \wedge t_k R_t\}$. So for $x = a_n^i, S^x = S_n^i$.

We can now simplify the fibring function $\mathbf{F}$. For each t, $\mathbf{F}(t)$ is a model $\mathbf{m}_n^i$. Since this model is characterised by a_n^i, we can let $\mathbf{F}(t)$ be a_n^i. Thus

$$\mathbf{F} : W \mapsto W_a.$$

$\mathbf{F}$ satisfies the same properties as before, namely it is one to one and always switching semantics, (i.e. for $t \in S^1, \mathbf{F}(t) \in W_2$ and for $t \in S^2, \mathbf{F}(t) \in W_1$).

The above considerations make $\mathbf{F}$ a jump function which corresponds to a unary operator $\mathbb{J}A$ of the form:

$- t \vDash \mathbb{J}A$ iff $\mathbf{F}(t) \vDash A$

We thus conclude that fibring two modalities is reduced to one modality with a jump connective $\mathbb{J}$.

We can now define the notion of SFM model (Simplified Fibred Models) for mult-modal logic, with modalities $\{\square_i \mid i \in I\}$.

Definition 2.3 (SFM-model) *An   SFM-model   has   the   form* $(W, W_i, W_a, R, w_0, h, \mathbf{F}_i), i \in I$, *where*

1. *W is a set of worlds for some $i_0, w_0 \in W_{i_0}$.*
2. *$W_i \subseteq W, i \in I$ are pairwise disjoint and nonempty, with $W_a = \bigcup_i W_i$.*
3. *For $t, s \in W_i$ let $S^t = \{x \mid \exists n \ tR^n x\}$.*
   Then

   $- t \neq s \rightarrow S^t \cap S^s = \varnothing.$

   $- W = \bigcup_{t \in W_a} S^t.$

4. *$\mathbf{F}_i$ is a function satisfying*

   $- x \in S^t$ *and* $t \in W_i \rightarrow \mathbf{F}_i(x) = x$

   $- x \in S^t$ *and* $t \notin W_i \rightarrow \mathbf{F}_i(x) \in W_i$

   $- x \neq y \rightarrow \mathbf{F}_i(x) \neq \mathbf{F}_i(y).$

5. *For each $t \in W_i$ the model $\mathbf{m}_t = (S^t, R \restriction S^t \times S^t, t, h \restriction S^t)$ is in the semantics $\mathcal{K}_i$ of $\mathbf{L}_i$.*
   In case $\mathbf{L}_i$ is complete for the models $\{\mathbf{m}_t \mid t \in W_i\}$ we say the model is a universal *model.*
6. *We can also assume that $\mathbf{F}$ generates W as follows.*
   Let $W^0 = S^{w_0}$
   $W^{n+1} = W^n \cup \bigcup_{y \in W^n, i \text{ arbitrary}} S^{\mathbf{F}_i(y)}$
   Then $W = \bigcup_n W^n$.

Condition (f) can be assumed because evaluation at a model is evaluation at w_0. Any point not in $\bigcup_n W^n$ is not reachable from w_0 by embedded modalities so cannot affect truth values. See Figure 2.

The figure shows a model which has possible worlds $W = S^t \cup S^s \cup \dots$. $t \in W_j$ means t is an actual world of a model of $\square_j$. $\mathbf{F}_i(x) = s$ means the model with actual world s (of $\square_i$) is fibred at the point x.

When dovetailing, the point s is identified with the point x. To distinguish between xRz in S^t and xRy in S^s, we write xR_jz and xR_iy.

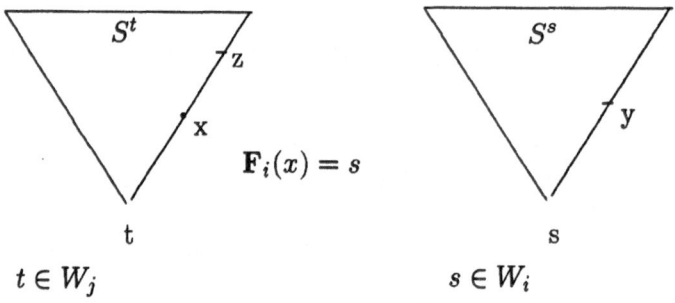

Figure 2.

We now take a new look at the fibring function $\mathbf{F}$. We can simplify the fibred semantics even further. Let us introduce new unary connectives of the form $\mathbf{J}_iA$ ($\mathbf{J}_i$ for a 'jump' operator) and modal operators $\square$ and $\lozenge$ with the table:

- $t \vDash \mathbf{J}_iA$ iff $\mathbf{F}_i(t) \vDash A$.
- $t \vDash \lozenge A$ iff for some $s(tRs$ and $s \vDash A)$.
- $t \vDash \square A$ iff for all $s(tRs \to s \vDash A)$.

According to this definition we can let $\square_i = \mathbf{J}_i\square$, $\lozenge_i = \mathbf{J}_i\lozenge$. We are referring to $\mathbf{J}_i$ as 'jump' operators because their truth table 'jumps' the evaluation from a world t to the world $\mathbf{F}_i(t)$.

Thus fibring several modalities together is like adding several unary jump operators to the modalities of the fibred semantics.

Conversely, assume we are given a modal logic with modalities $\Box$ and $\Diamond$ and several $\mathbb{J}_i$ and a class $\mathcal{K}$ of *SFM*-models of the form $(W, W_i, R, w_0, h, \mathbf{F}_i)$, as defined above.

We can define modal semantics classes $\mathcal{K}_i$ by letting $\mathcal{K}_i$ to be the set of all models of the form $(S^t, R \restriction S^t \times S^t, t, h \restriction S^t), t \in W_i$.

In the case of a finite number of modalities, one fibring function $\mathbf{F}$ and one jump operator $\mathbb{J}$ are sufficient. $\Box_1$ can be interpreted as $\Box$ and $\Box_k$ as $\mathbb{J}^k \Box$. $\mathbf{F}$ can be chosen to take any point x into $\mathbf{F}(x)$, an actual world of the next semantics in cyclical order.

Note that the jump operators are slightly more expressive. For q atomic, $\mathbb{J}q$ changes the index of evaluation of q from t to $\mathbf{F}(t)$. This cannot be done using modalities.

Example 2.4 (Dovetailing modal Kripke semantics) *Continuing the previous example, the case of dovetailing is when we require that for all atomic q*

$$h(t, q) = h(\mathbf{F}_i(t), q)$$

for all i. This means the actual world of the model fibred at t can be identified with t. We must be careful however. For each t, we must first introduce the notation tR_iy iff (definition) $\mathbf{F}_i(t)Ry$. Then we can identify the point $\mathbf{F}_i(t)$ with t and the dovetailed model can be equivalently represented as (S, R_i, a, h) with

$$t \vDash \Box_i A \quad \textit{iff} \quad \forall y(tR_iy \to y \vDash A)$$
$$t \vDash \Diamond_i A \quad \textit{iff} \quad \exists y(tR_iy \wedge y \vDash A)$$

The function $\mathbf{F}$ is no longer needed, since we identified t with $\mathbf{F}_i(t)$, so the dovetailed model reduces to (W, R_i, w_o, h).

It is interesting to see the role of the jump operators in the simplified dovetailed semantics.

We have in the fibred model $(W, R, w_0, h, \mathbf{F}_i)$

$$t \vDash \mathbb{J}_i \Box A \quad \textit{iff} \quad \mathbf{F}_i(t) \vDash \Box A$$
$$\textit{iff} \quad \forall x(\mathbf{F}_i(t)Rx \textit{ implies } x \vDash A)$$

This becomes

$$t \vDash \mathbb{J}_i \Box A \textit{ iff } \forall x(tR_ix \textit{ implies } x \vDash A).$$

$\mathbb{J}_i$ becomes a mode shifting operator, changing the mode of evaluation, namely of which R_i to use.

We can consider the two clauses for a satisfaction $\vDash_i$ dependent on mode i.

  $- \; t \vDash_i \Box A$ iff $\forall y(tR_iy$ implies $y \vDash_i A)$.

$- \ t \vDash_i J_j A \ \textit{iff} \ t \vDash_j A.$

2.5. STEP BY STEP SCENARIO FOR FIBRING TWO LOGICS

In view of the previous discussion, we need to describe the steps we take in fibring two arbitrary logics S_1 and S_2.

Step 1: Check the form in which these two logics are presented and see whether they can be recast into the format of subsection 2.2.

Step 2: If worst come to worst, extract the consequence relations $\mathrel{\vdash}_1$ and $\mathrel{\vdash}_2$ of these two logics.

Step 3: Using theorems of the next section, give $\mathrel{\vdash}_1, \mathrel{\vdash}_2$ basic relational semantics $\mathcal{K}_1$ and $\mathcal{K}_2$.

Step 4: Using definitions of next section, fibre the two logics.

Step 5: We can assume, one way or another, that we fibred the two systems, using some semantics.

Depending on what the logics are (e.g. modal logics, intuitionistic logics, fuzzy logics, etc.), use general theorems available in the following respective sections to assert transfer of decidability, finite model property, recursive axiomatisation, etc.

2.6. PLAN OF THE REST OF THIS OVERVIEW

So far we have described the fibring methodology in general. From now on we have to do case by case study. In each case special features of the logics involved will allow us to obtain:

1. A more simplified form of the fibring and additional properties of it and connections with other parts of logic.
2. Specific completeness and other transfer theorems.
3. Specific open problems for the logics involved.
4. Comparison with existing results in the literature.

For more details, see my papers (Gabbay, 1995a; Gabbay, 1993; Gabbay, 1995b; Gabbay, 1995d; Gabbay, 1995c; D'Agostino *et al.*, 1995; D'Agostino & Gabbay, 1996; Dörre *et al.*, to appear; Finger & Gabbay, 1992; Finger & Gabbay, to appear).

3. Logics and their Semantics

Our considerations in the general discussion of combining logics in the previous sections indicates that we may be presented with two logics that are completely incompatible in their style and formulation.

In such a case we cannot combine them directly. Our strategy in this case is to do the following:

Step 1: Extract from the given presentation of the logics two respective consequence relations $\mathrel{|\!\sim}_1$ and $\mathrel{|\!\sim}_2$ for them.

Step 2: Use the method of this section (as presented below) to give general basic point relational semantics for $\mathrel{|\!\sim}_1$ and $\mathrel{|\!\sim}_2$.

Step 3: Fibre the basic point relational semantics to obtain the combined system $\mathrel{|\!\sim}$.

Note that we offer here a technical solution. The basic point relational semantics may not mean much for the applications or motivations of these logics. We merely need it to bring the logic into some common background so that they can be combined. We can then use the semantics to prove transfer theorems and formulate an axiom system for the combined logic and possibly some confirmed properties of this system (e.g. decidability).

The purpose of this section is then to explain the following points.

- What is a logic and what is a basic relational semantics.
- Under what conditions on $\mathrel{|\!\sim}$ can we give it basic point relational semantics.
- How to fibre two such semantics.
- Transfer properties for the above.

Let us begin.

Definition 3.1 (Basic logic)   *1. A propositional language is a tuple* $\mathbf{L} = (Q, \mathbb{C})$ *where Q is a set of atomic propositions and $\mathbb{C}$ is a set of connectives. It may be convenient to include the connectives $\perp$ and $\top$.*

2. The set of well formed formulas of $\mathbf{L}$, $WFF_{\mathbf{L}}$ is defined inductively in the usual manner.

3. Let $\mathbf{L}_1 \cup \mathbf{L}_2$ denote the language $\mathbf{L} = (Q_1 \cup Q_2, \mathbb{C}_1 \cup \mathbb{C}_2)$.

4. A basic logic on a language $\mathbf{L}$ is a binary relation between wffs of $\mathbf{L}$ of the form $A \mathrel{|\!\sim} B$ satisfying the following conditions.[3]

---

[3]The general notion of a *logic* allows for relations between sets of wffs $\Delta \mathrel{|\!\sim} \Gamma$, and even lists or other structures. Most common is the form $\Delta \mathrel{|\!\sim} A$, where Δ is a set of wffs including the empty set $\varnothing$. In such a case we require $\mathrel{|\!\sim}$ to satisfy:

- *Reflexivity*
   $\Delta \mathrel{|\!\sim} A$, if $A \in \Delta$

- *Restricted monotonicity*
$$\frac{\Delta \mathrel{|\!\sim} A; \; \Delta \mathrel{|\!\sim} B}{\Delta, A \mathrel{|\!\sim} B}$$

– Identity

$$A \mathrel{\vertsim} A$$

– Logical equivalence

Let $A \equiv B$ abbreviate '$A\mathrel{\vertsim}B$ and $B\mathrel{\vertsim}A$' then for all A, B, C

$$\frac{A \equiv B}{(C\mathrel{\vertsim}A \text{ iff } C\mathrel{\vertsim}B)}$$

and

$$\frac{A \equiv B}{A\mathrel{\vertsim}C \text{ iff } B\mathrel{\vertsim}C} \; .$$

– Transitivity

$$\frac{A\mathrel{\vertsim}B;\, B\mathrel{\vertsim}C}{A\mathrel{\vertsim}C} \; .$$

– If $\top$ and $\bot$ are present we have

$$\bot\mathrel{\vertsim}A\mathrel{\vertsim}\top.$$

Definition 3.2 (Strengthening) *Let $\mathbf{L}$ be a language, let $\sharp(A_1, \ldots, A_n)$ be a connective in the language and $\mathrel{\vertsim}$ be a basic logic for $\mathbf{L}$.*

1. *We say that $\sharp$ satisfies upward (respectively, downward) strengthening in coordinate i, relative to $\mathrel{\vertsim}$, iff the following holds:*

$$\frac{\rule{0pt}{1.2em}X\mathrel{\vertsim}Y \,(\text{respectively } Y\mathrel{\vertsim}X)}{\sharp(A_1, \ldots, A_{i-1}, X, A_{i+1}, \ldots, A_n)\mathrel{\vertsim}\sharp(A_1, \ldots, A_{i-1}, Y, A_{i+1}, \ldots, A_n)}$$

2. *We say $\mathrel{\vertsim}$ satisfies strengthening iff each connective $\sharp$ satisfies either upward or downward (or both) strengthening in each coordinate.*
3. *A logic is said to be* isotonic *(I-Logic) iff it satisfies strengthening.*

Lemma 3.3 *Let $\mathbf{L}$ be a language and let $\mathrel{\vertsim}$ be a I-logic in $\mathbf{L}$. Let $C(q)$ be a wff of $\mathbf{L}$ with atomic q and let $C(q/X)$ denote the substitution of the wff X for q in C. Then the following holds*

$$\frac{A \equiv B}{C(q/A) \equiv C(q/B)}$$

– *Cut*

$$\frac{\Delta, A\mathrel{\vertsim}B;\, \Delta\mathrel{\vertsim}A}{\Delta\mathrel{\vertsim}B}$$

Proof. By structural induction on C, using strengthening.   ∎

Definition 3.4 (Rules) *Let* **L** *be a language and let* $\vdash\!\sim$ *be a logic.*

1. A logic rule ρ has the form

$$\frac{A_i \vdash\!\sim B_i, i = 1, \ldots, m}{A \vdash\!\sim B}.$$

2. $\vdash\!\sim$ *is said to satisfy the rule ρ iff whenever for all i, $A_i(q_j/C_j) \vdash\!\sim B(q_j/C_j)$ holds then $A(q_j/C_j) \vdash\!\sim B(q_j/C_j)$ holds.*

3. Let $\mathbb{R}$ *be a set of rules. Denote by* $\vdash\!\sim_{\mathbb{R}}$ *the smallest consequence relation satisfying all the rules in* $\mathbb{R}$.

Lemma 3.5 *Let* **L** *be a language and* $\vdash\!\sim$ *be an I-logic on* **L**. *Let* $\sharp$ *be a connective such that* $\sharp$ *satisfies both upward and downward strengthening in its first coordinate. Then* $\vdash\!\sim$ *is independent of its first coordinate.*

Proof. We will prove the lemma for the case either $\top$ or $\bot$ are present in the language. Since $\bot \vdash\!\sim A \vdash\!\sim \top$ for arbitrary A, we get $\sharp(\top) \equiv \sharp(\bot) \equiv \sharp(A)$ and hence $\sharp(A)$ is independent of A. If neither $\bot$ not $\top$ are present the proof is more complex, and we are not going to address it here. See Gabbay (Gabbay, 1994; Gabbay, 1995).   ∎

Definition 3.6   *1. Let* **L** *be a language and let* $\vdash\!\sim$ *be an I-logic. Let* $\sharp$ *be a connective. We can assume that each coordinate in* $\sharp$ *satisfies uniquely either downward or upward strengthening. We can therefore present* $\sharp$ *in the form* $(A_1, \ldots, A_r) \twoheadrightarrow_{\sharp} (B_1, \ldots, B_s)$ *where A_i are the wffs in the downward coordinates and B_j are the wffs in the upward coordinates.*

2. From now on we present all I-logics using connectives of the form $\{\twoheadrightarrow_1, \ldots, \twoheadrightarrow_k\}$, *where for each i, the arities of the connective $\twoheadrightarrow_i$ are written as (r_i, s_i).*

3. Let **L** *be a language with connectives* $\mathbb{C} = \{\twoheadrightarrow_i | i = 1, \ldots, k\}$ *with arities (r_i, s_i), respectively. We denote by* $\vdash\!\sim_{I(\mathbb{C})}$ *the smallest I-logic (i.e. smallest consequence relation) with the connectives* $\mathbb{C}$.

Definition 3.7 (Basic point relational structures) *Let* **L** *be a language and* $\vdash\!\sim$ *an I-logic for* **L**. *Assume its connectives are* $\{\twoheadrightarrow_1, \ldots, \twoheadrightarrow_k\}$ *with arities $m_i = r_i + s_i, i = 1, \ldots, k$. We present the connectives in the form* $(A_1, \ldots, A_{r_i}) \twoheadrightarrow_i (B_1, \ldots, B_{s_i})$.

1. A basic relational model for this language has the form

$$\mathbf{m} = (S, \prec, a, R_1, \ldots, R_k, h)$$

where S is a non-empty set, $a \in S$, $\prec$ is a reflexive and antisymmetric and transitive relation on S. We assume that for some $t_S \in S$ we have $\forall x \in S(t_S \prec x)$. Further, for $i = 1, \ldots, k, R_i$ is an $m_i + 1$ place relation on S satisfying the following conditions.

1.1. $\forall x_1, \ldots, x_{r_i}, y_1, \ldots, y_{s_i} \exists! t$
$[R_i(t, x_1, \ldots, y_1, \ldots) \land \forall u(R_i(u, x_1, \ldots, y_1, \ldots) \Leftrightarrow t \prec u)].$
We can denote this unique t by $t = f_i(x_1, \ldots, x_{r_i}, y_1, \ldots, y_{s_i})$, and call the function f_i the function associated with R_i. By 1.2 below, we have that f_i is monotonic upwards in $y_1, \ldots, y_{s_i}$ and downwards in $x_1, \ldots, x_{r_i}$ (relative to $\prec$).

1.2. $\forall x_1, \ldots, x_{r_i}, y_1, \ldots, y_{s_i}, x'_1, \ldots, x'_{r_i}, y_1, \ldots, y'_{s_i}$
$[R_i(t, x_1, \ldots, y_1, \ldots) \land \bigwedge_j x_j \prec x'_j \land \bigwedge_j y'_j \prec y_j \Rightarrow R_i(t, x'_1, \ldots, y'_1, \ldots s)].$

h is an assignment to the atoms. For atomic q, there exists a $t_q \in S$ such that $h(q) = \{x \mid t_q \prec x\}$.

2. *Satisfaction is defined as follows:*

   - *$t \vDash q$ iff $t \in h(q)$, q atomic.*
   - *$t \vDash (A_1, \ldots, A_{r_i}) \to_i (B_1, \ldots, B_{s_i})$ iff for all $x_j \vDash A_j, j = 1, \ldots, r_i$ there exist $y_j \vDash B_j, j = 1, \ldots, s_i$ such that $R_i(t, x_1, \ldots, x_{r_i}, y_1, \ldots, y_{s_i})$.*
   - *$\mathbf{m} \vDash B$ iff $a \vDash B$.*
   - *Let $\mathcal{K}$ be a class of models. We define $A \vDash_{\mathcal{K}} B$ iff for all $\mathbf{m} \in \mathcal{K}, \mathbf{m} \vDash A$ implies $\mathbf{m} \vDash B$.*

   We let $\vDash_{\mathcal{K}} B$ iff for all $\mathbf{m} \in \mathcal{K}, \vDash_{\mathbf{m}} B$.

3. *Note that there is another way of defining satisfaction in a model, which we denote by $\vDash^*$. Let $\hat{A} = \{t \mid t \vDash A\}$. Define $A \vDash^*_{\mathbf{m}} B$ iff for all $\prec$ minimal points $x \in \hat{A}$, we have $x \vDash_{\mathbf{m}} B$.*
   Similarly $A \vDash^*_{\mathcal{K}} B$ is defined. It is easy to see that if $\mathcal{K}$ is a class of models such that if $(S, \prec, a, R_i, h) \in \mathcal{K}$ then for all $x \in S$ we have also $(S, \prec, x, R_i, h)$ is in $\mathcal{K}$ then $A \vDash^*_{\mathcal{K}} B$ iff $A \vDash_{\mathcal{K}} B$.

4. *Consider the first-order language of classical logic with binary relation $\prec$ and relations $R_1, \ldots, R_k$ and possibly function symbols $f_1, \ldots, f_k$. Then the class of basic relational structures can be taken as the class of all models of some first-order theory β, comprised of all the conditions in (1) above.*

Lemma 3.8 *Let $\mathbf{m}$ be a basic relational structure. Let $\hat{B}$ be defined as $\hat{B} = \{t \mid t \vDash B\}$. Then there exists a $t_B \in S$ such that*

$$\hat{B} = \{t \mid t_B \prec t\}.$$

Since $\prec$ is transitive, we have $A \vDash_{\mathbf{m}} B$ iff $t_B \prec t_A$.

Proof. By definition, for q atomic we have:

$$h(q) = \{t \mid t_q \prec t\} = \hat{q},$$

where t_q is in S.

We show by induction that for any B there exists a t_B such that $\hat{B} = \{t \mid t_B \prec t\}$. We examine the sample inductive case of a binary $B \twoheadrightarrow C$. Let $R(t, x, y)$ be its relation.

Using the induction hypothesis we assume that t_B and t_C exist for $\hat{B}$ and $\hat{C}$ and that:

$$\widehat{B \twoheadrightarrow C} = \{x \mid \forall y(t_B \prec y \rightarrow \exists z(t_C \prec z \wedge R(x, y, z)))\}.$$

We are looking for $t_{B \rightarrow C}$. By condition (1.2) on R, there exists a t_0 such that $R(t_0, t_B, t_C)$ holds and for every u, $R(u, t_B, t_C)$ holds if $t_0 \prec u$. (I.e. $t_0 = f(t_B, t_C)$.). We claim t_0 is $t_{B \rightarrow C}$.

Let $t_0 \prec u$. Show $u \in h(B \twoheadrightarrow C)$. We know that $R(u, t_B, t_C)$ holds. Let $t_B \prec y$, and choose $z = t_C$. We need to show $R(u, y, t_C)$.

By condition (1.1) since $t_B \prec y$ and $t_C \prec t_C$ we get $R(u, y, t_C)$.

Assume $u \in \widehat{B \twoheadrightarrow C}$ and show $t_0 \prec u$. Choose $y = t_B$. Then for some z such that $t_C \prec z$ we have $R(u, t_B, z)$ and by (1) we have $R(u, t_B, t_C)$ and we have $t_0 \prec u$.

This concludes the proof of the lemma. ∎

Remark 3.9 *The proof of the previous theorem shows the basic point relational structures can be equivalently presented in the form*

$$\mathbf{m} = (S, \prec, a, f_1, \ldots, f_k, h)$$

where each $f_i(x_1, \ldots, x_{r_i}; y_1, \ldots, y_{s_i})$ is monotonic upwards in the y coordinates and downward in the x coordinates. The part without the h is called the frame *of the model and h, of course, is the assignment.*

The inductive truth definition is the following:

- $t \vDash (A_1, \ldots, A_{r_i}) \twoheadrightarrow_i (B_1, \ldots, B_{s_i})$ *iff for all $x_j \vDash A_j, j = 1, \ldots, r_i$, there exist $y_j \vDash B_j, j = 1, \ldots, s_i$ such that $f_i(x_1, \ldots, x_{r_i}; y_1, \ldots, y_{s_i}) \prec t$.*

The next definition and lemma shows how to calculate t_B, for any wff B in such a model.

Definition 3.10 *Let $\mathbf{m}$ be a basic point relational model for a language $\mathbf{L}$.*

With each formula $C(q_1, \ldots, q_n)$ with atoms $q_1, \ldots, q_n$ we can define a function $f_C(x_1, \ldots, x_n)$ on the domain of $\mathbf{m}$ as follows

- $f_{q_i}(x_i) = x_i$
- $f_C(x_1, \ldots, x_n) = f_i(f_{A_1}, \ldots, f_{A_{r_i}}, f_{B_1}, \ldots, f_{B_{s_i}})$ *where*
  $C = (A_1, \ldots, A_{r_i}) \twoheadrightarrow_i (B_1, \ldots, B_{s_i})$.

Lemma 3.11 *Let* **m** *be a basic point relational model and let* $C(q_1, \ldots, q_n)$ *be a wff built up from the atoms* $q_1, \ldots, q_n$. *Assume that* $h(q_i) = \{t \mid x_i \prec t\}$ *then*

$$\hat{C} = \{t \mid f_C(x_1, \ldots, x_n) \prec t\}.$$

Note that $\hat{\top}$ *is the smallest element of* $(S, \prec)$ *and* $\hat{\bot}$ *is the largest.*

Proof. Follows from the proof of a previous lemma. ∎

Theorem 3.12 (Strong completeness) *Let* **L** *be a language and* $\vdash\!\!\!\!\sim$ *be an I-logic. Let* $\mathbb{R}$ *be a set of rules such that* $\vdash\!\!\!\!\sim \; = \; \vdash\!\!\!\!\sim_{\mathbb{R}}$.

Then there exists a class $\mathcal{K}$ *of basic point relational structures such that* $\vdash\!\!\!\!\sim \; = \vDash_{\mathcal{K}}$, *and such that for each rule* ρ *of* $\mathbb{R}$ *of the form:*

$$\frac{A_i(q_1, \ldots, q_m) \vdash\!\!\!\!\sim B_i(q_1, \ldots, q_m)}{A(q_1, \ldots, q_m) \vdash\!\!\!\!\sim B(q_1, \ldots, q_m)}$$

the models satisfy in classical logic the following formula β_ρ:

$$\forall x_1, \ldots x_m [\bigwedge_i f_{B_i}(x_1, \ldots, x_m) \prec f_{A_i}(x_1, \ldots, x_n) \rightarrow$$
$$f_B(x_1, \ldots, x_n) \prec f_A(x_1, \ldots, x_m))].$$

Proof.

1. *Soundness*

   This can be verified.

2. *Completeness*

   Let S be the set of equivalence classes of WFF$_{\mathbf{L}}$ over $\equiv$.
   Let $A/ \equiv \prec B/ \equiv$ hold iff

   $$B \vdash\!\!\!\!\sim A.$$

For each connective $\twoheadrightarrow_i$ of **L**, let

$$f_i(A_1/\equiv, \ldots, A_{r_i}/\equiv, B_1/\equiv, \ldots, B_{s_i}/\equiv) =$$
$$((A_1, \ldots, A_{r_i}) \twoheadrightarrow_i (B_1, \ldots, B_{s_i}))/\equiv.$$

Let $R_i(t, x_1, \ldots, x_{r_i}, y_1, \ldots, y_{s_i})$ holds iff $f_i(x_1, \ldots, x_{r_i}, y_1, \ldots, y_{s_i}) \prec t$.
Let $h(q) = \{t \mid q/ \equiv \prec t\}$.
It is easy to show by induction that

- $\hat{A} = \{B/ \equiv \mid B \vdash\!\!\!\!\sim A\} = \{t \mid A/ \equiv \prec t\}$.

- $f_C(q_1/ \equiv, \ldots, q_n/ \equiv) = C(q_1, \ldots, q_n)/ \equiv$.

- $A \vDash B$ iff $A \vdash\!\!\!\!\sim B$ iff $f_B \prec f_A$

- Each formula β_ρ, for $\rho \in \mathbb{R}$ holds in the model because $\vdash\!\!\!\!\sim$ satisfies ρ

The above shows that the canonical model we have defined is the desired model. Assume $A \not\hspace{-0.1em}\vdash B$, then a countermodel would be $(S, \prec, A/ \equiv , R_i, h)$.

■

We shall fibre the basic relational semantics for the logics.

Remark 3.13 (Fibring of canonical models) *We are now going to be more specific about the classes of models $\mathcal{K}_1$ and $\mathcal{K}_2$ which we use. The completeness proof in Theorem 3.12 shows that the semantics of the logic $\vdash_i$ is obtained by taking one canonical frame, namely $\mathbf{m}^i_\alpha = (S^i, \prec^i, \alpha/ \equiv , f^i_1, \ldots, f^i_{k_i}, h^i)$, (where $S^i = WFF_{L_i}/ \equiv$ and $A/ \equiv \prec^i B/ \equiv$ is $B\vdash_i A$ etc.) and letting α run over all wffs. Whenever $A \not\hspace{-0.1em}\vdash B$ we let $\alpha = A$ and the model $\mathbf{m}^i_\alpha \vDash A$ but $\mathbf{m}^i_\alpha \not\vDash B$. Thus we can let $\mathcal{K}_i = \{\mathbf{m}^i_\alpha \mid \alpha \in WFF_{L_i}\}$. These models, however, are not pairwise disjoint because $S^i_\alpha = WFF_{L_i}/ \equiv$. To obtain pairwise disjoint models we can best base our WFF on new atoms. Let*

$$\mathbf{L}_1 = (Q, \mathbb{C}^1)$$
$$\mathbf{L}_2 = (Q, \mathbb{C}^2)$$

Assume $Q = \{q_1, q_2, q_3, \ldots\}$. Let us make more copies of the atoms. Let

$$Q^{(i,n)} = \{q_1^{i,n}, q_2^{i,n}, \ldots, \}.$$

Consider the language

$$\mathbf{L}_{i,n} = (q^{(i,n)}, \mathbb{C}^i).$$

Each formula $\alpha(q_1^*, \ldots, q_m^*) \in WFF_{L_i}$ built up from $Q_1^*, \ldots, q_m^* \in Q$ has its 'isomorphic' counterpart $\alpha^n \in WFF_{L_{i,n}}$ obtained by the simultaneous substitution of any atom $q_r \in Q$ by its corresponding $q_r^{i,n} \in Q^{(i,n)}$. Let $\{\alpha_1^i, \alpha_2^i, \alpha_3^i, \ldots\}$ be an enumeration of WFF_{L_i}.

We can now regard the semantics $\mathcal{K}_i$ as having the models

$$\mathbf{M}^i_{\alpha^i_n} = (WFF_{L_{i,n}}/ \equiv , \prec^i_n, \alpha^{i,n}_n, f^i_{r,n}, h^i_n).$$

Now all models in $\mathcal{K}_i$ are pairwise disjoint.

A fibring function $\mathbf{F}$ should associate with each $\beta^{i,n}_n/ \equiv \in WFF_{L_{i,n}}$ an actual world of a model of the other language, i.e. a formula $\alpha^{j,m}_m, j \neq i$. The persistence requirement can be observed by the following trick:

Let $\mathbf{f}$ and $\mathbf{g}$ be two functions $\mathbf{f} : WFF_{L_1} \mapsto WFF_{L_2}$ and $\mathbf{g} : WFF_{L_2} \mapsto WFF_{L_1}$ such that
$\alpha\vdash_1\beta$ implies $\mathbf{f}(\alpha)\vdash_2\mathbf{f}(\beta)$
$\gamma\vdash_2\delta$ implies $\mathbf{g}(\gamma)\vdash_1\mathbf{g}(\delta)$.
Now define a fibring function $\mathbf{F}$ as follows:

$$\mathbf{F}(\beta_n^{i,n}) = \alpha_m^{j,m}, m > n$$

satisfying the following

- if $i = 1$ then $\alpha_m^{2,m} = \mathbf{f}(\beta_n^{1,n})$
- if $i = 2$ then $\alpha_m^{1,n} = \mathbf{f}(\beta_n^{2,n})$.

The properties of $\mathbf{f}$ *and* $\mathbf{g}$ *will ensure that* $\mathbf{F}$ *satisfies persistence.*

Theorem 3.14 *Let* $\vdash_1$ *and* $\vdash_2$ *be two I-logics in* $\mathbf{L}_1$ *and* $\mathbf{L}_2$ *respectively. Let* $\vDash$ *be the fibred consequence relation arising from the fibred semantics for the combined language as defined in Remark 3.13. Then* $\vDash$ *is the smallest I-logic in the combined language which is a conservative extension of each pure component.*

Proof. Let $\vdash$ be the smallest I-logic in the combined language conservatively containing both $\vdash_1$ and $\vdash_2$. Such a logic exists as the intersection of all conservative extensions, $\vDash$ being one of them. Note that being an extension does not mean that $\vdash$ is closed under the rules defining $\vdash_1$ or $\vdash_2$, but only that $A \vdash_i B$ implies $A' \vdash B'$ where A', B' are substitution instances (in the combined language) of A and B respectively. We show $\vdash = \vDash$. We know that $\vdash \subseteq \vDash$, because $\vdash$ is the smallest conservative extension. We need to show $A_0 \nvdash B_0$ implies $A_0 \nvDash B_0$. Let $\mathbf{m}_{A_0}$ be a canonical model of $\vdash$. We know that the canonical models as constructed in Theorem 3.12 have the form

$$\mathbf{m}_\alpha = (\mathrm{WFF}/\equiv, \prec, \alpha/\equiv, f_i, g_j, h)$$

where $A/\equiv \prec B/\equiv$ iff $B \vdash A$, and f_i, g_j are the functions associated with the connectives f_i for $\mathbf{L}_1$ and g_j for $\mathbf{L}_2$.

We will turn $\mathbf{m}_\alpha$ into a fibred model and this will show that $A_0 \nvDash B_0$.

Each node $\beta/\equiv$ in the model is either atomic or has the form $(c_1, \ldots, c_k) \twoheadrightarrow (D_1, \ldots, D_m)$ where $\twoheadrightarrow$ is a connective in the language $\mathbf{L}_i, i = 1$ or $i = 2$.

$\mathbf{m}_\alpha$ can be viewed as a model of $\mathbf{L}_i$, where any β with main connective $\twoheadrightarrow$ not from $\mathbf{L}_i$ is regarded as atomic. We display this point of view by writing $\mathbf{m}_\alpha^i$. Thus $\mathbf{m}_\alpha^i$ is the same model $\mathbf{m}_\alpha$ but where any wff with main connective from $\mathbf{L}_j$ $j \neq i$ is regarded as atomic. We can now define a fibring function $\mathbf{F}$ on $\mathbf{m}_\alpha^i$. Take any $\beta/\equiv$ in the model $\mathbf{m}_\alpha^i$.

We can let $\mathbf{F}(\beta) = \mathbf{m}_\beta^j$ now regarded as a model of the $\mathbf{L}_j$ language, $j \neq i$. Thus $\mathbf{F}$ fibres with each node $\beta/\equiv$ a model of the other language. The assignment h_α^i in $\mathbf{m}_\alpha^i$ is defined as $h_\alpha^i(q) = \{t \mid q \prec t\}$.

We know that satisfaction in the canonical model is equal to $\vdash$ i.e. B holds at $A/\equiv$ iff $A \vdash B$. We now show that satisfaction in the fibred model is also the same i.e. $A/\equiv_{\vDash_{i,\beta}} B$ iff $A \vdash B$.

The proof is by induction.

1. *Case q atomic*
   $A/ \equiv\vDash_{i,\alpha} q$ iff $A/ \equiv \in h^i_\alpha(q)$ iff $A\!\!\mid\!\!\sim q$.
2. *Case the main connective of B is in the language* $\mathbf{L}_i$
   $A/ \equiv\vDash_{i,\alpha} (C_m) \twoheadrightarrow (D_n)$ iff $\forall x_m \vDash_{i,\alpha} C_m \ \exists y_n \vDash_{i,\alpha} D_n$ s.t. $f_{\twoheadrightarrow}(x_m, y_n) \prec t$. By the induction hypothesis $x_m \vDash_{i,\alpha} C_m$ is the same $x_m\!\!\mid\!\!\sim C_m$ and $y_n \vDash_{i,\alpha} D_n$ is the same as $y_n\!\!\mid\!\!\sim D_n$ and hence we continue, iff $A\!\!\mid\!\!\sim B$.
3. *Case B has main connective in the language* $\mathbf{L}_j \neq \mathbf{L}_i$
   $A/ \equiv\vDash_{i,\alpha} B$ iff $\mathbf{F}(A/ \equiv) \vDash_{j,A/\equiv} B$ iff, by the previous case, $A\!\!\mid\!\!\sim B$.
   Thus the model $\mathbf{m}_{A_0} \not\vDash B_0$.

This completes the proof of the theorem.   ∎

4. Combining Modal and Intuitionistic Logics

This section gives an overview of how the fibring methodology can be app-lied to combining traditional modal logics as well as modal and intermediate logics. The modal logics need not be normal. We show transfer of recursive axiomatizability, decidability and finite model property. Detailed proofs are given in (Gabbay, 1993).

Some results on combining logics (normal modal extensions of $\mathbf{K}$) have recently been introduced by Kracht and Wolter, Goranko and Passy and by Fine and Schurz as well as a multitude of special combined systems existing in the literature of the past 20-30 years. We hope our methodology will help organise the field systematically.

We already saw in section 2.2 item 7 what the fibred and dovetailed modal models look like. So proceeding directly from that point, we just quote here the theorems we get. For full details, see (Gabbay, 1993).

Theorem 4.1 (Completeness theorem for the fibred logic $\mathbf{L}_I^F$) *Let* $\mathbf{L}_i, i \in I$ *be modal logics with classes of structures* $\mathcal{K}_i$ *and set of theo-rems* $\mathbf{T}_i$. *(i.e.* $\mathbf{T}_i = \{A \text{ of } \mathbf{L}_i \mid A \text{ is valid in all } \mathcal{K}_i \text{ models }\}$). *Let* $\mathbf{T}_I^F$ *be the following set of wffs of* $\mathbf{L}_I^F$.

1. $\mathbf{T}_i \subseteq \mathbf{T}_I^F$
2. **Modal Fibring Rule:**[4]
   If $\square_i$ *is the modality of* $\mathbf{L}_i$ *and* $\square_j$ *of* $\mathbf{L}_j, i, j$ *arbitrary* $i \neq j$ *and* $C = \bigwedge_{k=1}^n \square_i A_k \to \bigvee_{k=1}^m \square_i B_k \in \mathbf{T}_I^F$ *then for all* d, $\square_j^d C \in \mathbf{T}_I^F$.

---

[4]The meaning of the modal fibring rule will be apparent from the proof below. Intui-tively, it has to do with substitutions of wffs of one language into a formula of the other language. If the substituted wffs are related (proof theoretically) we want to propagate this relation into the other language.

There are formal similarities between the modal fibring rule and necessitation. Consider a formula C built up only from 'atoms' of the form $\square_i B_k$. Then our special rule of necessitation says that from $\vdash C$ we can deduce $\vdash \square_j^d C$ for any modality $\square_j$ *other than* $\square_i$.

3. $\mathbf{T}_I^F$ *is the smallest set closed under (1), (2), modus ponens and substitution.*

Then $\mathbf{T}_I^F$ *is the set of all wffs of* $\mathbf{L}_I^F$ *valid in all the fibred structures of* $\mathbf{L}_I^F$.

Example 4.2 *Consider two logics* $\mathbf{K}_1$ *and* $\mathbf{K}_2$ *with two* $\mathbf{K}$ *modalities* $\square_1$ *and* $\square_2$. *Consider the fibred combination* $\mathbf{K}_{1,2}^F$. *By the previous theorem it can be axiomatised by taking all theorems of* $\mathbf{K}_1$ *and* $\mathbf{K}_2$ *together with the following modal fibring rule:*

$$\frac{\vdash \bigwedge_k \square_i A_k \to \bigvee_k \square_i B_k}{\vdash \square_j[\bigwedge_k \square_i A_k \to \bigvee_k \square_i B_k]}$$

where $i, j \in \{1, 2\}$, $i \neq j$.

Note that necessitation is not available. This example investigates under what conditions necessitation is admissible.

The logic $\mathbf{K}$ *has the following special properties:*

1. $\vdash \square\top$

2. $\vdash \square(A \wedge B) \leftrightarrow \square A \wedge \square B$

3. The disjunction property

$$\frac{\vdash \square A \to \square B \vee \square C}{\vdash (\square A \to \square B) \ or \ \vdash (\square A \to \square C)}$$

The above three properties can be used to prove that the modal fibring rule is equivalent to necessitation.

First the rule can be reduced to

$$\frac{\vdash \square_i A \to \square_i B}{\vdash \square_j(\square_i A \to \square_i B)}$$

using (2) and (3) above.

Second, since $(\top \to A) \leftrightarrow A$ *we get from (1) that* $(\square\top \to \square A) \leftrightarrow \square A$. *This can be used to derive*

$$\frac{\vdash \square_i A}{\vdash \square_j \square_i A}$$

Since necessitation is admissible in each component we get that

$$\frac{\vdash A}{\vdash \square_i A}$$

is admissible in the combination $\mathbf{K}_{1,2}$.

The key property which we have used is the disjunction property. This holds for $\mathbf{K}$, $\mathbf{K4}$, *and some other system and for all of these systems the above reduction of the modal fibring rule to necessitation stands.*

Theorem 4.3 *Assume* $\mathbf{L}_i, i \in I$ *has the finite model property then so does* $\mathbf{L}_I^F$.

Assume $\mathbf{L}_i, i \in I$ *are recursively axiomatisable, then so is* $\mathbf{L}_I^F$.

Dovetailing arises in many applications where the fibred model at t has the world t itself as its actual world. A major example is multimodal logics for action. In this example, assume we are at a given state t of a system. There are several possible actions α one can take. With each α we associate an accessibility relation R_α, where $tR_\alpha s$ means that s can nondeterministically arise after action α is applied to state s. We associate the modality $\square_\alpha A$ to read: A holds in all states arising from the application of α. Such a multimodal logic, with modalities $\{\square_\alpha\}$ is dovetailed, not fibred.

Theorem 4.4 (Completeness theorem for the dovetailed logic $\mathbf{L}_I^D$)
Let $\mathbf{L}_i, i \in I$ *be modal logics with semantical classes of structures* $\mathcal{K}_i$ *and set of theorems* $\mathbf{T}_i$. *Let* $\mathbf{T}_I^D$ *be the following set of wffs of* $\mathbf{L}_I^D$.

1. $\mathbf{T}_i \subseteq \mathbf{T}_I^D$
2. **Modal Dovetailing Rule**.[5]
   If $\square_i$ *is the modality of* $\mathbf{L}_i$ *and* $\square_j$ *of* $\mathbf{L}_j, i, j$ *arbitrary* $i \neq j$ *and*

$$C = \bigwedge_{k=1}^{n} \square_i A_k \wedge \bigwedge_{k=1}^{m} \Diamond_i \sim B_k \rightarrow \bigvee_{k=1}^{r} q_k \in \mathbf{T}_I^D$$

   then for all $d \; \square_j^d C \in \mathbf{T}_I^D$. *Where* q_k *are atoms or their negations, and* $q_1, \ldots, q_r$ *list all the atoms or their negations appearing in any* A_k *or* $B_k, k = 1, 2, \ldots$.
3. $\mathbf{T}_I^D$ *is the smallest set closed under (1), (2) modus ponens and substitution.*

Then $\mathbf{T}_I^D$ *is the set of all wffs of* $\mathbf{L}_I^D$ *valid in all the dovetailed structures of* $\mathbf{L}_I^D$.

Theorem 4.5 *Assume* $\mathbf{L}_i, i \in I$ *all are extensions of* $\mathbf{K}$ *formulated using traditional Hilbert axioms and the rule of necessitation, then* $\mathbf{L}_I^D$ *(the dovetailing of* $\mathbf{L}_i$*) can be axiomatised by taking the union of the axioms and the rules of necessitation for each modality* $\square_i$ *of each* $\mathbf{L}_i$

Theorem 4.6 *If* $\mathbf{L}_i, i \in I$ *admit necessitation and satisfy the disjunction property, then* $\mathbf{L}_I^F = \mathbf{L}_I^D$.

---

[5]The modal dovetailing rule is really a necessitation rule. It says that if C is a wff built up from 'atomic' units of the form $\square_i A_s$ and ordinary atoms q_k and $\square_j$ is a modality *different* from $\square_i$, then a limited necessitation rule holds: $\vdash C$ implies $\vdash \square_j^d C$ for any natural number d.

Theorem 4.7 *If $L_i, i \in I$ all have the finite model property (are finitely axiomatisable), so is L_I^D.*

Theorem 4.8 (Completeness theorem for fibring and dovetailing pure modalities) *Let $L_i, i \in I$ be modal logics in a language with connectives $\Box_i$ and/or $\Diamond_i$ together with atomic propositional variables. We assume for convenience that $\top, \bot$ are available.*

Let $\mathcal{K}_i$ be semantics (class of Kripke models) for which L_i is complete. Let L_I^F be the fibring (resp. Let L_I^D be the dovetailing) of $L_i, i \in I$. Let $\vdash_i$ be the consequence relation of L_i.
Let $\vdash$ be the consequence relation of L_I^F (resp. L_I^D) defined by the following rules.

1. *Any substitution instance of $\Delta_i \vdash_i \Gamma_i$, where $\Delta_i \vdash \Gamma_i$ is in the language of L_i.*
2. Pure Modalities fibring (resp. dovetailing) rules

$$\frac{A_1, \ldots, A_n \vdash C_1, \ldots, C_m}{\Box A_1, \ldots, \Box A_n, \Diamond B \vdash \Diamond C_1, \ldots, \Diamond C_m}$$

In case of fibring, A_i and C_j must begin with a modality.

$$\frac{A_1, \ldots, A_n \vdash C, B_1, \ldots, B_m}{\Box A_1, \ldots, \Box A_n, \vdash \Box C, \Diamond B_1, \ldots, \Diamond B_m}$$

In case of fibring, A_i, B_j and C must begin with a modality.

3. *$\vdash$ is the smallest consequence relation (monotonic, reflexive and transitive) closed under the above rules.*

The following holds.
Let Δ, Γ be two finite sets of wffs of L_I^F (resp. L_I^D). If $\Delta \not\vdash \Gamma$, then there exists a fibred model m of L_I^F (resp L_I^D) such that

- *$A \in \Delta$ implies $m \models A$*
- *$A \in \Gamma$ implies $m \not\models A$*

Theorem 4.9 *Assume $L_i, i \in I$ are all decidable then so are L_I^F and L_I^D.*

The next theorem studies the combination of pure modality $\Box$ and or $\Diamond$ into intuitionistic logic. This is the most general combination. The completeness theorem is rather involved, mainly because classical negation is not available, and we need to deal with negative information in a roundabout way, see (Gabbay, 1993) for details.

A lot of effort needs to be spent to ensure that the persistency condition

$$(*) \quad t \models A \text{ and } t \leq s \text{ implies } s \models A$$

holds in the fibred model. The above condition is the main difference between fibring of modalities alone and fibring of modalities with intuitionistic and intermediate logics.

We want a general fibring of an intuitionistic or intermediate logic connective $\Rightarrow$ with one or several modalities $\Box_1$ and/or $\Diamond_2$. Thus for example we would like to fibre the $\Rightarrow$ of Dummett's **LC** with the **K** modalities $\Box$ and $\Diamond$. The result of the fibring, as well as the axiomatisation, is very much dependent on the semantics we take for the modalities and on the additional connectives which happen to be present, such as $\wedge, \vee$ and $\neg$. Thus no general theorem can be easily given. We will therefore choose one type of modality, **K** modality and only $\Rightarrow$ and do the fibring and dovetailing for that.

We now fibre several intermediate logics $\Rightarrow_i$ with several **K** modalities $\Box_j, \Diamond_j$. Since we might not have $\neg, \wedge$ and $\vee$, we need to formulate the logics to be fibred using the notion of consequence relation. We choose the notion $\Delta \hspace{-2pt}\mid\hspace{-4pt}\sim\hspace{-2pt} \Gamma$ where both Δ and Γ are sets of wffs. Our proof works for the notion $\Delta \hspace{-2pt}\mid\hspace{-4pt}\sim\hspace{-2pt} B$, i.e. where Γ is a single formula. We therefore have to assume that the case of $\Delta \hspace{-2pt}\mid\hspace{-4pt}\sim\hspace{-2pt} \Gamma$ can be reduced to that of $\Delta \hspace{-2pt}\mid\hspace{-4pt}\sim\hspace{-2pt} B$. If the logic has disjunction then we can take B to be $\bigvee \Gamma$. Otherwise we an assume that $\hspace{-2pt}\mid\hspace{-4pt}\sim$ is *disjunctive*, namely that $\Delta \hspace{-2pt}\mid\hspace{-4pt}\sim\hspace{-2pt} \Gamma$ iff for some $B \in \Gamma, \Delta \hspace{-2pt}\mid\hspace{-4pt}\sim\hspace{-2pt} B$.

This condition is not too restrictive. The implicational fragment of intuitionistic logic is disjunctive i.e. if Δ, Γ are in the pure implicational fragment then $\Delta \hspace{-2pt}\mid\hspace{-4pt}\sim \bigvee \Gamma$ iff for some $B \in \Gamma, \Delta \hspace{-2pt}\mid\hspace{-4pt}\sim\hspace{-2pt} B$.

Dummett's **LC** does not satisfy the above. Its implicational fragment can be axiomatised by the additional axiom

$$((A \Rightarrow B) \Rightarrow C) \Rightarrow (((B \Rightarrow A) \Rightarrow C) \Rightarrow C)$$

Thus the conditions of Theorem 4.10 below do not apply to it. This does not mean, however, that we cannot prove the result for this case by some other means.

Theorem 4.10   (Completeness theorem for fibred (resp. dovetailed) intuitionistic modal logics).

1. *Let $\mathbf{L}_i$ be a logic with either an intermediate intuitionistic implication $\Rightarrow_i$ (and possibly $\wedge, \vee$ and $\bot$) or with one or both of the $\mathbf{L}_i$ modalities $\Box_i, \Diamond_i$. Let $\mathcal{K}_i$ be the semantics for $\mathbf{L}_i$. Let $\hspace{-2pt}\mid\hspace{-4pt}\sim_i$ be the consequence relation of $\mathbf{L}_i$. Then $\{A_1, \ldots, A_n\} \hspace{-2pt}\mid\hspace{-4pt}\sim_i \{B_1, \ldots, B_m\}$ means that in every Kripke model of $\mathbf{L}_i$ (modal or intuitionistic) of the form (S, R, a, h) whenever for all $j, a \vDash A_j$ then for some $k, a \vDash B_k$. We regard the consequence relation as a relation between finite sets of formulas. The*

completeness theorem characterises the consequence relation of $\mathbf{L}_I^F$ *(denoted by* $\hspace{0.05em}\vdash\hspace{-0.6em}\sim$ *) in terms of the consequence relations of* $\mathbf{L}_i$ *(denoted by* $\hspace{0.05em}\vdash\hspace{-0.6em}\sim_i$ *). We make the following assumptions.*

(*) For intuitionistic $\mathbf{L}_i$*, we assume that either disjunction is available or that it is disjunctive.*

2. *Let* $\hspace{0.05em}\vdash\hspace{-0.6em}\sim$ *be the smallest consequence relation containing* $\hspace{0.05em}\vdash\hspace{-0.6em}\sim_i$ *for* $i \in I$*, closed under substitution (and of course under reflexivity, monotonicity and cut) and closed under the following rules, where* $\Rightarrow$ *is a connective of an arbitrary* $\mathbf{L}_{i_1}$ *and* $\Box, \Diamond$ *of an arbitrary* $\mathbf{L}_{i_2}$*.*

(a)

$$\frac{A_1, \ldots, A_n \mathrel{\vdash\hspace{-0.6em}\sim} C_1, \ldots, C_m}{\Box A_1, \ldots, \Box A_n, \Diamond B \mathrel{\vdash\hspace{-0.6em}\sim} \Diamond C_1, \ldots, \Diamond C_m}$$

]for the case of fibring A_i *and* C_j *must begin with* $\Box, \Diamond$ *or* $\Rightarrow$*.*

$$\frac{A_1, \ldots, A_n \mathrel{\vdash\hspace{-0.6em}\sim} C, B_1, \ldots, B_m}{\Box A_1, \ldots, \Box A_n \mathrel{\vdash\hspace{-0.6em}\sim} \Box C, \Diamond B_1, \ldots, \Diamond B_m}$$

for the case of fibring A_i, B_j *and* C *must begin with* $\Box, \Diamond$ *or* $\Rightarrow$*.*

(b) Deduction theorem for each $\Rightarrow_i$*.*

Then $\hspace{0.05em}\vdash\hspace{-0.6em}\sim$ *is the consequence relation of all the fibred models of* $\mathbf{L}_I^F$ *(resp. dovetailed models of* $\mathbf{L}_I^D$*).*

Example 4.11 (Some dovetailed modal intuitionistic systems)

Note that Theorem 4.10 allows us to fibre or dovetail any intermediate logic with any modal logic provided the intermediate logic has disjunction in it or is disjunctive. The theorem gives us the axiom system for the combined logic. Especially for the case of dovetailing, the method is very powerful. Consider any modal logic where axioms are reduction of modalities, for example

- $\Box A \vdash A$
- $\Box A \vdash \Box\Box A$
- $\Diamond A \vdash \Box\Diamond A$
- $\Diamond\Box A \vdash \Box\Diamond A$
- $\Diamond^5\Box^6 A \vdash \Diamond^{18}\Box^{11}\Diamond^{12} A$
- $\Box\Diamond A \wedge \Diamond\Box B \vdash \Diamond\Diamond(A \wedge B)$

Some of these axioms you would recognize and some I have just invented. It does not matter. In the combined logic any $\bigwedge_i M_i A_i \vdash MA$ *becomes a theorem* $\bigwedge_i M_i A_i \Rightarrow MA$*, where* M_i, M *are strings of modalities. We can thus immediately have semantics and axioms for the dovetailed system of*

any intermediate logic with disjunction and any major modal logic (e.g. **K,
T, B, S4, S5,** *etc.).*

To be more specific, the above theorem allows us to fibre or dovetail the
following:

1. The intermediate logic **KC** obtained by adding the schema $\neg A \vee \neg\neg A$
   to intuitionistic logic with **S4** modality $\square$ and $\Diamond$.
2. Dummetts **LC** (obtained by adding the schema $(A \Rightarrow B) \vee (B \Rightarrow A)$
   with the **K** modality $\Diamond$.

*The axioms of the combined system are obtained from the completeness
theorem. For example, dovetailing the systems in (1) yields:*

1. **KC** axioms with Modus Ponens.
2. $\square A \wedge \square(A \Rightarrow B) \Rightarrow \square B$

   $\square(A \Rightarrow B) \wedge \Diamond A \Rightarrow \Diamond B$

   $\Diamond\Diamond A \Rightarrow \Diamond A$

   $\square A \vee \square\square A$

   $\square A \Rightarrow \Diamond A \vee \square B$

   $\Diamond\neg A \vee \square\neg\neg A$

   $\Diamond \neg A \vee \square\neg A$

   $\square A \Rightarrow A$

   $A \Rightarrow \Diamond A$

3. $\dfrac{\vdash A}{\vdash \square A}$

*Further note that the dovetailing method of adding modality to any in-
termediate logic is systematic. We check all modalities reductions of the
form* $\bigwedge_i M_i A_i \vdash MA$ *and turn them into an axiom*

$$M_1 A_1 \Rightarrow (\ldots \Rightarrow (M_n A_n \Rightarrow MA)\ldots).$$

There is nothing arbitrary here.

Another way to obtain intuitionistic modal logic is to take the modal logic
formulated axiomatically and change the underlying logic from classical to
intuitionistic. Thus modal **K**, when axiomatised by

- $\square(A \rightarrow B) \rightarrow (\square A \rightarrow \square B)$

- $\dfrac{\vdash A}{\vdash \square A}$

yields by chance, when '$\rightarrow$' *is changed to* '$\Rightarrow$', *the same system as dovetailing. but if we use an equivalent axiom*

- $\square(\neg A \vee B) \vee \neg\square A \vee \square B$

we get a different modal intuitionistic system.

Theorem 4.12 *Let* $\mathbf{L}_1, 1 \in I$, *be an intermediate logic and let* $\mathbf{L}_i, i \neq 1, i \in$
I *be modal logics . Let* $\mathbf{L}_I^D$ *and* $\mathbf{L}_I^F$ *be the dovetailing, resp. fibring of the
logics. Then if* $\mathbf{L}_i, i \in I$ *are all decidable, so are* $\mathbf{L}_I^D$ *and* $\mathbf{L}_I^F$.

COMPARISON WITH SOME OTHER WORK

The reader should compare our work with (Eiben *et al.*, 1992; Kracht & Wolter, 1991; Fine & Schurz, to appear) and (Goranko & Passy, 1992). (Eiben *et al.*, 1992) combines first-order theories and can combine logics by translation into classical logic. (Kracht & Wolter, 1991; Fine & Schurz, to appear; Goranko & Passy, 1992) combine Hilbert systems of normal modal logics. Our method is more general, is applicable to a wide variety of systems and yields transfer theorems on bulk. Their ideas may be most effective and transportable perhaps to combining theorem provers?

All in all, I think that given all the available papers in the literature, the subject is ready to take off. See also (Blackburn & de Rijke, to appear) for a discussion.

5. How to Make Your Logic Fuzzy

This section overviews results that show that fibring an arbitrary logic with a known fuzzy logic (e.g. Lukasiewicz infinite valued logic) is a general methodology for making ones logic fuzzy.

To appreciate the need for such a methodology, consider for example the modal propositional logic **K**, with one modality □, and let us examine our options for turning it into a fuzzy system. This logic is complete for the crisp (i.e. $\{0,1\}$ valued), Kripke semantics. Kripke models have the form $\mathbf{m} = (S, R, a, h)$, where $S \neq \varnothing$ is a set of possible worlds, $R \subseteq S \times S$ is a crisp binary relation, (of the form $R : S \times S \mapsto \{0,1\}$), $a \in S$ is the actual world, and h is a binary function assigning to each $t \in S$ and each atomic q a crisp value $h(t, q) \in \{0, 1\}$.

h can be extended to all wffs in the usual way with the inductive evaluation of $h(t, \Box A)$ being

$h(t, \Box A) = 1$ iff for all y such that tRy we have $h(y, A) = 1$.

or

$h(t, \Box A) = \text{Inf}\,\{h(y, A) \mid tRy\}$.

We say $\mathbf{m} \vDash A$ if $h(a, A) = 1$.

Let us try and turn this logic fuzzy!

Working intuitively, one may turn modal logic into a fuzzy modal logic in several ways (Fitting, 1991)):

1. changing the function $h(t, q)$ into a fuzzy function $h^{\sharp}(t, q) \in [0, 1]$ (obtaining real number values);
2. changing the crisp relation R into a fuzzy one $R^{\sharp} : S^2 \mapsto [0, 1]$;
3. making $a \in S$ fuzzy.
4. any combination of the above.

Is there a methodology involved to the above or do we just go from logic to logic and make fuzzy whatever semantical component we find?

What if we use a different semantics for **K** and make fuzzy the functions involved in that semantics? Do we get yet another batch of fuzzy modal logics?

In our particular example, let L_∞ be Lukasiewicz infinite valued logic (with values in $[0,1]$) and let us apply our fibring machinery to **K** and L_∞. We get the following:

1. The fibred semantics for $K(L_\infty)$ is the fuzzy semantics with h fuzzy, R crisp.
2. The fibred semantics for $L_\infty(K)$ is the semantics with R fuzzy and h crisp.
3. The semantics for $L_\infty(K(L_\infty))$ is the semantics where both R and h are fuzzy.

So in short, when you ask me how to make your logic L_1 fuzzy, I would answer—take a pure fuzzy logic L_2 (e.g. L_∞ or any other) and fibre it to L_1 in different ways.

Example 5.1 (Motivating fuzzy values) *We now give a concrete example of a fibred model of level 1. Figure 3 shows a* $\Box_1$ *Kripke model.*

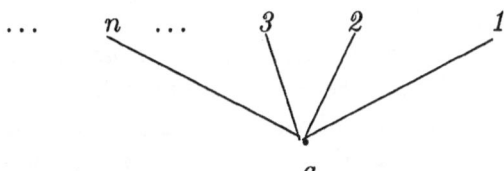

Figure 3.

Here $S = \{a\} \cup \{1, 2, 3, \ldots\}$ with aRn holding, for $n = 1, 2, \ldots$.

Assume $h(a, q) = 0$ and $h(n, q) = 1$ for $n = 1, 2, \ldots$. Try to evaluate $\Diamond_1 \Box_2 q$.

$a \vDash \Diamond_1 \Box_2 q$ iff for some n, $n \vDash \Box_2 q$. Since $\Box_2$ is in the L_2 language, we cannot continue to evaluate. We need an L_2 model to get a value at n. The fibring function $\mathbf{F}(n)$ gives an L_2 model (S_n, R_n, a_n, h_n). Let $S_m = \{a_m\} \cup \{(m, n) \mid n = 1, 2, 3, \ldots\}$. Let R_m be defined by

$$x R_m y \text{ iff} \begin{cases} x = a_m \\ or \\ x = (m, n_1) \text{ and } y = (m, n_2) \text{ and } n_2 \leq n_1 \end{cases}$$

and let

$$h_m(a_m, q) = 0$$

and

$$h_m((m,n), q) = 1 \text{ iff } m \leq n.$$

To complete the picture, let $\mathbf{F}(a) = \mathbf{F}(1)$. *Thus* $\Box_2 q$ *is false at* a_n *in all the models* $\mathbf{F}(n)$, *but we have*

$$(m, n) \vDash \Box_2 q \text{ iff } m \leq n.$$

This particular fibred model has a special feature which is important. All the models $\mathbf{F}(n)$, *have isomorphic frames; they are isomorphic to* $(T = \{0, 1, \frac{1}{2}, \frac{1}{3}, \ldots\}, \leq, 0)$, *through* π_m, *where* $\pi_m(a_m) = 0$, *and* $\pi_m(m, -n) = \frac{1}{n}$, *and they differ only in the assignment* $h_m(q)$. *The image of the truth set* $h_m(\Box_2 q) = \{y \mid y \vDash \Box_2 q\}$ *is projected on* $\{0, 1, \ldots, \frac{1}{n} \mid n = 1, 2, 3, \ldots\}$ *gets larger and larger as* m *increases. In the limit we have*

$$\bigcup_m \pi_m h_m(\Box_2 q) = \{\frac{1}{n} \mid n = 1, 2, \ldots\}.$$

Since we are interested in $a \vDash (\Diamond_1 \Box_2 q)$, *where the table for* $\Diamond_1$ *is existential, we can say that* $\Diamond_1$ *almost holds; it approaches the 'fuzzy' (or 'modal-*$\mathbf{L}_2$*') truth set* $\{\frac{1}{m} \mid m = 1, 2, \ldots\}$.

This is quite a conceptual jump. The model (S, R, a, h) *is a model of* $\mathbf{L}_1$ *and has no business getting set values from the set* T *via the mappings* π_m *of the models of* $\mathbf{L}_2$. *However, since all the fibred models* $\mathbf{F}(t), t \in S$ *are based on isomorphic frames, we can extend the evaluation from the fibred models back into the* $\mathbf{L}_1$ *language.*

It is important to note that the way we extended the evaluation from the fibred model to $\Diamond_1$ *of* $\mathbf{L}_1$ *was arbitrary. We chose a way of doing it which was reasonable, but nevertheless it was a choice. We could have said let us take as value for* $a \vDash \Diamond_1 \Box_2 q$, *not the union of* $\pi_m h_m(\Box_2 q)$ *but the maximum or some other reasonable definition.*

Having adopted a good definition, we now consider the expanded model $(S, R, a, h, \mathbf{F}, T, \pi)$.

We can define an $\mathbf{L}_2$*-fuzzy value* $\mu_t(A)$, *for* $t \in S$ *and any* A *as follows:*

- $\mu_t(A) = \pi_t h_t(A) = \{\pi_t(s) \mid s \in S_t \text{ and } s \vDash A\}$ *for* A *in* $\mathbf{L}_2$ *or* A *atomic.*
- $\mu_t(A \wedge B) = \mu_t(A) \cap \mu_t(B)$
- $\mu_t(\sim A) = T - \mu_t(A)$
- $\mu_t(\Diamond_1 A) = \bigcup_{\{s \mid tRs\}} \mu_s(A)$
- $\mu_t(\Box_1 A) = \bigcap_{\{s \mid tRs\}} \mu_s(A)$

What we have done can be best understood in algebraic terms. Let $\mathbb{B}$ be the Boolean algebra of the set $T = \{0, 1, \frac{1}{2}, \frac{1}{3}, \ldots\}$ with the interior operation $Q^{\Box_2}$, for $Q \subseteq T$ begin

$$Q^{\Box_2} = \{x \in T \mid \text{ for all } y \geqq x, y \in Q\}.$$

Assign to each atom q and $t \in S$ the 'fuzzy' algebraic subset $\mu_t(q) \subseteq T$. In our particular model we assign

$$\mu_n(q) = \{1, \tfrac{1}{2}, \ldots, \tfrac{1}{n}\}$$
$$\mu_a(q) = \mu_1(q)$$

We extend the assignment by

$-\ \mu_t(\Box_2 A) = (\mu_t(A))^{\Box_2}$
$-\ \mu_t(\Box_1 A) = \bigcap_{\{s \mid tRs\}} \mu_s(A)$

The next example brings the idea forward even more clearly.

Example 5.2 (Many valued modal logic) *This is an example of fibring semantical models (modal logic) with algebraic models (Lukasiewicz many-valued logic). We consider the modal language $\mathbf{L}_1$ with $\Box$ and the many valued language $\mathbf{L}_2$, with $\{\wedge, \vee, \rightarrow, \neg\}$ and with truth values at the real interval $[0, 1]$. We study $\mathbf{L}_1(\mathbf{L}_2)$. The algebraic models of $\mathbf{L}_2$ are linearly ordered Abelian groups which are embeddable in $[0, 1]$. So it is sufficient to consider assignments μ of values and truth table for values in $[0, 1]$. The following are the algebraic functions:*

$-$ *The domain is $[0, 1]$*
$-$ $\leq$ *is numerical $\leq$.*
$-$ $\top = \{0\}$ *(0 is truth).*
$-$ $\perp$ *is 1 (1 is falsity).*
$-$ $f_\wedge(x, y) = \ max\ (x, y)$.
$-$ $f_\vee(x, y) = \ min(x, y)$
$-$ $f_\neg(x) = 1 - x$
$-$ $f_\rightarrow(x, y) = \ max(0, y - x)$.

We now turn to fibring.

Let $\mathbf{m} = (S, R, a, h)$ be a Kripke model for $\Box$. The fibring function $\mathbf{F}$ associates with each $t \in S$ an algebraic model $\mathbf{a}_t = (A_t, \leq, f_\wedge, f_\vee, f_\rightarrow, f_\neg, \{0\}, \mu_t)$. Since $A_t = [0, 1]$, fibring algebras $\mathbf{a}_t$ to t is nothing more than associating with each t an arbitrary many-valued assignment μ_t to the atoms of the modal language.

Let us now evaluate $\Box(q \rightarrow p), q, p$ atomic, at the model $\mathbf{m}$.

$-$ $a \vDash \Box(q \rightarrow p)$ *iff for all $t \in S$ such that $aRt, t \vDash q \rightarrow p$.*

— *Since the main connective of $q \to p$ is many-valued, we have $t \vDash q \to p$ iff $\mathbf{a}_t \vDash q \to p$ iff $\mu_t(q \to p) = 0$ iff $max(0, \mu_t(p) - \mu_t(q)) = 0$ iff $\mu_t(p) \leq \mu_t(q)$.*

We would like to highlight a point which will be of importance later. Consider the above fibring. We start with $\mathbf{m} = (S, R, a, h)$. Then with each $t \in S$, we fibre an algebra $\mathbf{a}_t$. Since all the algebras have the same domain, the fibring reduces to μ_t, the assignment. Let us pause at this stage and consider the entity (S, R, a, h, μ) and let us try to evaluate $t \vDash \Box q$. Since $\Box q$ contains no many-valued connectives, we get $t \vDash \Box q$ holds iff $\forall s$ (tRs implies $s \vDash q$) iff $\forall s(tRs$ implies $h(t, q) = 1)$. Consider the wff $Iq = def (q \to q) \to q$. Really Iq is q but it is formally a many-valued wff. So we have to evaluate it at the algebra $\mathbf{a}_t$. We have $\mathbf{a}_t \vDash I(q)$ iff $\mu_t(q) = 0$. $t \vDash \Box I(q)$ iff for all $s(tRs$ implies $\mu_t(q) = 0)$.

To summarise, consider $t \vDash \Box q$; we have two ways of looking at it.

1. *Regard 'q' as an atom of the modal language, in which case*

$$t \vDash \Box q \text{ iff for all } S, tRs \text{ implies } h(s, q) = 1$$

2. *Regard 'q' as an atom of the many-valued language, in which case*

$$t \vDash \Box q \text{ iff for all } s, tRs \text{ implies } \mu_s(q) = 0.$$

The two evaluations need not give the same result.

We now have the opportunity to make $t \vDash \Box q$ fuzzy (i.e. 'fuzzle' the satisfaction $\vDash$, or in other words, 'fuzzle' the modal logic) by extending μ_t to $\Box q$:

($\natural$) $\mu_t(\Box q) = Sup_{\{s | tRs\}} \mu_s(q)$.

The reader should note that this definition is a chosen one and we could have chosen some other 'averaging' function.

Using ($\natural$) we can now fuzzle any wff of the modal logic and extend μ_t to all wffs, by taking the many-valued table for $\wedge, \vee, \neg$ and $\to$. We have thus by understandable intuitive definition, through ($\natural$), turned (S, R, a, μ) into a sort of modal many-valued logic by changing the crisp $\{0,1\}$ assignment h into a fuzzy μ. Note that what we are getting is not fibring, it is something new.

Example 5.3 (Persistence) *This example will fibre modal logic to the intermediate logic Dummett's* **LC**. *It will serve to prepare the ground for fibring in the presence of persistence. Let $\Rightarrow$ be intuitionistic implication.* **LC** *is the extension of intuitionistic logic with the axiom schema*

$$(p \Rightarrow q) \vee (q \Rightarrow p)$$

or if disjunction is not available, we can write an implicational axiom schema

$$(p \Rightarrow q) \Rightarrow (((q \Rightarrow p) \Rightarrow r) \Rightarrow r).$$

Let $\mathbf{L}_1$ be the language with $\{\Rightarrow, \wedge, \vee, \perp\}$ and let $\mathbf{L}_2$ be modal logic with $\square$. Consider the intuitionistic $\mathbf{LC}$ model with $U = [0,1]$ (unit real numbers interval) of the form $(U, \leq, 0, h)$. Since we are dealing with intuitionistic model, we must have persistence, i.e. for all atomic q and any $t, s \in U$.

(*) $t \leq s$ and $h(t, q) = 1$ imply $h(s, q) = 1$.

(**) We also require, for technical reasons, that for all q, $h(1, q) = 1$.

   Satisfaction is defined as follows:

 - $t \vDash A \wedge B$ iff $t \vDash A$ and $t \vDash B$
 - $t \vDash A \vee B$ iff $t \vDash A$ or $t \vDash B$
 - $t \vDash A \Rightarrow B$ iff $\forall s(t \leq s \wedge s \vDash A$ imply $s \vDash B)$.
 - $t \vDash \perp$ iff $t = 1$

   The reader familiar with **t**-conorms can view the above as follows: For each atomic q let

$$\mu(q) = \ Inf \ \{t \mid h(t, q) = 1\}$$

We have (because of persistence) that μ can be extended to all wffs as follows:

 - $\mu(A \wedge B) = \ max(\mu(A), \mu(B))$
 - $\mu(A \vee B) = \ min(\mu(A), \mu(B))$
 - $\mu(A \rightarrow B) = \ Inf \ \{t \mid \ max \ (t, \mu(A)) \geq \mu(B)\}$

   For each $t \in U$, let $\mathbf{F}(t) = (S_t, R_t, a_t, h_t)$ be a modal model of $\square$. Note that $\lozenge$ is not intuitionistically definable from $\square$ and so we have to explicitly include $\lozenge$ if we want. Here we assume we have $\square$ only.

   By general fibring principles, we must have persistence for modal formulas as well, for example, for $\square^k A$.

[*] $\mathbf{F}(t) \vDash \square^k A$ and $t \leq s$ imply $\mathbf{F}(s) \vDash \square^k A$

   This means that

$$t \leq s \rightarrow [\forall y[a_t R_t^k y \rightarrow y \vDash A] \rightarrow \forall y[a_s R_s^k y \rightarrow y \vDash A]]$$

   It is possible to show that we can assume without loss of generality (i.e. without changing the semantic consequence relation) that:

(†) $t \leq s \wedge x R_s y \rightarrow x R_t y.$[6]

---

[6]This condition is for $\square$. For $\lozenge$ we need $t \leq s \wedge x R_t y \rightarrow x R_s y$.

In fact if we let $S = \bigcup_t S_t$ we can assume that the fibred models are

$$\mathbf{F}(t) = (S, R_t, a_t, h_t).$$

We are going to assume the following additional properties: $a_t = a$ for some fixed a and $t \leq s$ and $h_t(x, q) = 1$ imply $h_s(x, q) = 1$ for all $x \in S$ and atomic q. We believe one can show that such assumptions can be made without loss of generality.

So the models differ only in their accessibility relation R_t which satisfies (†) above, and the assignment h_t.

Define functions $h^\sharp(x, q) \in U, q$ atomic, $x \in S$ and $R^\sharp : S^2 \mapsto U$ by letting

$$h^\sharp(x, q) = Inf\, \{t \mid h_t(x, q) = 1\}.$$
$$R^\sharp(x, y) = Sup\, \{t \mid xR_t y\}.$$

(Let us assume the Sup is attained.)

Consider the system $(U, \leq, 0, \mu, S, R^\sharp, a, h^\sharp)$. We can view this system in two ways:

1. *An **LC** model $(U, \leq, 0, \mu)$ with a fibring of modal models (S, R_t, a, h_t), where $xR_t y$ holds iff $R^\sharp(x, y) \geq t$, and $h_t(x, q) = 1$ iff $t \geq h^\sharp(x, q)$.*
2. *A fuzzy model $(S, R^\sharp, a, h^\sharp)$ where the accessibility relation $R^\sharp$ and the assignment $h^\sharp$ are fuzzy and where the fuzzy truth set is $(U, \leq, 0, \mu)$ and evaluation is done using the **t-conorm** max, as indicated above.[7]*

Let us explore further the fuzzy model $(S, R^\sharp, a, h^\sharp)$. Consider, for $x \in S$, the statement $x \vDash_t A$, i.e. $x \vDash A$ in the model $\mathbf{F}(t)$. Because of persistence, we can define

$$\mu^\sharp(x, A) = Inf\, \{t \mid x \vDash_t A\}.$$

Consider $\mu^\sharp(x, \Box A)$

$$\begin{aligned} \mu^\sharp(x, \Box A) &= Inf\{t \mid x \vDash_t \Box A\} \\ &= Inf\, \{t \mid \forall y(xR_t y \text{ implies } y \vDash_t A)\} \end{aligned}$$

but $xR_t y$ holds iff $t \leq R^\sharp(x, y)$ and $y \vDash_t A$ holds iff $\mu^\sharp(y, A) \leq t$.
Hence

$$\mu^\sharp(x, \Box A) = Inf\, \{t \mid \forall y(t \leq R^\sharp(x, y) \text{ implies } \mu^\sharp(y, A) \leq t)\}$$

---

[7]If we choose a different t-conorm, say

$$\begin{aligned} \mu(A \wedge B) &= min\,(1, \mu(A) + \mu(B)) \\ \mu(A \to B) &= max\,(0, \mu(B) - \mu(A)) \\ \mu(A \vee B) &= max\,(0, \mu(A) + \mu(B) - 1) \end{aligned}$$

we get evaluation which makes the accessibility relation Lukasiewicz fuzzy.

The previous two examples show that modal and many valued logic can be put together in two different ways. If we start with a modal model (S, R, a, h) then we can fuzzle (make fuzzy) h by changing it into a many valued assignment μ and extend to the entire modal language. If we start with a many valued model μ then we can fuzzle μ by changing it into a function into elements of a modal algebra. This turned out to be equivalent to looking at modal models where the possible world relation is fuzzy but the assignment is crisp. I.e. models of the form $(S, R^{\sharp}, a, \mu)$ where $R^{\sharp}(x, y) \in [0, 1]$, while μ is a $\{0, 1\}$ assignment. μ can be extended to all wffs, in which case it becomes a $[0, 1]$ valued function.

The obvious combination of the two approaches is to make both $R^{\sharp}$ and $\mu^{\sharp}$ fuzzy. This leads us to the following definition.

Definition 5.4 *An algebraic fuzzled many valued modal model has the form $(S, R^{\sharp}, a, \mu^{\sharp})$, where $R^{\sharp}; S^2 \mapsto [0, 1]$ is a fuzzy possible world relation and for each $s \in S$ and atomic q, $\mu^{\sharp}(q) \in [0, 1]$.*

$\mu^{\sharp}_s$ *can be extended to arbitrary formulas as follows:*

$$\mu^{\sharp}_s(A * B) = f_*(\mu^{\sharp}_s(A), \mu^{\sharp}_s(B))$$

where $* \in \{\wedge, \vee, \rightarrow, \neg\}$ *and f_* is the many valued truth table for $*$.*

$$\mu^{\sharp}_x(\Box A) = \mathrm{Inf}_t[\text{for all } y, R^{\sharp}(x, y) \geq x \text{ implies } \mu^{\sharp}_y(A) \leq t].$$

Summary 5.5 *We summarise the ideas of this section.*
 - *Making fuzzy is identical with fibring in a special way.*
 - *Any logic $\mathbf{L}_1$ can be 'made fuzzy' by fibring it with $\mathbf{L}_2$ as $\mathbf{L}_1(\mathbf{L}_2)$.*
 - *If $\mathbf{L}_1$ is the Lukasiewicz infinite valued logic and $\mathbf{L}_2$ is modal logic then $\mathbf{L}_1(\mathbf{L}_2)$ can be understood as modal logic with fuzzy accessibility but crisp assignment to atoms while $\mathbf{L}_2(\mathbf{L}_1)$ is modal logic with fuzzy assignment to atoms (but crisp accessibility).*
 In case of either $\mathbf{L}_1(\mathbf{L}_2(\mathbf{L}_1))$ or $\mathbf{L}_2(\mathbf{L}_1(\mathbf{L}_2))$ we get fuzzy accessibility and fuzzy assignment. Furthermore, any further iterations of the form $\mathbf{L}_1(\mathbf{L}_2(\mathbf{L}_2(\mathbf{L}_2\ldots)\ldots))$ can be given semantics which is comprised of fuzzy SFM models, (where both the accessibility and assignment are meny valued) of Definition 2.3 for the case of one modality and one jump operator.

More details are in the full paper (Gabbay, 1995b).

6. Self Fibring for Predicate Logics

Fibred semantics can be used to give meaning to expressions of the form $\varphi(t, A)$ or $P(A)$, where $\varphi(x, y)$ is a formula with x, y free and $P(x)$ is a unary predicate and A is a formula and t is a term. Such logical expressions arise extensively in logic programming, logical models of natural language and metalevel considerations. For example, 'John believes A', 'John said that A', '*Demo* (A_1, A_2)', '*Provable* (x, A)', '*Hold* (t, A)' and the like. See (Gabbay, 1994).

The basic idea of our approach for classical logic can easily be explained.

Suppose we start with an ordinary model **m** classical logic. This has a domain D. Let $P(x)$ be a unary predicate of the language. Then P is assigned a subset of D. See Figure 4.

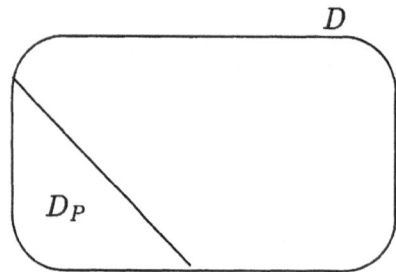

Figure 4.

For any element $a \in D$ we can evaluate the 'expression' $P(a)$. The evaluation gives us a truth value $\top$ or $\bot$. If $a \in D_P$ the value is $\top$. If $a \notin D_P$ the value is $\bot$. Now consider the expression $P(\sim P(a))$ or $P(c)$ where c is a non-denoting singular term.

We do not know how to evaluate it because the expression $y =\sim P(a)$, or $y = c$ inside the predicate is not assigned, (in our model theory), an element of the domain. However, we do notice that all we need to have is an answer to the question: is the truth value of $P(y)$ $\top$, or is it $\bot$? To get an answer we use t he 'fibring' methodology. We need a way of associating a value to the two 'parameters', P and y. We associate with D_P another model $\mathbf{F}(D_P)$, of the appropriate kind, which can help us get a value. The function $\mathbf{F}$ giving us the model is called the *fibring function*.

- For the case of a constant $y = c$, it is sufficient that the fibring function BF assigns a new model $\mathbf{n} = \mathbf{F}(D_P)$, for the language with c based on the do main $D' \supseteq D$. $\mathbf{n}$ of course assigns a value for c in D'. Thus $\mathbf{n} \models P(c)$ can be evaluated and so we can let
  $\mathbf{m} \models P(c)$ iff $\mathbf{F}(D_P) \models P(c)$.
- In the case of $y =\sim P(a)$, the model $\mathbf{n} = \mathbf{F}(D_P)$ is taken as another classical model, in which the expression y can be evaluated, because y

is a formula. So we can let

$\mathbf{m} \vDash P(\sim P(a))$ iff $\mathbf{F}(D_P) \vDash \sim P(a)$.

The reader should note the strong connection between the two cases, $y = c$ and y a formula. Indeed the connection is with Free logic. Free logic systems allow for non-denoting terms such as Pegasus, thus there are constants y in the language where $P(y)$ gets a truth value but y does not get assigned an element of the domain. For this kind of logic the rule $\forall x P(x) \rightarrow P(y)$ does hold but $P(y) \rightarrow \exists x P(x)$ does not hold. Depending on the system of free logic we might even not have $\forall x \forall y$ and $\forall y \forall x$ not being the same. See Bencivenga (Bencivenga, 1986), and also see (Lambert & Fraassen, 1972).

Our fibred semantics can also be of service to free logic, because we have the domain D of existing elements and also the non-existing 'elements' y, being the formulas of the language, or some other non-denoting constants.

The reader should note and observe that our method is not restricted to classical logic. We can equally study formulas of the $P(c), c$ non-denoting, or of the form $A(t, \varphi(s), f(B))$ in intuitionistic logic. The same approach applies. See my paper (Gabbay, 1995c) for full details.

We start with fibred semantics for self reference in monadic classical logic without function symbols and then gradually make the language more complicated. First add unary function symbols, then binary function symbols and then binary relations. In each case we develop a suitable fibred semantics and explore its properties. Further results can be found in a forthcoming paper, (Gabbay, 1995c).

We begin with our definitions and results.

Definition 6.1 (Fibred monadic syntax) *Let L be a language with monadic predicates, the classical connectives and quantifiers. We allow for two kinds of universal quantifiers, $(\forall x)$ and (x). $(\forall x)$ is intended as quantifying over terms and formulas and (x) quantifiers only over terms (i.e. it is a sort of 'free logic' quantifier). We can define $(\exists x) = \sim (\forall x) \sim$ and $(Ex) = \sim (x) \sim$.*

We have no function symbols, but we do allow individual constants. We define the notion of a fibred wff (f-wff) of the language as follows

1. *A is a f-wff of level 0 if A is an ordinary free classical wff of L (built up with $\forall x$ and (x)).*

2. *A is a f-wff of level $n + 1$ if there exists a free classical wff $\varphi(x_1, \ldots, x_n, y_1, \ldots, y_m)$ of L, with free variables $x_1, \ldots, x_n, y_1, \ldots, y_m$ and f-wffs $A_1, \ldots, A_n$ of level $\leq n$ such that $A = \varphi(x_1/A_1, \ldots, x_n/A_n, y_1, \ldots, y_m)$, where x_i/A_i denotes the simultaneous substitution of A_i for x_i in φ. The free variables of A are $y_1, \ldots, y_m$ together with all the*

*free variables of $A_1, \ldots, A_n$. We assume the substitution is carefully
made and does not result in incorrect binding of variables.*

The reader will know from context whether we assume **L** *to contain only
$(\forall x)$ or only (x) or both.*

Definition 6.2 (Fibred monadic models)   *1. A classical (free logic)
model for the language* **L** *has the form* $\mathbf{m} = (D, D', h, g)$ *where* $D \neq \varnothing$
and $D \cap D' = \varnothing$. D *is the domain of existing elements ((x) ranges over
D) and D' is the domain of non-existing elements ($(\forall x)$ ranges over
$D^* = D \cup D'$). h is an assignment function assigning to each predicate
P a subset $h(P) \subseteq D^*$. g is a function assigning to each variable x
a value $g(x) \in D$ and to each constant c value $g(c) \in D^*$ (thus $g(x)$
always exists in D but $g(c)$ may not exist and be in D'.*
  In our fibred semantics D' is intended to be the set of all wffs.
 *2. Traditional classical models are obtained by letting $D' = \varnothing$. In such a
  case we present the model as (D, h, g).*
  *A fibred model is obtained by letting $D' =$ set of all closed f-wffs of
  the language. In fact D' can be formally taken as all f-wffs—with free
  variables. The assignment g assigns elements of D to the variables
  which will fix the domain D' to be all wffs of the language based on the
  ' constants' $\{g(x_1), g(x_2), \ldots\}$. With this understanding we can define
  satisfaction for self fibred model later on.*
 *3. Satisfaction for classical models is defined using the mixed notation
  $\varphi(x)[x/\alpha]$ to mean the formula $\varphi(x)$ with the intended value α for x,
  where α is any value element of D^* (or as we shall see in the case of
  fibred models indeed a formula of the language).*
  With the above notation we let

  — $\mathbf{m} \vDash P(x)[x/d]$*, for* $d \in D^*$ *iff* $d \in g(P)$
  — $\mathbf{m} \vDash P(t)$ *iff* $g(t) \in h(P)$*.*
    The above can also be written as

$$\mathbf{m} \vDash P(t)[t/g(t)].$$

  — $\mathbf{m} \vDash (x)\varphi(x)$ *iff* $\mathbf{m} \vDash \varphi(x)[x/d]$ *for all* $d \in D$
  — $\mathbf{m} \vDash \forall x \varphi(x)$ *iff* $\mathbf{m} \vDash \varphi(x)[x/d]$*, for all* $d \in D^*$*.*
    *In the self fibred model, D' is the set of all f-wffs and so $\forall x$
    quantifies over real elements and formulas while (x) quantifies
    over real elements only.*

  — $\mathbf{m} \vDash t_1 = t_2$ *iff* $g(t_1) = g(t_2)$*.*
 *4. A fibred model has the form $\mathbf{n}_0 = (D, h_0, g, \mathbf{F})$, where (D, h_0, g) is a
  classical model and $\mathbf{F}$ is a function giving for each assignment h and
  each subset $X \subseteq D$ a new assignment $h' = \mathbf{F}(X, h)$.*

5. *A relation $\vDash$ is said to be a satisfaction relation between f-wffs A and fibred models $\mathbf{n} = (D, h, g, \mathbf{F})$ if the following holds:*

  - $\mathbf{n} \vDash P(t)$ iff $g(t) \in h(P)$, for t variable or constant and P atomic.
  - $\mathbf{n} \vDash P(x)[x/d], d \in D$ iff $d \in h(P)$.
  - $\mathbf{n} \vDash \forall x A(x)$ iff for all $\alpha \in D$ and all wffs $\alpha, \mathbf{n} \vDash A(x)[x/\alpha]$.
  - $\mathbf{n} \vDash (x)A(x)$ iff for all $\alpha \in D, \mathbf{n} \vDash A(x)[x/\alpha]$.
    Note that we obtain D' as the set of all f-wffs of the language.
  - $\mathbf{n} \vDash P(\varphi)$ iff $(D, h', g, \mathbf{F}) \vDash \varphi$ for $h' = \mathbf{F}(h(P), h)$.
  - $\mathbf{n} \vDash (A \wedge B)[x/\alpha]$ iff $\mathbf{n} \vDash A[x/\alpha]$ and $\mathbf{n} \vDash B[x/\alpha]$
  - $\mathbf{n} \vDash (\sim A)[x/\alpha]$ iff $\mathbf{n} \nvDash A[x/\alpha]$.

6. *The function $\mathbf{F}$ need not be defined for all subsets X of D. It is sufficient to have it defined for all subsets of the form $h(P)$, for atomic P of the language. This would make the model more effectively computable. Also note that we can relinquish the requirement that $\mathbf{F}$ is a function and let $\mathbf{F}$ be a relation. Thus $\mathbf{F}(X, h)$ would be a set of assignments and the truth condition for $P(\varphi)$ would be:*

  - $\mathbf{n} \vDash P(\varphi)$ iff $(D, h', g, \mathbf{F}) \vDash \varphi$ for all $h' \in \mathbf{F}(h(P), h)$.

This version of $\mathbf{F}$ is of particular interest for the interpretation of $P(\varphi)$ as a modality on φ. We shall discuss this later in the section.

Note that the above definition makes $\forall x P(x)$ range over all $x \in D$ and over all wffs φ as well.

Also note that for the case of $\forall$ the above definition of $\vDash$ is an implicit and not a recursive definition. Consider $\mathbf{m} \vDash \forall x P(x)$. We have to evaluate $\mathbf{m} \vDash P(\forall x P(x))$ and move to a new model $\mathbf{n} \vDash \forall x P(x)$. The variable x is instantiated also to the value $\forall x P(x)$. This causes an infinite recursive process. If we only have the quantifier (x) which ranges over $x \in D$, then $\vDash$ is recursively defined.

Further note that a more sophisticated satisfaction clause for $\mathbf{n} \vDash P(\varphi)$ can be the following

•• $\mathbf{n} \vDash P(\varphi)$ iff $(D, h', g, \mathbf{F}) \vDash \psi_P(\varphi)$ where $h' = \mathbf{F}(h(P), h)$ and $\psi_P(Q)$ is a classical wff involving a predicate Q and $\psi_P(\varphi)$ is the result of substituting φ for Q.

The previous satisfaction clause is obtained by letting $\psi_P(Q) = Q$.

The definition of the fibring function $\mathbf{F}$ makes it dependent on both assignments h and subsets of D. The dependence on h makes a difference. If $\mathbf{F}$ were independent of h and dependent only on subsets the following f-wff would be valid.

  - $(x)[P_1(x) \leftrightarrow Q(P_2(x))] \to [P_1(\varphi) \leftrightarrow Q(P_2(\varphi))]$

The f-wff says that if P_1 in $\mathbf{m}$ has the same extension as P_2 in $\mathbf{m}_Q$ then $\mathbf{m} \vDash P_1(\varphi)$ iff $\mathbf{m}_Q \vDash P_2(\varphi)$, for arbitrary φ.

Theorem 6.3 (Axiomatisation of the system with ∀ only)

Consider the logic with all classical axioms and rules together with the following additional axioms and rules:

- $P(\forall x A(x)) \leftrightarrow \forall x P(A(x))$
- $P(A \wedge B) \leftrightarrow P(A) \wedge P(B)$
- $\vdash A$ *implies* $\vdash P(A)$
- $\sim P(A) \leftrightarrow P(\sim A)$.

Then the logic is complete for satisfaction in fibred models where the universal quantifier ∀ ranges over all elements in the domain and all wffs as well.[8]

If we drop the last axiom ($\sim P(A) \leftrightarrow P(\sim A)$) we get completeness for the fibring of set function $\mathbf{F}$.

Proof. See (Gabbay, 1995c). ∎

Remark 6.4 (Connection with modal logic) *We already say that $P(\varphi)$ can be viewed as a modal operator $\Box_P \varphi$. This is reminiscent of dynamic logic where there are modalities associated with regular expressions σ of the form $\Box_\sigma \varphi$ and there are operations on the expressions such as*

$$(\sigma_1, \sigma_2) \mapsto \sigma_1 * \sigma_2$$

and various connected axioms such as

$$\Box_{\sigma_1} \Box_{\sigma_2} \varphi \leftrightarrow \Box_{\sigma_1 * \sigma_2} \varphi.$$

In our case the connections between $\Box_{P_1}, \Box_{P_2}$ etc. are done through a full quantified theory of monadic predicate logic. Furthermore the domain of the modal model is the same as the domain used to connect the modal operators. This is like doing dynamic logic with expressions of the form $\Box_{\sigma(x)} \varphi(x, y)$ with x appearing both in σ and in φ. In fact in our case we can get expressions like $\Box_{B(x,y)}(\varphi_1(x, y), \varphi_2(x, y))$; any complex formula $B(x, y)$ can be a sort of modality $B(\varphi_1, \varphi_2)$.

Let us see what we can express in this language: consider the well known irreflexivity rule in modal logic.

$$\frac{\vdash \sim q \wedge \Box q \to \varphi}{\vdash \varphi}$$

---

[8]Notice the similarity of the axioms with that of a modal operator. In this case the modal operator is a *next* operator. The axioms without $\sim P(A) \leftrightarrow P(\sim A)$ define a **K** modality and correspond to a fibring **F** which is a set function.

provided q is not in φ,
    This translates into

$$\forall \alpha [\sim (Ey(y = \alpha)] \wedge \sim \alpha \wedge P(\alpha) \rightarrow \varphi)] \rightarrow \varphi$$

$Ey(y = \alpha)$ is $\sim (y) \sim (y = \alpha)$ and it means α is a formula variable.

Remark 6.5 (Fibred binary relations) *Let us indicate how we might deal with the language of binary relations. We try to give fibred semantics for f-wffs of the form $R(A, B)$, where $R(x, y)$ is atomic binary predicate and A, B wffs. There are two main ways of dealing with $R(A, B)$.*

1. *Following the previous discussion, where we had a unary predicate $P(A)$ and a formula $\psi_P(Q)$, with $(D, h, \mathbf{F}) \models P(A)$ iff $(D, h', \mathbf{F}) \models \psi_P(A)$, we can define*

$$(D, h, \mathbf{F}) \models R(A, B) \text{ iff } (D, h', \mathbf{F}) \models \psi_R(A, B)$$

   where here $\mathbf{F}$ fibred a new model with each binary relation, i.e.

$$h' = \mathbf{F}(h(R), h), \text{ and } \psi_R(Q_1, Q_2)$$

   is a formula with two unary predicates Q_1, Q_2.

2. *We can adopt a more clever approach. Consider $R(x, t)$. For fixed t, we get a unary predicate in x. Let us use the notation $\lambda x R_t(x)$ for this predicate. Given a wff A, we can semantically evaluate $R_t(A)$ in the unary fibred models of the previous discussion.*

$$\mathbf{m} \models R_t(A) \text{ iff } \mathbf{F}(\lambda x R_t(x), h) \models A.$$

We can now consider $R_t(A)$, for A fixed, as a predicate in t, i.e. consider $\{t \mid \mathbf{m} \models R_t(A)\}$ for this predicate. Hence, for a wff B, we can evaluate $R_B(A)$

$$\mathbf{m} \models_1 R_B(A \mid \text{ iff } \mathbf{F}(\{t \mid \mathbf{m} \models R_t(A), h) \models B.$$

    Summarising, we get

$$\mathbf{m} \models_1 R(A, B) \text{ iff } \mathbf{F}(\{t \mid \mathbf{F}(\lambda x R(x, t), h) \models A\}, h) \models B.$$

We obtained $\models_1$ by looking at $\lambda x R(x, t)$ first. If we were to start with $\lambda t R(x, t)$ as our initial parameterised unary predicate, we get

$$\mathbf{m} \models_2 R(A, B) \text{ iff } \mathbf{F}(\{x \mid \mathbf{F}(\lambda t R(x, t), h) \models B\}, h) \models A.$$

The two definitions are not the same. $R(A, B)$ is syntactically symmetrical and we do not know whether A was substituted first and then B or the other way round, or perhaps both were substituted simultaneously.

We leave the investigation of these options to the full paper (Gabbay, 1995c).

We now turn to the case where function symbols are available in the language. We assume *unary* function symbols $e_i(x), i = 1, 2, 3, \ldots$ as well as unary predicates. Let $e(x)$ be a term. It has the form $e_1^{k_1}(e_2^{k_2} \ldots (e_n^{k_n}(x)) \ldots)$. The atomic expressions have the form $P(e(x))$. We thus have to give semantics to expressions of the form $e(\varphi)$ and $P(e(\varphi))$.

The idea is to treat $P(e(\varphi))$ as $P_1(\varphi)$ where $P_1(x)$ is defined as $P(e(x))$. We therefore have to let $e(\varphi)$ to be the formula $\varphi_1 = e(\varphi)$ where for any P

$$P(\varphi_1) = P(e(\varphi)) = P_1(\varphi).$$

Semantically we need to have

$$\mathbf{m} \vDash P(\varphi_1) \text{ iff } \mathbf{m}_P \vDash \varphi_1$$

which equals

$$\mathbf{m}_P \vDash e(\varphi) \text{ iff } \mathbf{m}_{e(P)} \vDash \varphi,$$

where e is now regarded as a function from predicates to predicates

$$e(P)(x) = P(e(x)).$$

Equivalently, for $X \subseteq D$, let:

$$e^{-1}(X) = \{y \mid e(y) \in X\}.$$

Definition 6.6 (Fibred syntax with unary functions) *Let* **L** *be the classical language of Definition 6.1. With additional unary function symbols* $\{e_1, e_2, e_3, \ldots\}$. *The notion of a well formed formula of level 0 is modified accordingly.*

The notion of f-wff of level $n + 1$ is defined similarly with the added clause that if A is of level n and $e(x)$ is a term then $e(A)$ is a wff of level $n + 1$.

Definition 6.7 (Semantics for fibred unary functions) *We* mo- *dify Definition 6.2. Since our language contains function symbols, the assignment g must assign to each function symbol e a function $g(e) : D^* \mapsto D^*$. We require that $g(e)$ takes D into D and D' into D'. With each set $X^* \subseteq D^*$ and each $g(e)$, we let $g(e)^{-1}(X^*) = \{y \mid g(e)(y) \in X^*\}$.*

Notice that all terms are rigid designators!

Satisfaction is defined inductively as before, with the following additional inductive clause for function symbol e.

- $(D, h, g, \mathbf{F}) \vDash e(\varphi)$ *iff for all $Y \subseteq D$ and all h_1 such that $h = \mathbf{F}(Y, h_1)$ we have that $(D, h', g, \mathbf{F}) \vDash \varphi$ where $h' = \mathbf{F}(g(e)^{-1}Y, h_1)$.*

The meaning of the clause is as follows:

We want to evaluate $h \vDash e(\varphi)$. We essentially look at all h_1 and Q such that $h = \mathbf{F}(h(Q), h_1)$ and evaluate $h_1 \vDash Q(e(\varphi))$.

We stop our discussions here. For more see the full paper (Gabbay, 1995c).

7. Conclusion and Discussion

We saw that the idea of fibred semantics is basically very simple. If we encounter a symbol of a foreign language while evaluating an expression in our own language, then we regard this symbol as atomic (in our language) and turn to the semantics of the foreign language to get a value for it.

The above methodology, when iterated, gives fibred semantics for the combined language. Once we have this fibred semantics, there are some natural general questions to ask about the combined system:

1. How do the properties of the combined system depend on the properties of the components (transfer theorems)?
2. What is the landscape of possible interaction axioms and their corresponding conditions on the fibred semantics?
3. Can the fibring process be mechanised?
4. Can we reformulate existing known systems as fibred systems?
5. Extend our methodology to quantified systems.
6. Check interpolation theorems for the combined language.

There are also specific problems to ask for each of the application areas. Here we list some currently active research problems.

Case of modal and intuitionistic logics
We saw that for this case decidability and finite model property transfer to the combined system. We can also axiomatise the combined system in terms of the axiomatisations of the components. What remains to be done is to investigate the landscape of natural interaction axioms and their corresponding semantic conditions. For example, when dovetailing two modalities $\Box_1$ and $\Box_2$, where both are extensions of modal $\mathbf{K}$, the basic models have the form (S, R_1, R_2, a, h). The interaction axiom $\Box_1 A \to \Box_2 A$ corresponds to the condition $R_2 \subseteq R_1$. This is easy to show. It is more difficult to find what condition corresponds to the commutativity axiom: $\Box_1 \Box_2 = \Box_2 \Box_1$. It is not clear whether we have completeness for models of the form $(S_1 \times S_2, R_1, R_2, (a_1, a_2))$ where R_i is a relation on S_i and where we have:

$$(x_1, x_2) R_1 (y_1, y_2) \text{ iff } x_2 = y_2 \wedge x_1 R_1 y_1$$
$$(x_1, x_2) R_2 (y_1, y_2) \text{ iff } x_1 y_1 \wedge x_2 R_2 y_2.$$

In general, charting the landscape for two modalities means looking at some natural interaction axioms and their corresponding semantic conditions.

Case of Fuzzy Logics
In this case the main immediate research problem is to axiomatise the combined system in terms of the axiomatisation of its components.

Case of Substructural Logics
We need to analyse the various ways of adding modality to a substructural implication and compare with the current systems (linear, relevant, etc.) in the literature. We should try to explain some existing peculiarities in terms of fibring. See (D'Agostino et al., 1995; D'Agostino & Gabbay, 1996; Dörre et al., to appear).

Case of Self Fibring of Predicate Logics
There is a lot to be done here:

- Apply the fibred semantics idea to intuitionistic logic. This should be straightforward.
- Check applications of the semantics to theories of self reference.
- Give fibred semantics to free logics and definite descriptions, and compare with existing solutions. Extend the method to intuitionistic logic.
- Investigate transfer theorems.

Case of the Conditional
The connection between the metalevel non-monotonic consequence relation and the object level subjunctive conditional are well-known. We know how to bring the consequence relation into the object level as a conditional, by using fibred semantics. We need to show systematically that the fibred semantics of the combined system is essentially the same as the traditional semantics of the conditional.

Case of Combining Metalevel with Object Level
Chart a methodology of bringing metalevel features into the object level in any reasonable system.

More General Fibring Problems
An alternative approach to combining logics and structures is in (Blackburn & de Rijke, to appear); the theory there evolves around the idea of explicitly controlling the communication between the structures (and logics) being combined through the use of modal operators to refer to the 'communication channels' between structures. The approach has clear links with out jump operator, but the precise connection remains to be determined.

For a discussion of combining general algebraic systems see (Baader & Schulz, 1995a; Baader & Schulz, 1995; Kepser & Schulz, 1996).

Also there is a connection with labelled deductive systems (Gabbay, 1995). It looks like the most natural environment for performing fibring and formulating the fibred semantics and proof system.

For topological fibring see (Pfalzgraf, 1987; Pfalzgraf, 1991; Pfalzgraf & Stokkermans, 1994).

D. M. GABBAY
Department of Computing
Imperial College
London, Great Britain

References

G. Amati and F. Pirri. Uniform tableaux methods for intuitionistic modal logic I, to appear *Studia Logica*.

A. R. Anderson and N. D. Belnap. *Entailment*, vol. 1. Princeton University Press, 1975.

F. Baader and K. Schulz. On the combination of symbolic constraints, solution domains and constraint solvers. In *Proc. CP-95*, 1995, Lecture notes in Computer Science, 976, pp. 380 – 397, Springer Verlag 1995.

F. Baader and K. Schulz. Combination of Constraints Solving Techniques: An Algebraic Point of View. Research report CIS-Rep-94-75, University of Munich, 1994. A short version has appeared in *Proc. RTA-95*, Lecture Notes in Computer Science 914, pp. 352–366, Springer-Verlag, 1995.

F. Baader and K. Schulz. Combination of constraint solvers for free and quasi-free structures. Research Report, CIS-Rep-95-120, University of Munich, 1995.

E. Bencivenga. Free logic. In D. M. Gabbay and F. Guenthner, eds, *Handbook of Philosophic al Logic, vol. 3*, Kluwer, 1986.

P. Blackburn and M. de Rijke. Why combine logics? To appear.

P. Blackburn and M. de Rijke. Zooming in, zooming out. *Journal of Logic, Language and Information*, to appear.

M. Božić and K. Došen. Models for normal intuitionistic modal logics. *Studia logica*, **43**, 217–245, 1984.

R. A. Bull. A modal extension of intuitionist logic. *Notre Dame Journal of Formal logic*, **6**, 1965.

M. D'Agostino, D. M. Gabbay and A. Russo. Grafting modality into substructural logics. Draft, Imperial College, 1995.

M. D'Agostino and D. M. Gabbay. Fibring labelled tableaux for substructrual logics. To appear Tableaux 96 conference.

J. P. Delgrande. An approach to default reasoning based on a first order conditional logic. *Artificial Intelligence*, **36** 63–90, 1988.

J. Dörre, D. Gabbay and E. König. Fibred semantics for feature based grammar logic. To appear in *Journal of Logic, Language and Information*.

K. Došen. Models for stronger normal intuitionistic modal logics. *Studia Logica*, **44**, 39–70, 1985.

A. E. Eiben, A. Jánossy and A. Kurucz. *Combining Logics*, draft presented at *Logic at Work*, Amsterdam, December 1992.

W. B. Ewald. Intuitionistic tense and modal logic. *JSL*,**51**, 166–179, 1986.

R. Fagin, J. Y. Halpern and M. Y. Vardi. A nonstandard approach to the logical omniscience problem in *Reasoning About Knowledge, TARK 1990*, R. Parikh, ed., pp. 41–55, Morgan Kaufmann, 1990.

L. F. del Cerro and A. Herzig. Combining classical and intuitionistic logic. This volume.

K. Fine and G . Schurz. Transfer theorems for stratified multimodal logics. To appear.

M. Finger and D. M. Gabbay. Adding a temporal dimension to a logic. *Journal of Logic, Language and Information*, 1, 203–233, 1992.

M. Finger and D. Gabbay. Combining temporal logic systems. To appear in *Notre Dame*

Journal of Formal Logic.

G. Fischer-Servi. On modal logic with an intuitionistic base. *Studia Logica*, **36**, 1977.

G. Fischer-Servi. Semantics for a class of intuitionistic modal calculi. In *Italian Studies in the Philosophy of Science*, M. L. Dalla Chiara, ed., pp. 59–71, D. Reidel, 1980.

G. Fischer-Servi. Axiomatizations for some intuitionistic modal logics. *Rend. Sem. Mat. Univ. Politecn. Torino*, **42**, 179–194, 1984.

F. B. Fitch. Intuitionistic modal logic with quantifiers. *Portugalia Mathematica*, **7**, 113–118, 1948.

M. Fitting. Logics with several modal operators. *Theoria*, **35**, 259–266, 1969.

M. Fitting. Many valued modal logics. *Fundamenta Informatica*, **15**, 235–254, 1991.

M. Fitting. Many valued modal logics II. *Fundamenta Informatica*, **17**, 55–73, 1992.

M. Fitting. Tableaux for many valued modal logic. Report Jan 25, 1994.

J. M. Font. Modality and possibility in some intuitionistic modal logic. *Notre Dame Journal of Formal Logic*, **27**, 533–546, 1986.

D. M. Gabbay. Theoretical foundations for non-monotonic reasoning in expert systems. In *Proceedings NATO Advanced Study Institute on Logics and Models of Concurrent Systems*, (ed. K. R. Apt), pp. 439–457. Springer-Verlag, Berlin, 1985.

D. M. Gabbay. *Fibred semantics and the weaving of logics, Part 1*, Lectures given at *Logic Colloquium 1992*, Veszprém, Hungary, August 1992. A version of the notes is published as a Technical Report No 36, by the University of Stuttgart, Sonderforschungbereich 340, Azenbergstr 12, 70174 Stuttgart, Germany, 1993. Full version to appear in *Journal of Symbolic Logic*.

D. M. Gabbay. *Labelled Deductive Systems, Part I*. Oxford University Press, 1995. First draft 1989. Preprint, Department of Computing, Imperial College, London SW7 2BZ, UK. Current draft February 1991, 265 pp. Published as CIS - Bericht-90-92, Centrum für Informations und Srachverabeitunt, Universität München, Germany. Third intermediate draft, Max Planck Institute, Saarbrucken, Technical Report, MPI-94-223, 460 pp, 1994.

D. M. Gabbay. Fibred semantics and the weaving of logics, part 2. In *Logic Colloquium 92*, L. Czermak, D. Gabbay and M. de Rijke, editors, pp. 95–113. SILLI/CUP, 1995.

D. M. Gabbay. Fibred semantics and the weaving of logic, part 3. How to make your logic fuzzy, draft, Imperial College, 1995.

D. Gabbay. Fibred semantics and the weaving of logics, part 4: Self-fibring of predicate logic. Draft, Imperial College, 1995.

D. M. Gabbay. Classical vs non-classical logic, *Handbook of Logic in AI*, volume 2, Oxford d University Press, 1994.

D. M. Gabbay. Conditional implication and nonmonotonic consequence. In *Views on Conditionals*, ed. L.F. del Cerro *et al.*, pp. 347–369, COUP, 1995.

V. Goranko and S. Passy. Using the Universal Modality: Gains and Questions. *Journal of Logic and Computation*, **2**, 5–30, 1992.

A. J. I. Jones. Towards a formal theory of defeasible deontic conditionals. To appear in *Annals of Mathematics and Artificial Intelligence.*

A. J. I. Jones and I. Pörn. Ideality, subideality and deontic logic. *Synthese*, **65**, 1985.

A. J. I. Jones and I. Pörn. 'Ought' and 'must'. *Synthese*, **66**, 1986.

S. Kepser and K. Schulz. Combination of constraint systems II: Rational amalgamation. To appear in *Proc. CP-96.*

M. Kracht and F. Wolter. Properties of independently axiomatizable bimodal logics. *Journal of Symbolic Logic*, **56** 1469–1485, 1991.

S. Kraus, D. Lehmann, and M. Magidor. Nonmonotonic reasoning, preferential models and cumulative logics, *Artificial Intelligence*, **44**, 167–208, 1990.

K. Lambert and Bas C. van Fraassen. *Derivation and Counterexample*, Dickinson Publishing Co, 1972.

R. E. Meyer and E. D. Mares. Semantics of Entailment 0. In K. Dosen and P. Schroeder-Heister, eds, *Substructural Logics*, pp. 239–258, Oxford University Press, 1993.

H. Ono. On some intuitionistic modal logic. *Publications Research Institute of Mathema-*

tical Science, **13**, 687–722, 1977.

H. Ono. Some problems in intermediate predicate logics. *Reports on Mathematical Logic*, **21**, 55–68, 1987.

J. Pfalzgraf. A note on simplexes as geometric configurations. *Archiv der Mathematik*, **49**, 134–140, 1987.

J. Pfalzgraf. Logical fiberings and polycontextural systems. In *Fundamentals of Artificial Intelligence Resarch*, Ph. Jorrand and J. Kelemen, eds. LNCS 535, Springer-Verlag, 1991.

J. Pfalzgraf and K. Stokkermans. On robotics scenarios and modeling with fibered structures. In *Springer series texts and Monographs in Symbolic Computation, Automated Practical Reasoning: Algebraic Approaches*, ed. J. Pfalzgraf and D. Wang. Springer-Verlag, 1994.

A. K. Simpson. Proof theory and semantics of intuitionistic modal logic, Technical report CTS-114-94, Univ of Edinburgh, 1994.

N. Y. Suzuki. An algebraic aporach to intuitionistic modal logics in connection with intermediate predicate logics. *Studia Logica*, **48**, 141–155, 1988.

N. Y. Suzuki. Kripke bundles for intermediate predicate logics and Kripke frames for intuitionistic modal logics. *Studia Logica*, **49**, 289–306, 1990.

D. Wijesekera. Constructive modal logic 1, *Annals of Pure and Applied Logic*, **50**, 271–301, 1990.

ALESSANDRA RUSSO

GENERALISING PROPOSITIONAL MODAL LOGIC USING LABELLED DEDUCTIVE SYSTEMS

Abstract. A family of labelled deductive systems called *Propositional Modal Labelled Deductive* (PMLD) systems is described. These logics combine the standard syntax of propositional modal logic with a simple subset of first–order predicate logic, called a *labelling algebra*, to allow syntactic reference to a Kripke–like structure of possible worlds. PMLD systems are a generalisation of normal propositional modal logic in that they facilitate reasoning about what is true at different points in a (possibly singleton) structure of actual worlds, called a *configuration*. A model–theoretic semantics (based on first–order logic) is provided and its equivalence to Kripke semantics for normal propositional modal logics is shown whenever the initial configuration is a single point. A sound and complete natural deduction style proof system is also described. Unlike traditional proof systems for modal logics, this system is uniform in that every deduction rule is applicable to (the generalisation of) each normal modal logic extension of K obtained by adding combinations of the axiom schemas $T, 4, 5, D$ and B.

1. Introduction

This paper builds upon the methodology in (Gabbay, 1994) to develop generalisations of normal propositional modal logics, called *Propositional Modal Labelled Deductive* (PMLD) systems. The idea is to define a logical framework in which explicit reference can be made to particular possible worlds and to the relationships between possible worlds, while retaining the conventional syntax of modal logic. This is partly to provide a logic, based on modal reasoning, able to fulfil the increasing need in some application areas for using not only sets, but *structures* of information (in the Discussion section an example application in distributed database systems is given). The combined approach described in this paper shows how explicit references to structures of arbitrary modal theories can be made, whilst

F. Baader and K.U. Schulz (eds.), Frontiers of Combining Systems, 57–73.
© 1996 *Kluwer Academic Publishers.*

retaining a concise proof–theoretical presentation of modal logic. This is achieved by combining a modal logic with a *labelling algebra*. The labelling algebra is defined as a binary first–order theory which axiomatises the Kripke semantic notion of an accessibility relation. The modal language is given in a traditional way as a modal extension of a propositional language.

The two languages (modal language and labelling language) are combined via the notion of *declarative unit*. The declarative unit $\lambda : \alpha$ expresses that the modal formula α is true at the label (i.e. possible world) λ. This combined approach retains the advantages of both implicit (e.g. (Fitting, 1983)) and explicit (e.g. (Ohlbach, 1991)) traditional formalisations for modal logic. As in the implicit approach, statements such as "necessary α" can be captured succinctly using the modal operator $\Box$. This statement is simply written as the single declarative unit $W_a : \Box \alpha$ (where W_a is the labelling algebra representation of the actual world). Like the explicit approach, the language is rich enough to allow explicit syntactic reference to the accessibility relation between possible worlds.

PMLD systems are in fact generalisations of normal propositional modal logics in that they facilitate reasoning about what is true at different points in a (possibly singleton) structure of actual worlds. Information like "A holds at a possible world W_1 accessible from W_0 and $\neg A$ holds at a possible world W_2 not accessible from W_0", which could not be formalised at all in the implicit approach, can be directly represented as information about the relation between the possible worlds, together with modal formulae holding locally at W_1 and W_2. Loosely speaking, the PMLD theory would in this case be the set $\{R(W_0, W_1), \neg R(W_0, W_2), W_1 : A, W_2 : \neg A\}$. Of course, the same kind of generalisation could be achieved using explicit (first–order translation) methods, although less succinctly.

Furthermore, the explicit syntactic reference to a Kripke–like accessibility relation allows a natural deduction system to be developed for PMLD systems which is uniform, in that every deduction rule can be applied to each propositional modal logic. The difference between one modal logic and another is captured entirely by the labelling algebra[1].

The paper is organized as follows. In Section 2 the language and syntax of PMLD systems is defined together with the notion of a *configuration* – a PMLD system's equivalent to a modal theory. In Section 3, a natural deduction style proof system for PMLD systems is given in which inference rules are applied to configurations. In Section 4, a model–theoretic semantics, based on a translation method into classical logic, is described together with a notion of semantic entailment. Its equivalence to Kripke semantics for normal propositional modal logics is also shown whenever the initial

---

[1]The term "algebra" has been chosen here to reflect the terminology adopted in (Gabbay, 1994).

configuration is a single point. Soundness and completeness results of the proof system described in Section 3 are stated with respect to this semantics. The paper ends with a general discussion in Section 5. (Russo, 1995) is a longer version of this paper which includes detailed definitions, proofs of theorems and further examples.

Some remarks may be helpful regarding notation. Throughout the paper constant and predicate symbols begin with an upper-case letter, whereas variables and function symbols begin with a lower-case letter. Greek-letter meta-variables are used to refer in general to terms and expressions in the system. Larger entities such as structures, sets, theories and languages are symbolised in calligraphic font, $\mathcal{A}, \mathcal{B}, \mathcal{C}$, etc.. The power set of a given set $\mathcal{A}$ is denoted by $PW(\mathcal{A})$. A wide family of PMLD systems, corresponding to each of the normal modal logic extensions of K obtained using combinations of the axiom schemas $T, 4, 5, D$ and B, (e.g. $K, T, K4, K45, KD, KD4, KB$, etc.), will be considered. These systems will be referred to as the K–PMLD system, the T–PMLD system, etc. respectively. The symbol S will generally refer to an arbitrary PMLD system within this family.

2. Language and Syntax

In this section PMLD systems are described formally. Basic definitions of the PMLD language and syntax are given together with the notion of a *configuration* – a PMLD system's equivalent to a modal theory.

2.1. LANGUAGES

A PMLD language is defined as an ordered pair $\langle \mathcal{L}_L, \mathcal{L}_M \rangle$, where $\mathcal{L}_L$ is a *labelling language* and $\mathcal{L}_M$ is a *propositional modal language*. As in the implicit approach, $\mathcal{L}_M$ is composed of a countable set of propositional letters, $\{p, q, r, \ldots\}$, the logical connectives $\neg, \vee, \wedge, \rightarrow$, and the two modal operators $\square$ and $\diamond$. The labelling language $\mathcal{L}_L$ is a binary fragment of a first–order language composed of a countable set of constant symbols $\{W_0, W_1, W_2, \ldots\}$, a countable set of variables, $\{x, y, z, \ldots\}$, a unary function symbol $succ$, a binary predicate R, the set of logical connectives, $\{\neg, \wedge, \vee, \rightarrow, \equiv\}$, and the quantifiers $\forall$ and $\exists$. The first–order language $Func(\mathcal{L}_L, \mathcal{L}_M)$ is an extension of $\mathcal{L}_L$ defined as follows.

Definition 2.1 Let $\mathcal{L}_M$ be a modal propositional language and $\{\alpha_1, \alpha_2, \ldots\}$ be the set of all wffs of $\mathcal{L}_M$. The first–order language $Func(\mathcal{L}_L, \mathcal{L}_M)$ is defined as the language $\mathcal{L}_L$ extended with the sets of unary function symbols $\{f_{\alpha_1}, f_{\alpha_2}, \ldots\}$ and $\{box_{\alpha_1}, box_{\alpha_2}, \ldots\}$.

The binary predicate R represents the accessibility relation between possible worlds, whereas the ground terms of $Func(\mathcal{L}_L, \mathcal{L}_M)$, also called *labels*, refer to the possible worlds. Intuitively, for each possible world (label) λ and modal formula α, the term $f_\alpha(\lambda)$ names a *particular* world specifically associated with α. Such terms will be used whenever Kripke semantic notions of the form "there exists a possible world ..." need to be formalised. In contrast, ground terms of the form $box_\alpha(\lambda)$ can be thought of as referring to any *arbitrary* world specifically associated with α. These terms will be used whenever Kripke semantic notions of the form "for all possible worlds ..." need to be expressed. However, formally speaking, $f_\alpha(\lambda)$ and $box_\beta(\lambda)$ are both just labels – within a particular model they might even refer to the same possible world.

Different classes of Kripke frames are formalised in a PMLD system by a first–order axiomatisation, called *labelling algebra*, written in the language $\mathcal{L}_L$. For example, for the T–PMLD system, the labelling algebra $\mathcal{A}_T$ is the set $\{\forall x R(x,x)\}$ and for the $S4$–PMLD system, $\mathcal{A}_{S4}$ is the set of axioms $\{\forall x R(x,x), \forall x,y,z((R(x,y) \land R(y,z)) \rightarrow R(x,z))\}$. Table 1 shows the labelling algebra associated with some example PMLD systems. Each

TABLE 1. Labelling algebrae

PMLD system S	Labelling algebra $\mathcal{A}_S$
K	$\{\}$
T	$\{\forall x R(x,x)\}$
$K4$	$\{\forall x,y,z((R(x,y) \land R(y,z)) \rightarrow R(x,z))\}$
$K45$	$\{\forall x,y,z((R(x,y) \land R(y,z)) \rightarrow R(x,z)),$ $\forall x,y(R(x,y) \rightarrow R(y,x))\}$
KD	$\{\forall x R(x, succ(x))\}$
KB	$\{\forall x,y(R(x,y) \rightarrow R(y,x))\}$

set $\mathcal{A}_S$ of axioms reflects van Benthem's correspondence theory shown in (van Benthem, 1983). However, to reflect the "pre–skolemized" approach given by the use of the function symbols f_{α_i} and box_{α_i}, the seriality property of the D class of systems is formalised by the axiom $\forall x R(x, succ(x))$ instead of the axiom $\forall x \exists y R(x,y)$. It is clear that for each possible world λ the term $succ(\lambda)$ names one accessible world.

2.2. SYNTAX

The PMLD language facilitates the formalisation of two types of information, (i) what holds at particular possible worlds and (ii) which worlds are in relation with each other and which are not. As exemplified above, (i) and (ii) are captured within the syntax of a PMLD system by two different types of syntactic entity, the *declarative unit* and the *R–literal*. A declarative unit is defined as a pair separated by a colon – "*label:modal formula*" – expressing that a modal formula is true at a possible world. The label component is a ground term of the language $Func(\mathcal{L}_L, \mathcal{L}_M)$ and the modal formula is a wff of the modal language $\mathcal{L}_M$. Other examples of declarative units are $W_0:p$, $f_{\diamond q}(W_3):p$ and $f_q(W_1):p \rightarrow r$. An R–literal is any ground literal in the language $Func(\mathcal{L}_L, \mathcal{L}_M)$ of the form $R(\lambda_1, \lambda_2)$ or $\neg R(\lambda_1, \lambda_2)$, where λ_1 and λ_2 are labels, expressing that λ_2 is or is not accessible from λ_1. Examples of R–literals are $R(W_1, W_2)$, $\neg R(W_2, box_p(W_2))$ and $R(f_p(W_0), box_q(f_p(W_0)))$. For each R-literal Δ, the *conjugate* of Δ, written $\overline{\Delta}$, is the opposite in sign of Δ (i.e $\neg R(\lambda_1, \lambda_2)$, if $\Delta = R(\lambda_1, \lambda_2)$ and $R(\lambda_1, \lambda_2)$ if $\Delta = \neg R(\lambda_1, \lambda_2)$).

As mentioned above, this combined aspect of the PMLD syntax yields a definition of a PMLD theory more general than the traditional notion of a modal theory ((Huges & Cresswell, 1989), (Fitting, 1983)). Informally, a PMLD theory, called a *configuration*, is composed of two sets, a set of R–literals and a set of declarative units. An example of a PMLD theory is the pair of sets $\{R(W_0, W_1), R(W_0, W_2), R(W_1, f_p(W_1))\}$ and $\{W_0 : \square(p \rightarrow q), W_0 : \square r, W_1 : \diamond p, f_p(W_1) : p, W_2 : q\}$. The formal definition of a configuration is as follows.

Definition 2.2 Given a PMLD language, a configuration is a tuple $\langle \mathcal{D}, \mathcal{F} \rangle$ where $\mathcal{D}$ is a set of R–literals and $\mathcal{F}$ is a function from the set of ground terms of $Func(\mathcal{L}_L, \mathcal{L}_M)$ to the set $\mathrm{PW}(\mathrm{wff}(\mathcal{L}_M))$ of sets of wffs of $\mathcal{L}_M$.

The $\mathcal{D}$ component of a configuration $\mathcal{C} = \langle \mathcal{D}, \mathcal{F} \rangle$ will sometimes be referred to as a *diagram* and set membership statements of the form $\alpha \in \mathcal{F}(\lambda)$ will sometimes be written as $\lambda : \alpha \in \mathcal{C}$. Configurations can be infinite. This occurs either when the diagram $\mathcal{D}$ is infinite (i.e. the configuration contains an infinite number of R–literals), or when for some label λ of the language $Func(\mathcal{L}_L, \mathcal{L}_M)$, the value of $\mathcal{F}(\lambda)$ is an infinite set of formulae of $\mathcal{L}_M$ (i.e. the configuration contains an infinite number of declarative units referring to λ), or when $\mathcal{F}(\lambda) \neq \emptyset$ for an infinite number of terms λ of $Func(\mathcal{L}_L, \mathcal{L}_M)$.

In the next section, a natural deduction style proof system for an arbitrary PMLD system S is given, in which inference rules and a derivability relation are defined between configurations. A set $\mathcal{R}$ of such deduction rules together with a PMLD language $\langle \mathcal{L}_L, \mathcal{L}_M \rangle$ and a labelling algebra $\mathcal{A}$ uniquely define a PMLD system (i.e. for any PMLD system S, $S = \langle \langle \mathcal{L}_L, \mathcal{L}_M \rangle, \mathcal{A}, \mathcal{R} \rangle$).

3. A Natural Deduction System

As mentioned above, explicit syntactic reference to a Kripke–like accessibility relation facilitates reasoning about structures of actual worlds (i.e.

configurations). A natural deduction style proof system is described in this section whose inference rules and derivability relation are defined between configurations. This system allows reasoning about both the declarative units and the R–literals within a given configuration. Elimination and introduction rules are given for each classical connective and modal operator. For the R–literals, three additional rules are defined. One of these (the $\mathcal{I}_{R-A}$ rule) facilitates first–order derivations of relationships between possible worlds using the labelling algebra $\mathcal{A}$. This rule captures entirely the difference between one modal system and another, allowing all other deduction rules, including those for the modal operators, to be equally applicable to any PMLD system S. It is in this sense that the PMLD systems described here are *uniform*. The basic notion of an inference rule is defined as follows.

Definition 3.1 An inference rule $\mathcal{I}$ is a set of pairs of configurations, where each such pair is written as C/C'. If $C/C' \in \mathcal{I}$ then we say C is an *antecedent configuration* of $\mathcal{I}$, and C' is an *inferred (or consequence) configuration* of $\mathcal{I}$ with respect to C.

All the rules except one have the effect of expanding the antecedent configuration. These rules can extend an antecedent configuration C with either a declarative unit, or with an R–literal or with both. However, configurations equal or smaller than the antecedent one can also be inferred. This is facilitated by an inference rule called the C–Reduction ($\mathcal{I}_{C-R}$) rule (see Table 4). Informally, a *proof* is a non empty sequence of configurations, $C_0, \ldots, C_n$, where, for each $0 < i \leq n$, C_i is obtained from C_{i-1} by the application of an inference rule. A configuration C' is said to be *derivable* from a configuration C, written $C \vdash_S C'$, if and only if there exists a proof $C, \ldots, C'$.

Because of space limitations, only a schematic representation of the inference rules is given here. A complete mathematical formalisation of the proof system can be found in (Russo, 1995). Table 2 summarises the set of natural deduction rules for the classical connectives, Table 3 describes the natural deduction rules for the modal operators and in Table 4 additional rules for R–literals are given. However, some remarks are essential to clarify this schematic representation. For any configuration C, the informal notation $C\langle \lambda : \alpha \rangle$ (respectively $C\langle \Delta \rangle$) denotes that C includes a declarative unit $\lambda : \alpha$ (respectively R–literal Δ). Declarative units and R–literals contained in square brackets (see e.g. the $\mathcal{I}_{\vee E}$ rule) are assumptions introduced within a derivation that are subsequently discharged. The notation $C'\langle \psi \rangle$ represents that the inferred configuration C' is C extended with the declarative unit or R–literal ψ. In the $\mathcal{I}_{\vee E}$ rule, $\tilde{C}$ and $\overline{C}$ are the configurations derived in subderivations after adding to the antecedent configuration C the assumptions $\lambda : \alpha$ and $\lambda : \beta$ respectively. For the $\mathcal{I}_{\rightarrow I}$, $\tilde{C}$ is the configuration

TABLE 2. Natural deduction rules for classical connectives

$$\frac{\mathcal{C}\langle\lambda:\alpha\wedge\beta\rangle}{\mathcal{C}'\langle\lambda:\alpha\rangle}\quad\mathcal{I}_{\wedge E}\qquad\qquad\frac{\mathcal{C}\langle\lambda:\alpha,\lambda:\beta\rangle}{\mathcal{C}'\langle\lambda:\alpha\wedge\beta\rangle}\quad\mathcal{I}_{\wedge I}$$

$$\frac{\mathcal{C}\langle\lambda:\alpha\vee\beta\rangle\quad\overset{\mathcal{C}\langle[\lambda:\alpha]\rangle}{\underset{\tilde{\mathcal{C}}\langle\lambda:\gamma\rangle}{\vdots}}\quad\overset{\mathcal{C}\langle[\lambda:\beta]\rangle}{\underset{\overline{\mathcal{C}}\langle\lambda:\gamma\rangle}{\vdots}}}{\mathcal{C}'\langle\lambda:\gamma\rangle}\quad\mathcal{I}_{\vee E}\qquad\frac{\mathcal{C}\langle\lambda:\alpha\rangle}{\mathcal{C}'\langle\lambda:\alpha\vee\beta\rangle}\quad\mathcal{I}_{\vee I}$$

$$\frac{\mathcal{C}\langle\lambda:\alpha\rightarrow\beta,\lambda:\alpha\rangle}{\mathcal{C}'\langle\lambda:\beta\rangle}\quad\mathcal{I}_{\rightarrow E}\qquad\frac{\overset{\mathcal{C}\langle[\lambda:\alpha]\rangle}{\underset{\tilde{\mathcal{C}}\langle\lambda:\beta\rangle}{\vdots}}}{\mathcal{C}'\langle\lambda:\alpha\rightarrow\beta\rangle}\quad\mathcal{I}_{\rightarrow I}$$

$$\frac{\mathcal{C}\langle\lambda:\neg\neg\alpha\rangle}{\mathcal{C}'\langle\lambda:\alpha\rangle}\quad\mathcal{I}_{\neg E}\qquad\frac{\overset{\mathcal{C}\langle[\lambda:\alpha]\rangle}{\underset{\tilde{\mathcal{C}}\langle\lambda':\perp\rangle}{\vdots}}}{\mathcal{C}'\langle\lambda:\neg\alpha\rangle}\quad\mathcal{I}_{\neg I}$$

derived after adding the assumption $\lambda:\alpha$ to the antecedent configuration, to get the declarative unit $\lambda:\beta$. In the $\mathcal{I}_{\neg I}$ rule, the symbol $\perp$ is a short–hand for any wff of $\mathcal{L}_M$ of the form $\alpha\wedge\neg\alpha$. $\tilde{\mathcal{C}}$ is, in this case, the configuration derived after adding the assumption $\lambda:\alpha$ to $\mathcal{C}$ to get a contradiction – the declarative unit $\lambda':\perp$. In this case the label λ' can be different from λ because contradictions might occur in any possible world of a configuration[2]. When $\lambda'=\lambda$, the rules in Table 2 faithfully reflect the natural deduction proof system for propositional logic defined in (Prawitz, 1965). In fact, since these rules do not define new labels, they can simply be considered as "local natural deduction rules" for propositional logic[3].

---

[2]This is because a configuration might already have declarative units of the form $\lambda':\alpha$ and $\lambda':\neg\alpha$, or because applications of rules for modal operators might generate contradictions in some accessible world.

[3]Of course, strictly speaking, two $\wedge$–elimination rules and two $\vee$–introduction rules are included in any such system. For example, from $p\wedge q$, both p and q can be classically

The natural deduction rules for modal operators are somewhat different. As shown in Table 3, these rules involve declarative units with different labels. They express the interaction between the local modal theories within a configuration. They refer explicitly to relationships between points (R–literals) to infer new formulae in accessible worlds, and allow the introduction of "new particular accessible worlds". The $\mathcal{I}_{\Box E}$ and $\mathcal{I}_{\Diamond I}$ rules clearly reflect the Kripke semantic truth definitions of the two modal operators. If at a local actual point λ_1, $\Box\alpha$ holds, this implies that for any point accessible from λ_1, the formula α must hold. Therefore, if a given configuration $\mathcal{C}$ contains a declarative unit $\lambda_1 : \Box\alpha$ and a relationship between λ_1 and another local actual point λ_2 (i.e. $R(\lambda_1, \lambda_2)$), then it is obvious to extend this configuration with the declarative unit $\lambda_2 : \alpha$. Similarly for the $\mathcal{I}_{\Diamond I}$ rule. The $\mathcal{I}_{\Diamond E}$ and $\mathcal{I}_{\Box I}$ rules show how ground terms of the form $f_\alpha(\lambda)$

TABLE 3. Natural deduction rules for modal operators

$$
\frac{\mathcal{C}\langle\lambda : \Diamond\alpha\rangle}{\mathcal{C}'\langle f_\alpha(\lambda) : \alpha, R(\lambda, f_\alpha(\lambda))\rangle} \; \mathcal{I}_{\Diamond E}
\qquad
\frac{\mathcal{C}\langle\lambda_2 : \alpha, R(\lambda_1, \lambda_2)\rangle}{\mathcal{C}'\langle\lambda_1 : \Diamond\alpha\rangle} \; \mathcal{I}_{\Diamond I}
$$

$$
\frac{\mathcal{C}\langle\lambda_1 : \Box\alpha, R(\lambda_1, \lambda_2)\rangle}{\mathcal{C}'\langle\lambda_2 : \alpha\rangle} \; \mathcal{I}_{\Box E}
\qquad
\frac{\begin{array}{c}\mathcal{C}\langle[R(\lambda, box_\alpha(\lambda))]\rangle \\ \vdots \\ \tilde{\mathcal{C}}\langle box_\alpha(\lambda) : \alpha\rangle\end{array}}{\mathcal{C}'\langle\lambda : \Box\alpha\rangle} \; \mathcal{I}_{\Box I}
$$

and $box_\alpha(\lambda)$ are used within this proof system[4]. The consequence of the $\mathcal{I}_{\Diamond E}$ rule is equivalent to "there exists an accessible point where α holds", whereas the subderivation of the $\mathcal{I}_{\Box I}$ rule corresponds to checking whether "at any accessible world α holds".

The following table completes the description of the PMLD deduction system. The $\perp$–Introduction rule allows the inference of falsity (i.e. $\lambda : \perp$) whenever an R–literal and its negation are present in a configuration. This is necessary because since no compound classical formulae with R–literals can be inferred in a configuration, inconsistency of this form would not otherwise be captured. The R–Assertion rule, $\mathcal{I}_{R-A}$, expands the diagram inferred.

---

[4]Intuitively, $f_\alpha(\lambda)$ represents a "new" possible world when introduced within a proof. Indeed, in the completeness proof described in the next section, a canonical model is defined over a Herbrand universe of such terms.

of the antecedent configuration using the labelling algebra $\mathcal{A}$. Thus different R–literals are inferred for different PMLD systems. The R–Introduction rule, $\mathcal{I}_{R-I}$, is the equivalent of a $\neg$–Introduction rule for R-literals. Finally, with the C–reduction rule, $\mathcal{I}_{C-R}$, it is possible to infer any smaller configuration contained in an existing one. To guarantee completeness, this rule must be included because, as mentioned above, all the other rules have the effect of expanding their antecedent configurations.

TABLE 4. Other rules

$$
\mathcal{I}_{\perp I} \quad \frac{\mathcal{C}\langle \Delta, \overline{\Delta}\rangle}{\mathcal{C}'\langle \lambda : \alpha\rangle} \qquad\qquad \mathcal{I}_{C-R} \quad \frac{\mathcal{C}}{\mathcal{C}'} \\[2pt]
\text{where } \mathcal{C}' \subseteq \mathcal{C}
$$

$$
\mathcal{I}_{R-I} \quad \frac{\begin{array}{c}\mathcal{C}\langle[\,\overline{\Delta}\,]\rangle \\ \vdots \\ \bar{\mathcal{C}}\langle \lambda' : \perp\rangle\end{array}}{\mathcal{C}'\langle \Delta\rangle} \qquad\qquad \mathcal{I}_{R-A} \quad \frac{\mathcal{C}}{\mathcal{C}'\langle \Delta\rangle} \\[2pt]
\text{if } \mathcal{A} \cup \mathcal{D} \vdash_{FOL} \Delta
$$

Unfortunately, space limitations do not allow example derivations to be given in detail here (see (Russo, 1995) for these). However, the reader is invited to use the rules in Tables 2, 3 and 4 to check the following. (i) Any configuration of any PMLD system which contains the declarative units $W_0 : \Box(p \to q)$ and $W_0 : \Box p$ can be expanded to a configuration containing $W_0 : \Box q$ (use $\mathcal{I}_{\Box I}$, $\mathcal{I}_{\Box E}$ and $\mathcal{I}_{\to E}$). (ii) Any configuration of the $K4$–PMLD system containing the declarative units $W_0 : \Box p$, $W_2 : \Box(p \to q)$ and $W_3 : \neg q$, and the R–literals $R(W_0, W_1)$ and $R(W_1, W_3)$, can be expanded to a configuration containing the R-literal $\neg R(W_2, W_3)$ (use $\mathcal{I}_{R-I}$, $\mathcal{I}_{\Box E}$, $\mathcal{I}_{R-A}$, $\mathcal{I}_{\to E}$ and $\mathcal{I}_{\wedge I}$). Note that derivations such as (ii) have no obvious counterpart in traditional modal logic.

4. Semantics, Soundness and Completeness

A PMLD system can be considered to be a "semi–translated" approach to modal logic — a Kripke–like accessibility relation is syntactically expressed, but without requiring the full translation of modal formulae into first–order sentences. Therefore, a model–theoretic semantics could be equally given in terms of the traditional Kripke possible worlds semantics (as discussed

in (Gabbay, 1994)) or in terms of a first–order semantics using some translation method. The second approach has been chosen here. In this section,
a translation method of PMLD systems into first–order logic is defined
and the notions of model, satisfiability of a configuration and semantic
entailment are then given in terms of classical semantics. Soundness and
completeness results are stated which show that the derivability relation
introduced in the previous section corresponds to the semantic entailment
relation defined below. Finally, results on correspondences between PMLD
systems and traditional axiomatisations of modal logic are given.

4.1. SEMANTICS

As pointed out above, a declarative unit $\lambda:\alpha$ represents that the formula α
holds at the possible world λ. In what follows, such Kripke–style semantic
notions are expressed in terms of first–order statements of the form $[\alpha]^*(\lambda)$,
where $[\alpha]^*$ is a predicate symbol. The relationships between these predicates are constrained by a set of first–order axiom schemas which capture the
satisfiability conditions of each type[5] of formula α. The *extended labelling
language* $Mon(\mathcal{L}_L, \mathcal{L}_M)$ is an extension of the language $Func(\mathcal{L}_L, \mathcal{L}_M)$ given by adding a monadic predicate symbol $[\alpha]^*$ for each wff α of $\mathcal{L}_M$. The
extended algebra $\mathcal{A}^+$ is a first–order theory written in $Mon(\mathcal{L}_L, \mathcal{L}_M)$ which
extends the labelling algebra $\mathcal{A}$ as follows.

Definition 4.1 Given an extended labelling language $Mon(\mathcal{L}_L, \mathcal{L}_M)$ and
a labelling algebra $\mathcal{A}$ written in $Func(\mathcal{L}_L, \mathcal{L}_M)$, the extended algebra $\mathcal{A}^+$
is the first–order theory in $Mon(\mathcal{L}_L, \mathcal{L}_M)$ consisting of the following axiom
schemas (Ax1)–(Ax8), together with the axioms of $\mathcal{A}$. For any wffs α and
β of $\mathcal{L}_M$:

$$\forall x([\alpha \wedge \beta]^*(x) \equiv ([\alpha]^*(x) \wedge [\beta]^*(x))) \tag{Ax1}$$

$$\forall x([\neg\alpha]^*(x) \equiv \neg[\alpha]^*(x)) \tag{Ax2}$$

$$\forall x([\alpha \vee \beta]^*(x) \equiv ([\alpha]^*(x) \vee [\beta]^*(x))) \tag{Ax3}$$

$$\forall x([\alpha \rightarrow \beta]^*(x) \equiv ([\alpha]^*(x) \rightarrow [\beta]^*(x))) \tag{Ax4}$$

$$\forall x([\Diamond\alpha]^*(x) \rightarrow (R(x, f_\alpha(x)) \wedge [\alpha]^*(f_\alpha(x)))) \tag{Ax5}$$

---

[5]The type of a wff is given by the main connective of the wff itself, e.g. the wff
$\Diamond(p \rightarrow q)$ is a $\Diamond$–formula.

$$\forall x (\exists y (R(x,y) \land [\alpha]^*(y)) \to [\Diamond\alpha]^*(x)) \tag{Ax6}$$

$$\forall x ((R(x, box_\alpha(x)) \to [\alpha]^*(box_\alpha(x))) \to [\Box\alpha]^*(x)) \tag{Ax7}$$

$$\forall x ([\Box\alpha]^*(x) \to (\forall y (R(x,y) \to [\alpha]^*(y)))) \tag{Ax8}$$

The first four axiom schemas express the distributive properties of the logical connectives among the monadic predicates of $Mon(\mathcal{L}_L, \mathcal{L}_M)$. (Ax5) and (Ax8) force the accessibility relation R on labels generated by the application of the function symbols f_α and box_α of $Mon(\mathcal{L}_L, \mathcal{L}_M)$ whenever their respective antecedents hold. Clearly, all the axiom schemas (Ax1)–(Ax8) reflect the traditional Kripke semantic definition of satisfiability of modal wffs[6]. Note that a unique extended algebra $\mathcal{A}^+$ is associated with each PMLD system.

A translation method is defined next. It associates syntactical expressions of a PMLD language with sentences of the language $Mon(\mathcal{L}_L, \mathcal{L}_M)$, and PMLD theories (configurations) with first–order theories in the language $Mon(\mathcal{L}_L, \mathcal{L}_M)$. Each declarative unit $\lambda : \alpha$ is associated with the sentence $[\alpha]^*(\lambda)$, and each R–literal Δ is associated with itself. The first–order translation of a configuration is defined as follows.

Definition 4.2 Let $\mathcal{C} = \langle \mathcal{D}, \mathcal{F} \rangle$ be a configuration written in $\langle \mathcal{L}_L, \mathcal{L}_M \rangle$. The first order translation of $\mathcal{C}$, written $FOT(\mathcal{C})$, is the theory $\mathcal{D} \cup \mathcal{DU}$, where $\mathcal{DU} = \{[\alpha]^*(\lambda) \mid \alpha \in \mathcal{F}(\lambda), \lambda$ is a ground term of $Func(\mathcal{L}_L, \mathcal{L}_M)\}$.

Here a first–order translation of a given configuration is a set of ground literals of the language $Mon(\mathcal{L}_L, \mathcal{L}_M)$. Notions of model, satisfiability and semantic entailment are given in terms of classical semantics using the above definitions, as follows (where "$\mathcal{M} \Vdash_{FOL} \psi$" signifies that the classical formula ψ is true in the classical model $\mathcal{M}$, according to the standard definition). Given a PMLD system S, the associated extended algebra $\mathcal{A}^+$, a declarative unit $\lambda : \alpha$ and a R–literal Δ,

$$\mathcal{M} \text{ is a semantic structure of } S \quad \Leftrightarrow_{\text{def}} \quad \mathcal{M} \text{ is a model of } \mathcal{A}^+ \tag{1}$$

$$\mathcal{M} \Vdash_S \lambda : \alpha \quad \Leftrightarrow_{\text{def}} \quad \mathcal{M} \Vdash_{FOL} [\alpha]^*(\lambda) \tag{2}$$

$$\mathcal{M} \Vdash_S \Delta \quad \Leftrightarrow_{\text{def}} \quad \mathcal{M} \Vdash_{FOL} \Delta \tag{3}$$

(1) defines the class of models of a PMLD system S in terms of models of the extended algebra $\mathcal{A}^+$ associated with S. (2) and (3) define the satisfiability of declarative units and R–literals in terms of classical satisfiability of

---

[6]This is easily seen by interpreting the truth of $[\alpha]^*(x)$ as the truth of the modal formula α in the possible world x.

their associated first–order translations. A semantic structure $\mathcal{M}$ satisfies a configuration $\mathcal{C}$, written $\mathcal{M} \Vdash_S \mathcal{C}$, if and only if for each $\pi \in \mathcal{C}$ (where π may be a declarative unit or an R–literal), $\mathcal{M} \Vdash_S \pi$. A notion of semantic entailment in a PMLD system is given here as a relation between configurations. It is formally defined as follows.

Definition 4.3 Let $S = \langle\langle \mathcal{L}_L, \mathcal{L}_M \rangle, \mathcal{A}, \mathcal{R}\rangle$ be a MLDS and let $\mathcal{A}^+$ be the extended algebra of S. Let $\mathcal{C} = \langle \mathcal{D}, \mathcal{F} \rangle$ and $\mathcal{C}' = \langle \mathcal{D}', \mathcal{F}' \rangle$ be two configurations of S and $FOT(\mathcal{C}) = \mathcal{D} \cup \mathcal{DU}$ and $FOT(\mathcal{C}') = \mathcal{D}' \cup \mathcal{DU}'$ their respective first order translations. The configuration $\mathcal{C}$ semantically entails $\mathcal{C}'$, written $\mathcal{C} \models_S \mathcal{C}'$, iff for each $\Delta \in \mathcal{D}'$, $\mathcal{A}^+ \cup FOT(\mathcal{C}) \models_{FOL} \Delta$, and for each $[\alpha]^*(\lambda) \in \mathcal{DU}'$, $\mathcal{A}^+ \cup FOT(\mathcal{C}) \models_{FOL} [\alpha]^*(\lambda)$.

4.2. SOUNDNESS, COMPLETENESS AND CORRESPONDENCE

Soundness and completeness results are stated below. Their proofs, given in full in (Russo, 1995), take advantage of the soundness and completeness of first–order classical logic. The soundness result, Theorem 4.1, states that if there exists a natural deduction proof of a configuration $\mathcal{C}'$ from a configuration $\mathcal{C}$, then $\mathcal{C}$ semantically entails $\mathcal{C}'$.

Theorem 4.1 [Soundness] Let $S = \langle\langle \mathcal{L}_L, \mathcal{L}_M \rangle, \mathcal{A}, \mathcal{R}\rangle$ be a MLDS and let $\mathcal{C}, \mathcal{C}'$ be two configurations of S. If $\mathcal{C} \vdash_S \mathcal{C}'$ then $\mathcal{C} \models_S \mathcal{C}'$.
    Proof: See (Russo, 1995).

The proof of Theorem 4.1 is represented diagrammatically in Figure 1.

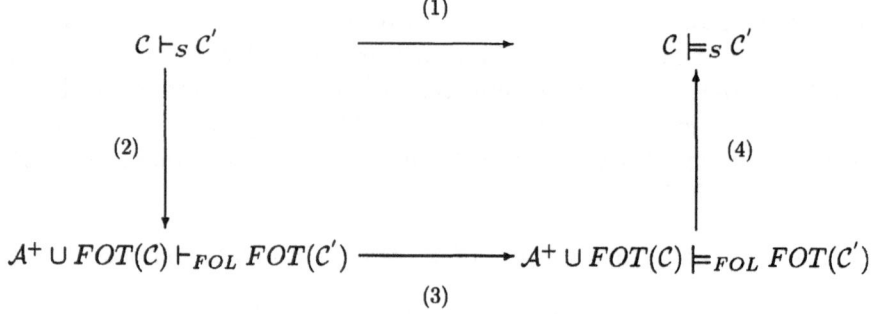

Figure 1. Proof of the Soundness Theorem

The soundness statement, which corresponds to the arrow labelled with (1), is proved by the composition of three main steps, arrows (2), (3) and (4) respectively. The first step (arrow (2)) proves that the hypothesis, $\mathcal{C} \vdash_S \mathcal{C}'$, implies that $\mathcal{A}^+ \cup FOT(\mathcal{C}) \vdash_{FOL} FOT(\mathcal{C}')$ (see (Russo, 1995), Lemma 3.1). This trivially implies (by soundness of first–order logic) that $\mathcal{A}^+ \cup$

$FOT(C) \models_{FOL} FOT(C')$, which gives the second step of the proof (arrow (3)). Arrow (4) is given by the definition of the semantic entailment between configurations.

The completeness result, Theorem 4.2, states that, given a MLDS S and two configurations C and C' of S, such that their configuration difference $C' - C$ is finite, if C' is semantically entailed from C then C' is also derived from C. $C' - C$ (defined formally in (Russo, 1995)) is basically the set of declarative units and R–literals in C' but not in C.[7] The proof of Theorem 4.2, given in (Russo, 1995), uses a Henkin–style methodology (Huges & Cresswell, 1989).

Theorem 4.2 [Completeness] Let $S = \langle \langle \mathcal{L}_L, \mathcal{L}_M \rangle, \mathcal{A}, \mathcal{R} \rangle$ be a MLDS. Let $\mathcal{A}^+$ be the associated extended algebra. Let C and C' be two configurations of S such that the configuration difference $C' - C$ is finite. If $C \models_S C'$ then $C \vdash_S C'$.

Proof: See (Russo, 1995).

An informal description of the completeness proof, referring to its diagrammatical representation given in Figure 2, is as follows.

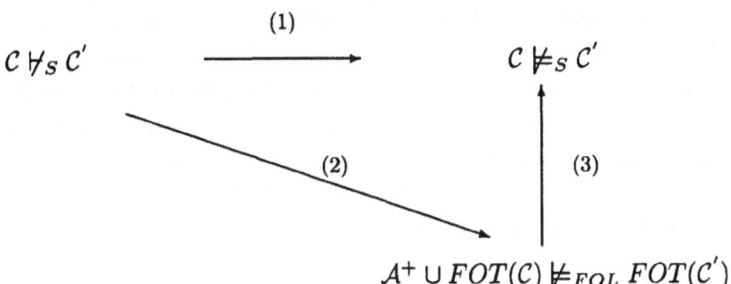

Figure 2. Proof of the Completeness Theorem

The proof is of the contrapositive statement of Theorem 4.2 (arrow (1)). It is proved by the composition of two main steps, arrows (2) and (3). Arrow (3) is already given by Definition 4.3, while arrow (2) represents the main part of the completeness theorem. Its proof is based on the statement *if C is a consistent configuration then C is satisfiable*, known as the "Model Existence Lemma". The hypothesis that C' is not derivable from C, $C \not\vdash_S C'$, implies (see (Russo, 1995), Theorem 4.1) that there exists a $\pi \in C' - C$ (where π is a declarative unit or an R–literal) such that $C \not\vdash_S \pi$.

---

[7]Obviously, if the configuration difference $(C' - C)$ were infinite, an infinite proof sequence would be required to prove C' from C.

So $\mathcal{C}$ extended with $\neg\pi$ (written $\mathcal{C}+[\neg\pi]$) is a consistent configuration[8]. This implies that the configuration $\mathcal{C}+[\neg\pi]$ is satisfiable (see (Russo, 1995), *Model Existence Lemma* and Corollary 4.2). Therefore, there exists a semantic structure $\mathcal{M}$ of S which satisfies $\mathcal{C}$ and that also satisfies $\neg\pi$. Hence, $\mathcal{M}$ does not satisfy π. Thus, since $\pi \in \mathcal{C}'$, by the definition of satisfiability of a configuration, $\mathcal{M}$ does not satisfy $\mathcal{C}'$. Hence $\mathcal{A}^+ \cup FOT(\mathcal{C}) \not\models_{FOL} FOT(\mathcal{C}')$.

It was claimed in the introduction that a PMLD system is a generalisation of normal propositional modal logic, in that it facilitates reasoning about structures of actual worlds, which may or may not be single points. This claim is substantiated in Theorem 4.3 below. It states that under the restriction of considering only singleton configurations, a PMLD system is equivalent to the corresponding normal propositional modal logic. The equivalence is stated with respect to traditional axiomatisations of normal propositional modal logics.

Theorem 4.3 Let N be any normal propositional modal logic obtained by extending K with combinations of the axiom schemas $T, 4, 5, D$ and B. Let $\langle N_{Ax}, \vdash_{N_{Ax}}\rangle$ be an axiomatic system for N which is sound and complete with respect to Kripke semantics. Let N-PMLD be the corresponding PMLD system. Let $\Psi = \{\alpha_1 \ldots, \alpha_n\}$ $(n \geq 0)$ be an arbitrary (possibly empty) set of wffs of N, and let $\mathcal{C}$ be the configuration of N-PMLD consisting only of all declarative units of the form $W_0 : \alpha_i$, $\alpha_i \in \Psi$, and containing no R–literals. Then for any formula γ of N, $\mathcal{C} \vdash_N W_0 : \gamma$ if and only if $\Psi \vdash_{N_{Ax}} \gamma$, and $\mathcal{C} \models_N W_0 : \gamma$ if and only if $\Psi \models_{N_{Ax}} \gamma$.
    Proof: See (Russo, 1995).

Hence the semantics given here is equivalent to Kripke semantics whenever the initial configuration is a single point. However, PMLD systems allow for a more general form of initial configuration – wffs associated with different local actual worlds, and explicit relationships between possible worlds. These configurations cannot be represented within traditional modal logic, where the initial set of formulae must always refer to the same initial actual world.

5. Discussion

The research described above can be summarised as follows. In Section 2 a family of labelled deductive systems has been defined in which modal logic and first–order logic are combined to generalise the notion of a modal

---

[8]A configuration $\mathcal{C}$ is inconsistent if it derives a configuration $\mathcal{C}'$ which contains a declarative unit of the form $\lambda : \bot$. A configuration is consistent if it is not inconsistent.

theory. In Section 3 a natural deduction style proof system has been described which reflects this more general notion (the derivability relation is defined between configurations), and which is uniform for a wide family of modal logics. In Section 4 a model–theoretic semantics has been given, based on a translation method from the combined formalism into first–order logic. The soundness and completeness of the system with respect to this semantics has been stated (Theorems 4.1 and 4.2), as well as the sense in which normal propositional modal logics can be regarded as special cases of PMLD systems (Theorem 4.3).

The PMLD approach is similar to that of Fitting's prefixed tableaux (Fitting, 1983), in that, in both cases, modal formulae are labelled. However, whereas in Fitting's system the notion of an accessibility relation is embedded in the definition of "accessible prefixes", in a PMLD system it is made explicit. Consequently, the different side conditions imposed on the prefixed tableaux modal rules by different modal logics are replaced here by the R–Assertion rule, allowing the PMLD system to be uniform. In (Sympson, 1993) Sympson has developed a natural deduction system for propositional intuitionistic modal logic. His work is similar to the approach described in this paper in that it is also built on the framework of LDS. Formulae are labelled with variables (lwffs), which denote possible worlds, and relationships between possible worlds are also made explicit (rwffs). However, rwffs are only used as assumptions in "geometric" rules. Consequently, different geometric rules are given for different intuitionistic modal logics, whereas the PMLD natural deduction rules are applicable to any normal modal logic. Other work related to the approach in this paper can be found in (Frisch & Scherl, 1991), (Gent, 1993), (Massacci, 1994) and (Borghuis, 1993).

The translation method described in Section 3 is comparable with other existing relational translation approaches (e.g. (Moore, 1980) and (Ohlbach, 1991)). It is hoped that the PMLD translation will result in a more compact first–order theory than that of Ohlbach (where an exponential number of clauses are generated) – this is the subject of further investigation. This could give an advantage when using existing classical theorem provers for modal logics. Moreover, in this case the need for expensive unification steps would be eliminated by use of the function symbols f_α and box_α.

The framework described in this paper is presently being extended to the predicate case. It is expected that this extension will provide a uniform proof system which is sound and complete with respect to varying domain semantics.

One reason for developing the PMLD framework has been to provide a logic based on modal reasoning which is closer to the needs of computing and A.I. applications (see (Gabbay, 1992) for a general discussion of this issue). Although the focus of this paper has not been on applications, it is perhaps worth briefly discussing an example. A distributed database system can be represented as a configuration, where each label is associated with a local database and connections between databases are described by R-literals. Here, modal operators would impose constraints which might affect the evolution of several local databases when an update is made at a particular node. This idea has also been discussed in (Benevides, 1995).

Finally, this paper illustrates an important general point regarding work in Combining Logics. Most work in this area is concerned with showing how entirely *new* logics can be generated as hybrids of two or more existing logics. This paper makes a contribution in a slightly different direction. It shows how an existing class of logics can be *generalised* by combining them with another. It could be that such generalisations are possible on a much wider scale.

Acknowledgements

Many thanks to Krysia Broda, Melvin Fitting, Dov Gabbay, Rajeev Gore, Ian Hodkinson, Rob Miller and Steve Vickers for useful discussions on the subject of this paper, and for careful reading of earlier drafts. This research has been funded by the Italian National Research Council (CNR).

ALESSANDRA RUSSO
Imperial College of Science, Technology and Medicine,
Department of Computing,
London, United Kingdom

References

M.R.F. Benevides. Multiple Database Logic. *ECSQARU'95, Lecture Notes in Artificial Intelligence,* 946, 1995.
T. Borghuis. Interpreting Modal Natural Deduction in Type Theory. In Maarten de Rijke editor, *Diamonds and Defaults,* Kluwer Academic Publishers, 67-102, 1993.
M. Fitting. *Proof Methods for Modal and Intuitionistic Logics.* D. Reidel, Dordrecht, 1983.
A. M. Frisch and R. B. Scherl. A General Framework for Modal Deduction. In *Proceedings of the 2nd Conference on Principles of Knowledge Representation and Reasoning.* Morgan-Kaufmann, 1991.
D.M. Gabbay. How to Construct a Logic for Your Application. *GWAI'92, Lecture Note in Artificial Intelligence,* 671:1-30, 1992.
D.M. Gabbay. LDS - Labelled Deductive Systems, Volume 1 - Foundations. Technical Report MPI-I-94/223, Max-Planck-Institut Für Informatik, 1994.
I. Gent. Theory Matrices (for Modal Logics) using Alphabetical Monotonicity. *Studia Logica,* 52(2):233-257, 1993.
G.E. Hughes and M.J. Cresswell. *An Introduction to Modal Logic.* Methuen and Co. Ltd, 1968. Reprinted by Routledge, 1989.

F. Massacci. Strong Analytic Tableaux for Normal Modal Logics. *In Proceedings of CADE-12, LNAI 814*, Springer, 1994.

R.C. Moore. *Reasoning About Knowledge and Action*. MIT, Cambridge, 1980.

H.J. Ohlbach. Semantics–based Translation Methods for Modal Logics. *Journal of Logic and Computation*, 1(5):691-746, 1991.

D. Prawitz. *Natural Deduction: a Proof-Theoretical Study*. Almqvist and Wiksell, 1965.

A. Russo. Modal Labelled Deductive Systems. Technical Report 95/7, Imperial College, Department of Computing, 1995.
Available at http://theory.doc.ic.ac.uk/~ar3/PMLDSreport.ps.

A. Sympson. The Proof Theory and Semantics of Intuitionistic Modal Logics. PhD Thesis, University of Edinburgh, 1993.

J. van Benthem. Correspondence Theory. In D. Gabbay and F. Guenthner, editors, *Handbook of Philosophical Logic*, volume II, Extensions of Classical Logics. D. Reidel Publishing Company, 1983.

Kanger, Stig: *A Note on Partial Postulate Sets for Normal Modal Logics*, in *Theoria*, vol. 23, pp. 49–64, Stockholm, 1957.

Montague, Richard: *Universal Grammar*, in *Theoria*, vol. 36, pp. 373–398, 1970.

[...] *Pragmatics and Intensional Logic*, in *Semantics of Natural Language* (ed. Davidson and Harman), D. Reidel, 1972.

[...]

Scott, Dana: *Advice on Modal Logic*, in *Philosophical Problems in Logic* (ed. K. Lambert), D. Reidel, 1970.

Van Benthem, J.: *Modal Logic and Classical Logic*, Bibliopolis, Naples, 1983.

DAVID BASIN, SEÁN MATTHEWS, AND LUCA VIGANÒ

A TOPOGRAPHY OF LABELLED MODAL LOGICS

Abstract. Labelled Deductive Systems provide a general method for representing logics in a modular and transparent way. A Labelled Deductive System consists of two parts, a base logic and a labelling algebra, which interact through a fixed interface. The labelling algebra can be viewed as an independent parameter: the base logic stays fixed for a given class of related logics from which we can generate the one we want by plugging in the appropriate algebra. Our work identifies an important property of the structured presentation of logics, their combination, and extension. Namely, there is tension between modularity and extensibility: a narrow interface between the base logic and labelling algebra can limit the degree to which we can make use of extensions to the labelling algebra. We illustrate this in the case of modal logics and apply simple results from proof theory to give examples.

1. Introduction

The idea of a Labelled Deductive System (LDS) has been proposed as a general technique for representing logics and their consequence relations. In the LDS approach, instead of defining a consequence relation over formulae $(\ldots A \vdash B \ldots)$, we define it over pairs consisting of a label and a formula $(\ldots x : A \vdash y : B \ldots)$. The labels then allow us to track the information we need in order to formalize the more subtle aspects of the relation. For modal logic, for instance, we might want to distinguish between local and global consequence, so the label could keep track of which world (in the Kripke sense) the formula lives in. Or for a substructural logic, where the consequence relation should be sensitive to operations like weakening and contraction, the labels could track resources and their use (D'Agostino & Gabbay, 1994).

Labelled formulae have been implicitly used in proof systems (e.g. (Anderson & Belnap, 1975; Fitting, 1983)) for some time. More recently, howe-

F. Baader and K.U. Schulz (eds.), Frontiers of Combining Systems, 75–92.
© 1996 *Kluwer Academic Publishers.*

ver, people have begun to study Labelled Deductive Systems systematically as a general tool, in which the labelling is explicitly considered as part of the logic itself (Gabbay, 1994a; Gabbay, 1994b; Simpson, 1993). Along with the 'fine control' that explicit labelling offers it also supports modular theory development. Let us fix some terminology which will be useful for discussing this. We view a proof system for an LDS as consisting of two parts: a *base logic* for manipulating labelled formulae, and a separate *labelling algebra* for reasoning about properties of the labels. In the base logic (and possibly also the labelling algebra) there may be rules which combine formulae from the two parts; these rules form an *interface* for communication between the otherwise separate parts. This construction supports *modularity* in that we can view the labelling algebra as an independent parameter: the base stays fixed for a given class of related logics from which we can choose the one we want by plugging in the appropriate labelling algebra. D'Agostino and Gabbay (D'Agostino & Gabbay, 1994, p.244) explain such modularity as one of the important features of the LDS approach:

> The labelling algebra represents this metalevel information as a *separate* component of a standard derivation system and can be treated as an independent parameter. In the LDS approach, logical systems are not studied statically, in isolation, but dynamically, observing the process of their generation and their interaction (via modifications of the labelling algebras) on the basis of a fixed proof-theoretical hard core (the underlying system of deduction).   [their emphasis]

We give a concrete example of this kind of modular development in (Basin *et al.*, 1997), where we formalize a class of propositional modal logics. Our base logic there is a labelled natural deduction presentation of propositional calculus extended with introduction and elimination rules for $\Box$ (i.e. the logic K). Our labelling algebras are *relational theories* comprised of Horn clause axioms expressing accessibility of worlds in Kripke frames. That is, labels are possible worlds in the frame and the labelling theory provides a binary relation R, modelling the accessibility relation. Then the labelled formula $x : A$ specifies that A holds at the world indicated by its label x, and A is a theorem if it holds at an arbitrary world. The interface is unidirectional, from the labelling algebra to the base theory, and is given by the rules for $\Box$, e.g. the elimination rule says how we can combine a labelled formula $x : \Box A$ and a relation $x \, R \, y$ to conclude a new labelled formula $y : A$. This combination is enough to formalize, among others, the large and well-known Geach hierarchy, simply by varying the behavior of R.

In our work we used a generic proof system (Isabelle (Paulson, 1994)) supporting a metalogic to axiomatize the rules of the base and labelling theories. Our interface between the two parts was strictly enforced. Unlike other approaches based on semantic embedding (e.g. (Ohlbach, 1993)), in which both parts are combined in a single first-order theory, we used different judgements in the metalogic (as in (Harper, 1993)) to separate the two theories and control interaction between them. We showed that our formalization yields a modular presentation of modal logics that is correct (sound and complete with respect to the intended Kripke semantics) for modal consequence for any logic where R can be axiomatized using Horn clause axioms.

This work raises an interesting question. Given that separation between parts of an LDS is a philosophically attractive way to study classes of logics, is there any limit to which classes of modal logics we can capture? In (Basin et al., 1997) we show correctness for modal logics where the labelling algebra is axiomatized using Horn clause rules, but why stop there? After all, there are modal logics whose relational theories require full first-order, or even higher-order, axiomatizations (van Benthem, 1984; van Benthem, 1985). Shouldn't it be possible to represent them soundly and completely simply by axiomatizing their corresponding labelling algebras?

Our contribution in this paper is to show, perhaps surprisingly, that the issue is not that simple. There are several natural candidates for the 'hard core' of the base logic, which differ in their interfaces, essentially in the way that the rules for *falsum*, $\perp$, make use of labelling information. We show that the way the two LDS parts are separated has not only theoretical but also practical significance.

local base *falsum*: If we enforce a strong separation, by using only the $\square$ rules as an interface and making the rules for $\perp$ purely 'local' (so labels in labelled formulae can *only* be manipulated with the $\square$ rules), then we get a class of logics that does not correspond to any modal logic, although it has independently interesting paraconsistent properties.

global base *falsum*: If we allow slightly more communication, adopting 'global' rules for $\perp$ and again retaining the $\square$ rules as our interface, then we get the system we describe in (Basin et al., 1997), which allows us to formalize modal logics with Horn clause axiomatizable Kripke semantics.

universal *falsum*: In order to formalize all modal logics with first-order axiomatizable frames, we must allow even a wider interface between the two parts; in particular we must identify *falsum* in the base logic and labelling algebra by adding rules to propagate $\perp$ between the parts. The result is, effectively, a loss of separation similar to the approaches based on semantic embedding.

In the remainder of this paper we investigate these three possibilities and give a proof theoretic analysis and supporting examples to illustrate how tradeoffs arise. In Section 2 we present a labelled version of K and extensions which build on the second possibility above. After, in Section 3, we present proof normalization results for these theories. These are based on standard results for natural deduction proofs, e.g. (Prawitz, 1965; Prawitz, 1971; van Dalen, 1994), which we adapt to our natural deduction LDS presentations. These results are interesting in their own right (e.g. for automated theorem proving based on search for normal-form proofs) but our main application of them is to show tradeoffs in the interface of an LDS and which class of logics it captures. In particular, in Section 4, we show that a narrow interface, while leading to a clean separation of the base logic and labelling algebra, strongly restricts the class of representable logics. In the final section we draw conclusions and compare and apply our results to related work on labelled deduction and modal logics.

2. Labelled Modal Logics

We begin by introducing a labelled natural deduction (ND) system for the base modal logic K and its labelling algebra. Our primary focus is on this system, corresponding to 'global base *falsum*' above, and we will later consider changes of the base logic corresponding to the other possibilities listed above. This system was introduced in (Basin *et al.*, 1997) and we briefly review relevant parts. We assume the reader is familiar with modal logics (see, e.g., (Chellas, 1980)) and natural deduction (e.g. major and minor premises in an inference, discharged and open assumptions, see (Prawitz, 1965; Prawitz, 1971; van Dalen, 1994)).

2.1. LABELLED K

Let W be a set of *labels* and R a binary relation over W. If x and y are labels, then $x \, R \, y$ is a *relational formula* (*rwff*). We associate labels with formulae: if A is a propositional modal formula built from $\perp$, $\rightarrow$, $\Box$, then $x : A$ is a *labelled formula* (*lwff*). Hence, if p is a sentence letter, and A, B are propositional modal formulae, then $x : p$, $x : \perp$, $x : A \rightarrow B$, $x : \Box A$ are all lwffs. Lwffs containing other propositional and modal connectives (e.g. $\neg$, $\wedge$, $\vee$, $\Diamond$) can be defined in the usual manner, e.g. $x : \Diamond A \equiv x : (\Box(A \rightarrow \perp) \rightarrow \perp)$.

We henceforth assume that the possibly subscripted variables x, y, z range over labels, the possibly subscripted variables A, B, C range over propositional modal formulae, and $\Gamma = \{x_1 : A_1, \ldots, x_n : A_n\}$ and $\Delta = \{x_1 \, R \, y_1, \ldots, x_m \, R \, y_m\}$ are arbitrary sets of lwffs and rwffs.

The rules given in Figure 1 determine K, the base ND system which formalizes a labelled version of the modal logic K. We also give there the two derived rules for possibility, since we will explicitly use them in the latter. Notice that $\perp E$ is what we call a 'global' rule, and its restriction is

$$\begin{array}{c|c|c}
\begin{array}{c} [x:A \to \bot] \\ \vdots \\ \dfrac{y:\bot}{x:A} \ \bot E \end{array} \quad \dfrac{\begin{array}{c}[x:A]\\ \vdots \\ x:B \end{array}}{x:A \to B} \to I \qquad \dfrac{x:A \to B \quad x:A}{x:B} \to E
&
\dfrac{\begin{array}{c}[x\,R\,y]\\ \vdots \\ y:A \end{array}}{x:\Box A} \ \Box I \qquad\qquad \dfrac{x:\Box A \quad x\,R\,y}{y:A} \ \Box E
&
\begin{array}{c} \dfrac{y:A \quad x\,R\,y}{x:\Diamond A} \ \Diamond I \\[2em] \dfrac{x:\Diamond A \quad \begin{array}{c}[y:A]\ [x\,R\,y]\\ \vdots \\ z:B \end{array}}{z:B} \ \Diamond E \end{array}
\end{array}$$

Where, in $\bot E$, A is atomic and different from $\bot$; in $\Box I$ (respectively $\Diamond E$), y is different from x (respectively x and z) and does not occur in those assumptions on which $y : A$ (respectively $z : B$) depends, except those of the form $x\,R\,y$ (respectively $y:A$ and $x\,R\,y$), which are discharged by the inference.

Figure 1. The rules of K (with derived rules for $\Diamond$)

inessential but technically convenient (van Dalen, 1994, p.194). We discuss this rule and its 'global nature' in detail in Section 4.

2.2. EXTENSIONS OF K: RELATIONAL THEORIES

In K no assumptions are made on the relations between possible worlds in the Kripke frame $\langle W, R \rangle$, hence there are no assumptions about R in the labelling algebra. Other modal logics are obtained from K by placing conditions on the accessibility relation in the frame; e.g. we get the well-known logic T from K by adding that R is reflexive, and then $S4$ from T by further adding transitivity. In (Basin *et al.*, 1997) we show that a large subclass of the modal logics with first-order axiomatizable frames, including those usually of actual interest ($K, D, T, B, S4, S4.2, KD45, S5, \dots$) can be captured by extending K with Horn relational theories.

But there are properties of R which fall outside the Horn fragment, since they can only be expressed in a full first-order or higher-order setting. We briefly review Horn relational theories, and then introduce extensions of K by first-order theories.

2.3. HORN RELATIONAL THEORIES

A *Horn formula (over the binary relation R)* is a closed formula of the form $\forall x_1 \dots \forall x_n ((t_1\ R\ s_1 \wedge \dots \wedge t_m\ R\ s_m) \to t_0\ R\ s_0)$, where the t_i and s_i are terms built from the labels $x_1, \dots, x_n$ and function symbols. Corresponding to each such Horn formula is a *Horn rule*

$$\frac{t_1\ R\ s_1 \quad \dots \quad t_m\ R\ s_m}{t_0\ R\ s_0}$$

A *Horn (relational) theory* $\mathcal{T}_H$ is then a theory generated by a set of such Horn rules. Since the addition of a Horn formula to a theory is equivalent (in first-order theories) to adding the corresponding rule, we will talk about additions based on either formulae or rules as is convenient.

Let i, j, m, and n be natural numbers, and let $\Box^n$ (respectively $\Diamond^n$) stand for a sequence of n consecutive $\Box$s (respectively $\Diamond$s); for example $\Diamond^2\Box^3\Diamond^0 A$ is $\Diamond\Diamond\Box\Box\Box A$. Among all the properties of the accessibility relation that can be expressed as Horn rules in the theory of one binary predicate R, it is possible to identify the subclass of *restricted* (i, j, m, n) *convergency axioms*, as the class of properties of the accessibility relation which correspond to instances of the *generalized Geach axiom schema* $\Diamond^i\Box^m A \to \Box^j\Diamond^n A$ (see (Chellas, 1980)), and yield, among others, the logics K, D, T, B, $S4$, $S4.2$, $KD45$ and $S5$. Restricted (i, j, m, n) convergency axioms are closed formulae of the form

$$(1) \qquad \forall x \forall y \forall z ((x\ R^i\ y \wedge x\ R^j\ z) \to \exists u(y\ R^m\ u \wedge z\ R^n\ u)),$$

where $x\ R^0\ y$ means $x = y$ and $x\ R^{i+1}\ y$ means $\exists v(x\ R\ v \wedge v\ R^i\ y)$, and $m = n = 0$ implies $i = j = 0$. Although there are instances of (1) which explicitly require identity, the restriction that $m = n = 0$ implies $i = j = 0$ is a necessary and sufficient condition for identity to be inessential. In this paper we consider only theories without identity.

We point out in (Basin *et al.*, 1997, Prop. 3) that for any theory defined by a collection of instances of (1) there is a Horn relational theory conservatively extending it. As an example, consider the properties given in Figure 2, all of which are instances of (1); e.g. transitivity and convergency are given by $(0, 2, 1, 0)$ and $(1, 1, 1, 1)$. We also present there the corresponding Horn rules and characteristic modal axioms.

Various combinations of Horn rules define labelled equivalents of standard modal logics: the logic $L = K + \mathcal{T}_H$ is obtained by extending K with a given Horn relational theory $\mathcal{T}_H$.

We adopt the convention of naming the logic $K + \mathcal{T}$ as KAx, where $\mathcal{T}$ is a relational theory and Ax is a string consisting of the standard names of the characteristic axioms corresponding to the relational rules generating $\mathcal{T}$; e.g. $KT4$ identifies the logic also known as $S4$. Moreover, we distinguish between relational theories generated by Horn or first-order rules by appending subscripts H and F to $\mathcal{T}$, e.g. $KT4_H$, $KT4_F$. For example, $KT4_H$ is obtained by extending K with the Horn rules R_refl and R_trans, or alternatively by extending either KT_H with R_trans or $K4_H$ with R_refl.

2.4. FIRST-ORDER RELATIONAL THEORIES

In addition to Horn relational theories, we also consider a full first-order labelling language, i.e. we extend the previous definition of rwffs as follows:

Seriality, $D: \Box A \to \Diamond A$	Reflexivity, $T: \Box A \to A$
$$\frac{}{x\,R\,f(x)}\ R\_ser$$	$$\frac{}{x\,R\,x}\ R\_refl$$
Transitivity, $4: \Box A \to \Box\Box A$	Euclideaness, $5: \Diamond A \to \Box\Diamond A$
$$\frac{x\,R\,y\quad y\,R\,z}{x\,R\,z}\ R\_trans$$	$$\frac{x\,R\,y\quad x\,R\,z}{y\,R\,z}\ R\_eucl$$

Convergency, $2: \Diamond\Box A \to \Box\Diamond A$	
$$\frac{x\,R\,y\quad x\,R\,z}{y\,R\,g(x,y,z)}\ R\_conv1$$	$$\frac{x\,R\,y\quad x\,R\,z}{z\,R\,g(x,y,z)}\ R\_conv2$$

Where $f: W \to W$ and $g: (W \times W \times W) \to W$ are function constants.

Figure 2. Some properties of R, characteristic axioms, and Horn rules

if x, y are labels and ρ_1, ρ_2 are rwffs, then $x\,R\,y$, $\emptyset$ (relational *falsum*), $\rho_1 \supset \rho_2$ (implication), $\forall x.\rho_1$ are all rwffs. Notice that we use algebraic symbols ($\emptyset$, $\supset$) to avoid confusion with the symbols in the base logic. As notation, we henceforth assume that the possibly subscripted variable ρ ranges over rwffs, and we will call $x\,R\,y$ a *basic rwff*.

The rules of the first-order ND system of the accessibility relation, ND_R, are given in Figure 3. We also give there the three derived rules for disjunction ($\sqcup$); rwffs over other connectives (e.g. negation $\sim$, conjunction $\sqcap$, $\exists$) and corresponding rules are defined as usual, and we will explicitly use them in the following.

The axioms representing first-order properties of R can now be added directly in their full form, and the relational theory $\mathcal{T}_F$ (and hence the logic $K + \mathcal{T}_F$) is obtained by extending ND_R with a collection of such rules. As an example, for instances of restricted (i, j, m, n) convergency and for irreflexivity we add the rules (*rconv* is schematic):

$$\frac{}{\forall x.y.z.\ (x\,R^i\,y \sqcap x\,R^j\,z) \supset \exists u(y\,R^m\,u \sqcap z\,R^n\,u)}\ rconv$$

$$\frac{}{\forall x.\ \sim(x\,R\,x)}\ irrefl$$

2.5. DERIVATIONS

Consider an arbitrary modal logic $L = K + \mathcal{T}$, where $\mathcal{T}$ is a Horn or first-order relational theory, and let φ be either an lwff or an rwff. An *L-derivation* of φ from a set of lwffs Γ and a set of rwffs Δ is a tree formed using the rules in L, ending with φ and depending only on Γ, Δ. We write

$$
\cfrac{\begin{array}{c}[\rho_1]\\ \vdots\\ \rho_2\end{array}}{\rho_1 \supset \rho_2}\;\supset I
\qquad
\cfrac{\rho}{\forall x.\,\rho}\;\forall I
\qquad
\cfrac{\rho_2}{\rho_1 \cup \rho_2}\;\cup Il
\qquad
\cfrac{\rho_1}{\rho_1 \cup \rho_2}\;\cup Ir
$$

$$
\cfrac{\begin{array}{c}[\rho \supset \emptyset]\\ \vdots\\ \emptyset\end{array}}{\rho}\;\emptyset E
\qquad
\cfrac{\rho_1 \supset \rho_2 \quad \rho_1}{\rho_2}\;\supset E
\qquad
\cfrac{\forall x.\,\rho}{\rho[t/x]}\;\forall E
\qquad
\cfrac{\rho_1 \cup \rho_2 \quad \begin{array}{cc}[\rho_1] & [\rho_2]\\ \vdots & \vdots\\ \rho & \rho\end{array}}{\rho}\;\cup E
$$

Where, in $\forall I$, x must not occur free in any open assumption on which ρ depends.

Figure 3.   The rules of ND_R (with derived rules for $\cup$)

$\Gamma, \Delta \vdash_L \varphi$ when φ can be so derived. φ is an *L-theorem*, $\vdash_L \varphi$, if it is *L*-derivable from empty Γ and Δ. In the following we will often omit the '*L*'. We systematically use Π, with or without indices, to range over derivations, and we write $\overset{\Pi}{\varphi}$ to specify that the formula φ is the conclusion of the derivation Π. Similarly, we write $\overset{\varphi}{\Pi}$ (respectively $\overset{[\varphi]}{\Pi}$) to distinguish a possibly empty set of occurrences of the open (respectively discharged) assumption φ in Π. Furthermore, we write $\Pi[z/y]$ for the systematic substitution of z for y in Π, with a suitable renaming of the variables to avoid clashes. We use superscripts to associate discharged assumptions with rule applications. Also we combine multiple applications of the same rule into one, and mark the rule with an asterisk.

As an example, consider the following $K4$-derivation of $x : \Diamond \Diamond B$ from the assumptions $x : \Box A$, $y : \Diamond (A \to B)$, and $x\,R\,y$:

$$
\cfrac{y:\Diamond(A\to B) \quad \cfrac{\cfrac{[z:A\to B]^1 \quad \cfrac{x:\Box A \quad \cfrac{x\,R\,y \quad [y\,R\,z]^1}{\overset{\Pi}{x\,R\,z}}}{z:A}\;\Box E}{z:B}\;\to E}{y:\Diamond B}}{\cfrac{y:\Diamond B \quad \cfrac{[y\,R\,z]^1}{} \quad \cdots}{}}
$$

$$
\cfrac{x:\Diamond\Diamond B}{}
$$

The subderivation Π depends on the relational theory: for $K4_H$ we simply apply R_trans; for $K4_F$ we replace Π with:

$$
\cfrac{\cfrac{\cfrac{}{\forall x.y.z.\,(x\,R\,y \cap y\,R\,z) \supset x\,R\,z}\;trans}{(x\,R\,y \cap y\,R\,z) \supset x\,R\,z}\;\forall E^* \qquad \cfrac{x\,R\,y \quad [y\,R\,z]^1}{x\,R\,y \cap y\,R\,z}\;\cap I}{x\,R\,z}\;\supset E
$$

3. Normalization

Our goal in this paper is to show that there are design decisions in modularizing LDS presentations: the interface between the labelling algebra and the base logic effects which class of logics it correctly captures. To show this we need to show that with particular base logics (e.g. K) certain desired formulae (e.g. characteristic modal axioms) are not provable. We will establish this using syntactic methods based on adapting standard results of ND (Prawitz, 1965; Prawitz, 1971; van Dalen, 1994) to our labelled systems. To begin with, we show that each L-derivation can be transformed into a normal form which does not contain unnecessary detours or redundancies. We first discuss normalization for K, introducing the necessary terminology and pointing out the modifications with respect to standard results. After, we consider extensions with relational theories.

3.1. NORMALIZATION FOR K

Normalization need not be considered for rwffs in K, since rwffs only appear as assumptions, or as results of 'trivial' derivations, i.e.

Fact 1 $\Gamma, \Delta \vdash_K x\,R\,y$ iff $\Delta \vdash_K x\,R\,y$ iff $x\,R\,y \in \Delta$.

An immediate consequence is that rwffs appear only in the fringes of K-derivations of lwffs. This important characteristic of the shape of derivations is preserved in extensions of the base modal logic K with relational theories, where rwffs can be derived and not simply assumed (c.f. Lemma 6 below).

An occurrence of an lwff in a derivation that stands at the same time as the conclusion of an introduction rule and as the major premise of an elimination rule is called a *maximum lwff*. A maximum lwff is a detour in a derivation which can be removed to obtain a simpler derivation. This is achieved by applying the two following *proper reductions*, corresponding to the two possible forms of a maximum lwff (one for $\to$ and one for $\square$). We only show the part of the derivation where the reduction actually takes place; the missing parts remain unchanged. As notation, we write $\Pi \succ_1 \Pi'$ when Π (properly) reduces to Π', and we write $\succ$ (respectively $\succeq$) for the transitive (respectively transitive-reflexive) closure of $\succ_1$.

$$
\dfrac{\dfrac{\begin{array}{c}[x:A]\\ \Pi_2\\ x:B\end{array}}{x:A \to B}\to I \quad \dfrac{\Pi_1}{x:A}}{x:B}\to E \quad\succ_1\quad \begin{array}{c}\Pi_1\\ x:A\\ \Pi_2\\ x:B\end{array}
\qquad\qquad
\dfrac{\dfrac{\begin{array}{c}[x\,R\,y]\\ \Pi\\ y:A\end{array}}{x:\square A}\square I \quad x\,R\,z}{z:A}\square E \quad\succ_1\quad \begin{array}{c}x\,R\,z\\ \Pi[z/y]\\ z:A\end{array}
$$

A derivation Π is *irreducible* if there is no Π' such that Π reduces to Π'. If Π is irreducible, then we say that Π is a *normal derivation*, or, equivalently, that Π is in *normal form*, and if $\Pi \succeq \Pi'$ where Π' is normal, then we say that Π *normalizes* to Π'.

Theorem 2 *All K-derivations normalize.*

The theorem follows by a straightforward adaptation of the proof for standard ND systems, where it is shown that proper reductions always reduce a well-founded measure on derivations.

3.2. NORMALIZATION FOR $L = K + \mathcal{T}$

Fact 1 generalizes to:

Fact 3 $\Gamma, \Delta \vdash_{K+\mathcal{T}} \rho$ *iff* $\Delta \vdash_{\mathcal{T}} \rho$.

We therefore distinguish between $\mathcal{T}$-derivations of an rwff ρ from a set of rwffs Δ, and L-derivations of an lwff $x : A$ from sets Γ, Δ.

All the rwffs in a $\mathcal{T}_H$-derivation are basic rwffs, and $\mathcal{T}_H$ does not contain introduction or elimination rules. Hence, every $\mathcal{T}_H$-derivation is already normal. Moreover, definitions and results for K-derivations generalize to $K + \mathcal{T}_H$-derivations under some minor modifications reflecting the fact that in the latter basic rwffs can be also derived and not only assumed (e.g. substitute $\underset{x R z}{\overset{\Pi}{}}$ for the assumption $x \, R \, z$ in the proper reduction for $\square$). ¿From Theorem 2 we immediately have that:

Theorem 4 *All $K + \mathcal{T}_H$-derivations normalize.*

Let us now turn to first-order relational theories. Since $\mathcal{T}_F$ is a first-order ND system, it is straightforward to show that it possesses all the standard normalization properties for these systems (c.f. (van Dalen, 1994)). In particular, we can define normal $\mathcal{T}_F$-derivations and show that all $\mathcal{T}_F$-derivations normalize, and then combine the results for K and $\mathcal{T}_F$ to prove that:

Theorem 5 *All $K + \mathcal{T}_F$-derivations normalize.*

Normal derivations have a number of convenient properties, which follow from their structure. To discuss them, we further adapt some standard terminology.

A *path* in an L-derivation of an lwff $x : A$ is a sequence of formulae $\varphi_0, \ldots, \varphi_n, x : A$, such that φ_0 is an hypothesis, and φ_i is a premise immediately above φ_{i+1}, for $0 \leq i \leq n - 1$. Each φ_i can be either an lwff or an rwff, depending on L. An *rpath* in an L-derivation of an rwff ρ_n is a sequence of rwffs $\rho_0, \ldots, \rho_n$, such that ρ_0 is an hypothesis, and ρ_i is a premise immediately above ρ_{i+1}, for $0 \leq i \leq n - 1$. A *track* is an initial part of a path which stops at the first minor premise of an elimination rule of K or at the conclusion. In other words, a track can only pass through the major premises of elimination rules of K (we do not consider singular tracks consisting of one single formula, nor do we define tracks for rwffs). A track that is also a path, and thus ends at the conclusion of the derivation, is called a *main track*.

As an example, consider the following $K4_H$-derivation, together with its underlying tree (with numbers substituted for the formulae): the rpaths (dashed lines) are $(7,6)$ and $(8,6)$; the only, and thus main, track (highlighted in bold) is $(5,4,3,2,1)$.

$$\cfrac{[x:\Box A]^3 \quad \cfrac{\cfrac{[x\,R\,y]^2 \quad [y\,R\,z]^1}{x\,R\,z}\,R\_trans}{\cfrac{\cfrac{\cfrac{z:A}{y:\Box A}\,\Box I^1}{x:\Box\Box A}\,\Box I^2}{\qquad}\,\Box E}}{x:\Box A \to \Box\Box A}\,\to I^3$$

Lemma 6 *Let Π be a normal $K + \mathcal{T}$-derivation of an lwff. As shown in Figure 4, a track in Π is divided into two parts: a central subderivation in the base logic, and a $\mathcal{T}$ part. The central K-subderivation is divided into (at most) three subparts, in which only rules of K are applied: an elimination part, in which only elimination rules are applied; a $\bot$-part, in which $\bot E$ is applied; an introduction part, in which only introduction rules are applied. The $\mathcal{T}$ part is attached to the (elimination part of the) central K-subderivation by instances of $\Box E$, in which only basic rwffs appear. No lwffs appear in the $\mathcal{T}$ part, which consists of a fringe of subderivations purely in the labelling algebra. In particular, for K (Figure 4.a), it consists only of assumed basic rwffs; for $K + \mathcal{T}_H$ (Figure 4.b), it consists of $\mathcal{T}_H$-derivations of basic rwffs; for $K + \mathcal{T}_F$ (Figure 4.c), it consists of normal $\mathcal{T}_F$-derivations of basic rwffs (hence, the last steps in normal $\mathcal{T}_F$-derivations in a normal $K + \mathcal{T}_F$-derivation are either eliminations or applications of $\emptyset E$).*

The lemma follows by a straightforward adaptation of (van Dalen, 1994, Fact 6.2.10). Notice that the observations about the $\mathcal{T}$ part (its content and the way it is connected to the central K-subderivation) hold for all, not just normal $K + \mathcal{T}$-derivations. Moreover, if $\Diamond I$ is explicitly added, then $\mathcal{T}$-derivations of basic rwffs appear also at the fringes of the introduction part (and the lemma can be suitably modified).

Finally, it is possible to introduce a suitable notion of subformula for lwffs (by the standard syntactic analysis of the propositional modal formula), and derive a subformula principle. The notion of *sublwff* is defined inductively by: $y:A$ is a sublwff of $x:A$ for all x, y; if $y:B \to C$ is a sublwff of $y:A$, then so are $y:B$ and $y:C$; if $y:\Box B$ is a sublwff of $y:A$, then so is $y:B$.

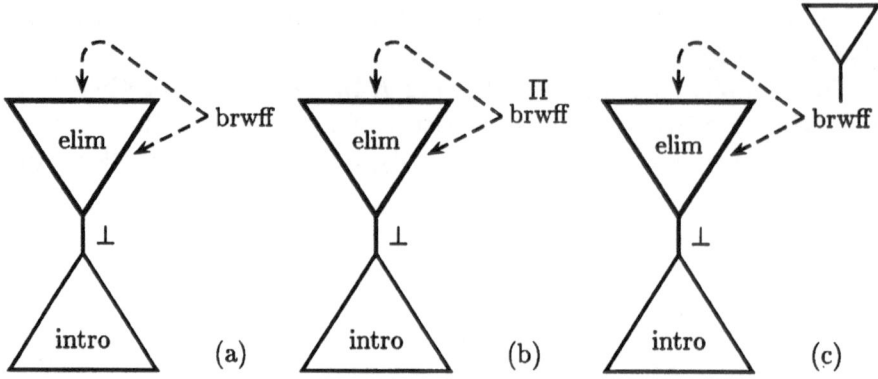

Figure 4.   The form of tracks in a normal derivation of an lwff in K, $K + \mathcal{T}_H$, $K + \mathcal{T}_F$
(brwff stands for basic rwff)

Lemma 7 (Sublwff principle for $K + \mathcal{T}$) *Every lwff in a normal $K +$
$\mathcal{T}$-derivation of $x : A$ from Γ, Δ is either a sublwff of $x : A$ or of some lwff
of Γ, or is an assumption $y : B \rightarrow \bot$ discharged by an application of $\bot E$
and $y : B$ is a sublwff of the kind mentioned.*

Although we cannot define a notion of subformula for basic rwffs in the logics $K + \mathcal{T}_H$, we remark that in some Horn relational rules with non-empty
premises, the labels in the conclusion already occur in the premises (in the
case of functions, the arguments of the function already occur in the premises, but not the function itself). On the other hand, a subformula principle
for rwffs in $\mathcal{T}_F$ can be easily shown by adapting standard argumentations
for first-order logic.

4. The Role of *falsum* and Theory Separation

The technical results developed above allow us to analyze modularity in
our LDS presentations of modal logics. We describe and compare the three
candidates for a base logic given in the introduction. We start with the
second, corresponding to the base logic K described in Section 2.1.

4.1. GLOBAL BASE *FALSUM*

We have indicated in Section 2.3 that K is an expressive enough base logic
to capture, using Horn theories, the modal logics usually of interest. Also,
we have shown that we have a precisely defined and restricted interaction
between the base logic and any algebra with which we extend it: any track
in a normal derivation of an lwff consists of a central subderivation purely in
the base logic and a fringe of subderivations purely in the labelling algebra,
attached to the central subderivation by instances of $\Box E$ (Lemma 6). This

is in the philosophical spirit of LDSs, and it is also pragmatically useful: since derivations of rwffs use only the resources of the labelling algebra, we may be able to employ theory specific reasoners successfully to automate proof construction.

However, clearly not all first-order Kripke frames correspond to some algebra $\mathcal{T}_H$. Thus we have to extend the range of possible algebras we allow to all those definable with first-order logic ($\mathcal{T}_F$). In this case we can again apply previous results (Lemma 6) to conclude that the same kind of controlled interaction we observed between K and $\mathcal{T}_H$ holds between K and $\mathcal{T}_F$. Unfortunately with the same interface we loose completeness.

Theorem 8 *There are extensions of K with first-order relational theories which are incomplete with respect to their intended Kripke semantics.*

Proof Consider the logic $L = K + \mathcal{T}_F$, where $\mathcal{T}_F$ is obtained by adding to ND_R the rule

$$\frac{\rule{6cm}{0.4pt}}{\forall x.y.z.\ (x\ R\ y \cap x\ R\ z) \supset (y\ R\ z \cup z\ R\ y)}\ conn.$$

This rule corresponds to the modal axiom $\Box(\Box A \to B) \vee \Box(\Box B \to A)$ (van Benthem, 1985, p.104), which is equivalent to $\Diamond(\Box A \wedge \neg B) \to \Box(\Box B \to A)$. We show[1] that if A and B are different sentence letters, then $\nvdash_L x : \Diamond(\Box A \wedge \neg B) \to \Box(\Box B \to A)$. It follows immediately that since the characteristic axiom is not derivable, the modal logic L is incomplete. To show non-provability, assume $\vdash_L x : \Diamond(\Box A \wedge \neg B) \to \Box(\Box B \to A)$ where A and B are different sentence letters. By Lemma 6, there exists a normal derivation where the last inferences are introductions, and, starting from the goal, we build the subderivation:

$$[x : \Diamond(\Box A \wedge \neg B)]^1\ [x\ R\ z]^2\ [z : \Box B]^3$$
$$\Pi_1$$
$$\cfrac{\cfrac{\cfrac{z : A}{z : \Box B \to A} \to I^3}{x : \Box(\Box B \to A)} \Box I^2}{x : \Diamond(\Box A \wedge \neg B) \to \Box(\Box B \to A)} \to I^1$$

Since A is a sentence letter, by Lemma 6, $z : A$ is the result of an application of an elimination rule; by permutative reductions (c.f. Footnote 1), we can assume it to be an application of $\Diamond E$:

---

[1] We omit many tedious details: the proof is obtained by considering all the possible inferences with respect to normalization and the sublwff principle. To improve readability, we explicitly use derived rules for K and $\mathcal{T}_F$. Strictly speaking, by analogy with the disjunction and existential elimination rules in first-order intuitionistic logic, the explicit presence of $\Diamond E$ and $\cup E$ requires us to consider other reductions (the so-called *permutative reductions*), which can be easily adapted from the ones presented in (van Dalen, 1994).

$$\cfrac{[x:\Diamond(\Box A \wedge \neg B)]^1 \qquad \cfrac{\cfrac{\cfrac{[x\,R\,z]^2 \ [z:\Box B]^3 \ [y:\Box A \wedge \neg B]^4 \ [x\,R\,y]^4}{\begin{array}{c}\Pi_2 \\ z:A\end{array}} \quad }{z:A}\Diamond E^4}{\cfrac{z:\Box B \to A}{x:\Box(\Box B \to A)}\Box I^2} \to I^3}{x:\Diamond(\Box A \wedge \neg B) \to \Box(\Box B \to A)} \to I^1$$

Now $z:A$ can only be the result of an application of $\bot E$ or of $\Box E$, since we can exclude the other elimination rules by Lemmata 6 and 7. Suppose that we apply $\bot E$. Then we can replace Π_2 with either one of two similar derivations; we consider only one, and point out that analogous reasoning applies for the other:

$$\cfrac{\cfrac{\cfrac{[y:\Box A \wedge \neg B]^4}{y:\neg B}\wedge Er \qquad \cfrac{[z:\Box B]^3 \quad \cfrac{\begin{array}{c}[x\,R\,z]^2 \ [x\,R\,y]^4 \ [z:\neg A]^5 \\ \Pi_3 \\ z\,R\,y\end{array}}{}\Box E}{y:B}}{y:\bot}\neg E}{z:A}\bot E^5$$

But, given $conn$ and $x\,R\,z$ and $x\,R\,y$, we can only derive $y\,R\,z \cup z\,R\,y$:

$$\cfrac{\cfrac{\cfrac{}{\forall x.y.z.\ (x\,R\,y \cap x\,R\,z) \supset (y\,R\,z \cup z\,R\,y)}conn}{(x\,R\,y \cap x\,R\,z) \supset (y\,R\,z \cup z\,R\,y)}\forall E^* \qquad \cfrac{[x\,R\,y]^4 \quad [x\,R\,z]^2}{x\,R\,y \cap x\,R\,z}\cap I}{y\,R\,z \cup z\,R\,y}\supset E$$

and $z\,R\,y$ cannot be a consequence of $y\,R\,z \cup z\,R\,y$.[2] Hence, $z:A$ cannot be the result of an application of $\bot E$. A similar reasoning applies in the case of $\Box E$, and we conclude that $\nvdash_L x:\Diamond(\Box A \wedge \neg B) \to \Box(\Box B \to A)$.

Analogous problems arise when trying to derive other modal formulae. However, the above is the simplest counter-example to the completeness of general $K + \mathcal{T}_F$ logics that we have been able to identify. Moreover, the provability of the corresponding modal axiom is philosophically the first requirement to be fulfilled by the addition of a relational rule. Other counter-examples can be devised in a similar way; e.g. it is possible to show that $x:\Box A \to \Diamond A$ does not follow from the rule for seriality, $\forall x. \exists y.\ x\,R\,y$.

---

[2]This follows by a straightforward semantic argument, with the observation that there are no further (relational) assumptions. Note that if R were symmetric, then $z\,R\,y$ would follow from $y\,R\,z \cup z\,R\,y$ by $\cup E$. Hence, this particular counter-example does not hold for extensions of the logic KB, for which, however, other counter-examples can be devised in a similar way.

4.2. UNIVERSAL *FALSUM*

We can isolate the reason for incompleteness in the way that $\perp$ is treated in K: $\perp E$ allows *falsum* to propagate across worlds, but in fact we have to allow more than this. *falsum* needs to be able to propagate also between the base logic and the algebra; i.e. we have to add rules equivalent to:

$$\frac{x:\perp}{\emptyset} \qquad \frac{\emptyset}{x:\perp}$$

Effectively, we identify the constants for *falsum* together in the two theories. In practice, perhaps the most sensible way to do this is to redefine the base logic and the labelling algebra by replacing the rule $\perp E$ with the rules $\perp E_L$ and $\perp E_R$, and the rule $\emptyset E$ with the rules $\emptyset E_L$ and $\emptyset E_R$:

$$\begin{array}{cccc}
[x:A \to \perp] & [\rho \supset \emptyset] & [x:A \to \perp] & [\rho \supset \emptyset] \\
\vdots & \vdots & \vdots & \vdots \\
\dfrac{y:\perp}{x:A}\ \perp E_L & \dfrac{y:\perp}{\rho}\ \perp E_R & \dfrac{\emptyset}{x:A}\ \emptyset E_L & \dfrac{\emptyset}{\rho}\ \emptyset E_R
\end{array}$$

Deriving the characteristic axiom for *conn* given these rules is now a simple exercise. However, with this wider interface we loose the controlled interaction between the base logic and the labelling algebra in derivations: Lemma 6 does not hold in these systems, since we can now derive lwffs from rwffs and rwffs from lwffs. For example, we can infer an arbitrary rwff from a modal contradiction, or an arbitrary lwff from a contradiction in the relational theory. As indicated in the introduction, by doing this we move much closer to a traditional semantic embedding in which there is only one theory and we cannot distinguish between their connectives and, in particular, *falsum*.

4.3. LOCAL BASE *FALSUM*

We have shown that K is not enough to formalize all the modal logics with first order semantics, but we have not yet considered what might be beneath it. And in fact there is a further obvious narrowing of the base logic, and we should ask what is, or is not, possible there.

We can easily formalize the consequence relation for the logic of $\to$ and $\perp$ without labels; it is only when we introduce $\square$ that we have a problem, because this involves (in the usual semantics) a notion of truth relative to a location. We might, therefore expect that we need to make use of the labels *only* in formalizing the rules for $\square$: after all we interpret $\to$ and $\perp$ purely locally in the Kripke semantics. And indeed in K the rules we give for $\to$ are local; $\perp E$, however, is not, since it relates two worlds (x and y): if we can derive *falsum* at an arbitrary y from $x:A \to \perp$, then we are able to derive A at x. A natural question to ask is what happens if we insist that x and y be the same; i.e.

$$\frac{\begin{array}{c}[x:A \to \bot] \\ \vdots \\ x:\bot\end{array}}{x:A} \bot E$$

It turns out that this base logic is not quite suitable, but the reasons are subtle. We lack space to go into the details, so we will confine ourselves to a simple example, showing that it is not suitable for any serial modal logic, since by the same style of analysis as in the proof of Theorem 8 we can show that:

$$x \, R \, y \not\vdash_K x:\Box A \to \neg\Box\neg A$$

However the systems we can obtain from local base *falsum* do have interesting paraconsistent properties: $\bot$ can only be propagated 'forwards' from a world to an accessible one, i.e. we have that:[3]

$$\frac{\dfrac{x:\bot}{x:\Box\bot} \bot E \quad x \, R \, y}{y:\bot} \Box E$$

In fact, this is not enough to capture the consequence relation of any standard modal logic, since we loose the duality between $\Box$ and $\Diamond$.

5. Discussion and Comparison

Our work identifies an important property of the structured presentation of logics, their combination, and extension. Namely, there is tension between modularity and extensibility: a narrow interface between the base logic and labelling algebra can limit the degree to which we can make use of extensions to the labelling algebra. As a consequence, there are important design decisions in implementing LDSs whose resolution requires predicting the range of possible applications.

This problem is a practical one which arose in our own work in encoding propositional modal logics while trying to take the idea of modular development seriously. Our approach has been inspired by the LDS methodology proposed by Gabbay, and further developed for modal logics, in parallel with our work, by Russo (Russo, 1995). However, our work differs from theirs in several important respects. First, Gabbay's proposed LDS methodology is very general and is based on a notion of diagrams and logic

---

[3] Clearly, when R is symmetric we can propagate $\bot$ also 'backwards'.

data-bases, manipulated by rules with multiple premises and conclusions. For example, in (Gabbay, 1994a, p.57) the rule for $\Diamond E$ is presented as:

$$\frac{s : \Diamond B}{create \; r, \; s < r \; and \; r : B}$$

the application of which updates a modal data-base with the two new conclusions (a rule to the same effect is given in (Russo, 1995)). The formal details are different from the kind of pure natural deduction encoding that we have presented, e.g. Figure 1. Second, although we have not provided details here, in our work we use a metalogic in which different judgements (e.g. in the sense of (Harper, 1993)) serve to separate the logic of the base core and the labelling algebra. In this paper, this difference is reflected by using different syntax for these theories. This separation is critical: it is only when we attempt to modularize and separate these two theories formally and define a precise interface between them that we see that only limited modularity (i.e. there are limits to the relational theories) is actually possible.

Of course, in implementing particular LDSs, Gabbay and Russo can similarly separate theories. The precise nature of this would be reflected in the rules they choose for propagating results between data-bases. It should be the case that if their rules enforce a similar separation, then they will encounter similar limitations to those reported here. That is, the problems we identify have some generality and should appear in other frameworks where theories are separated and results are communicated in a limited way between them.

The kind of labelled natural deduction encoding we employ is closest to the work of Simpson (Simpson, 1993). However his focus, proof techniques, and applications are based on using LDSs to investigate intuitionistic versions of modal logics, and his correctness considerations are quite different. Moreover, his relations have no independent theory with which one can work.

Notice that the universal *falsum* approach is adopted explicitly in (Russo, 1995). Simpson's approach is different, and difficult to compare: he treats rwffs only as assumptions in inferences of lwffs via his 'geometric' rules, which are derivable in our systems. An example of an approach in which, like with local *falsum*, local inconsistency does not imply global inconsistency, is the work of Giunchiglia and Serafini (Giunchiglia & Serafini, 1994), who show that particular 'multicontext systems', where (indexed) formulae are translated between contexts using 'bridge rules', define the same classes of provable formulae as certain standard modal logics. However their approach is, in general, radically different from ours, and not comparable.

In contrast, theorem provers for non-classical logic based on semantic embedding in first-order logic, e.g. (Ohlbach, 1993), also manipulate labelled formula and have no limitations on their labelling algebra. But then, there is no separation either, i.e. there is precisely one $\perp$ from which one can conclude arbitrary relational or labelled formulae.

DAVID BASIN, SEÁN MATTHEWS, LUCA VIGANÒ
Max-Planck-Institut für Informatik
Saarbrücken, Germany

References

A. Anderson and N. Belnap Jr. *Entailment, The Logic of Relevance and Necessity*, volume 1. Princeton University Press, 1975.

D. Basin, S. Matthews, and L. Viganò. A Modular Presentation of Modal Logics in a Logical Framework. In *Proc. of the Tbilisi Symposium on Language, Logic and Computation*, CSLI. To appear. Available at http://www.mpi-sb.mpg.de/~luca.

B. Chellas. *Modal Logic*. Cambridge University Press, New York, 1980.

M. D'Agostino and D. Gabbay. A Generalization of Analytic Deduction via Labelled Deductive Systems. Part I : Basic Substructural Logics. *JAR 13*, 1994.

M. Fitting. *Proof Methods for Modal and Intuitionistic Logics*. Kluwer, Dordrecht, 1983.

D. Gabbay. LDS - Labelled Deductive Systems, Volume 1 - Foundations. Technical report, MPI für Informatik, Saarbrücken, 1994.

D. Gabbay. What is a Logical System? In D. Gabbay, editor, *What is a Logical System?* Clarendon Press, Oxford, 1994.

F. Giunchiglia and L. Serafini. Multilanguage Hierarchical Logics (or: how we can do without modal logics). *Artificial Intelligence 65*, 1994.

R. Harper, F. Honsell, and G. Plotkin. A Framework for Defining Logics. *JACM*, 40, 1993.

H.-J. Ohlbach. Translation methods for non-classical logics: an overview. In *Bulletin of the IGPL*, volume 1, Saarbrücken, 1993.

L. Paulson. *Isabelle: a generic theorem prover*. LNCS-828. Springer, Berlin, 1994.

D. Prawitz. *Natural Deduction, a Proof-Theoretical Study*. Almqvist and Wiksell, Stockholm, 1965.

D. Prawitz. Ideas and Results in Proof Theory. In J. E. Fensted, editor, *Proc. of 2nd Scandinavian Logic Symposium*, Amsterdam, 1971. North-Holland.

A. Russo. Modal Labelled Deductive Systems. Technical Report 95/7, Department of Computing, Imperial College, London, UK, 1995.

A. Simpson. *The Proof Theory and Semantics of Intuitionistic Modal Logic*. PhD thesis, University of Edinburgh, Edinburgh, 1993.

J. van Benthem. Correspondence Theory. In D.M. Gabbay and F. Guenthner, editors, *Handbook of Philosophical Logic II*. D. Reidel Publishing Company, 1984.

J. van Benthem. *Modal Logic and Classical Logic*. Bibliopolis, Napoli, 1985.

D. van Dalen. *Logic and Structure*. Springer, Berlin, 1994.

LUIS FARIÑAS DEL CERRO AND ANDREAS HERZIG

COMBINING CLASSICAL AND INTUITIONISTIC LOGIC

Or: Intuitionistic Implication as a Conditional

Abstract. We study how a logic $C+J$ conbining classical logic C and intuitionistic logic J can be defined. We show that its Hilbert axiomatization *cannot* be attained by simply extending the union of the axiomatizations of C and J by so called interaction axioms. Such a logic would collapse into classical logic.

1. Introduction

In this short note we study how a logic $C+J$ containing classical logic C as well as intuitionistic logic J can be defined. Precisely, we shall study intuitionistic implication in the style of conditional logic. There, building on classical logic a new implicational connective is introduced.

Our language is based on a set of atoms $ATOMS = \{p, q, r \ldots\}$. The set of formulas $FOR = \{A, B, C, \ldots\}$ will constructed from atoms and the connectives that we shall give below.

We denote classical implication, equivalence and negation by $\rightarrow$, $\leftrightarrow$ and $\sim$, and intuitionistic implication and negation by $\Rightarrow$ and $\neg$. We suppose that there is a 0-ary connective $\perp$ denoting falsehood.

We suppose classical implication $A \rightarrow B$ to be an abbreviation of $\sim A \vee B$, and intuitionistic negation $\neg A$ to be that of $A \Rightarrow \perp$.

The semantics of $C+J$ is as expected: starting from the standard possible worlds semantics for J, we add the usual interpretation of classical implication and negation within a possible world.

There are more surprises on the axiomatic side: We show that the Hilbert axiomatization of that semantics *cannot* be attained by simply joining the axiomatizations of C and J and then adding some supplementary interaction axioms (simultaneously involving connectives from both). Such a logic would collapse into classical logic. Basically, what we must do is to modify the intuitionistic weakening axiom schema $A \Rightarrow (B \Rightarrow A)$ to

$$A \rightarrow (B \Rightarrow A) \text{ if } A \text{ is a persistent formula,}$$

93

F. Baader and K.U. Schulz (eds.), Frontiers of Combining Systems, 93–102.

where a formula of $C+J$ is said to be persistent if every occurrence of classical implication or negation is in the scope of some intuitionistic implication or negation.

In the rest of the paper, after a detour on conservative extensions we give semantics and axiomatics of $C+J$, and state soundness and completeness theorems. Finally, completeness will be proved using semantic tableaux, which give us at the same time a proof procedure for $C+J$.

2. Conservative extensions and the combination of logics

We say that some logic L_2 is the conservative extension of some other logic L_1 iff the language of L_1 is contained in that of L_2, and every theorem schema A that is valid in L_1 is valid in L_2 as well. The crucial point here is that we speak in terms of schemas and not instances: As the language of L_2 may contain more connectives as that of L_1, the validity test in L_2 involves more instances of A than before. In our case, the validity test of the schema $A \Rightarrow (B \Rightarrow A)$ in L_2 involves e.g. that of $\sim p \Rightarrow (q \Rightarrow \sim p)$ (which turns out to be not valid). [1] Therefore, $C+J$ cannot be axiomatized as a conservative extension of J.

Now suppose we want to combine two logics L_1 and L_2 (whose semantics and axiomatics are known) to a new logic $L_1 + L_2$. Suppose we already know the semantics of $L_1 + L_2$ (which we might have obtained by fibering (Gabbay, 1996) or some other technique). What we would like to have is an axiomatics of $L_1 + L_2$.

It is natural to try to in some sense combine the axiomatics of L_1 and L_2. Clearly, the most favourable case is when $L_1 + L_2$ is a conservative extension of both L_1 and L_2. This is the case e.g. if we combine modal logics of time and knowledge, and more generally when we combine several modal logics into a multimodal logic ((Catach, 1988), (Kracht, 1993)). A necessary and sufficient condition for $L_1 + L_2$ to be a conservative extension of L_i $(i = 1, 2)$ is that every axiom schema of L_i is valid in $L_1 + L_2$, and that every rule schema of L_i preserves validity in $L_1 + L_2$.

In the case of $C+J$, we shall see that we must weaken the axiom schemas of J in order to get a sound axiomatization of the natural semantics.

3. Semantics

A $C+J$-**model** is a triple $M = < W, R, I >$ where

- W is a set (called the set of possible worlds),

---

[1] If we tested validity of instances (and not of schemas) we would get another sense of conservative extension. In this weaker sense, $C+J$ would be a conservative extension of J.

- R is a partial preorder on W, i.e. a reflexive and transitive relation (called the accessibility relation),
- I is a mapping from W into 2^{ATOMS} such that $I(w_1) \subseteq I(w_2)$ whenever $w_1 R w_2$.

The **forcing relation** $\models$ is such that

- $w \models A$ if $A \in ATOMS$ and $A \in I(w)$.
- $w \models A \land B$ if $w \models A$ and $w \models B$.
- $w \models A \lor B$ if $w \models A$ or $w \models B$.
- $w \models {\sim}A$ if $w \not\models A$.
- $w \models A \Rightarrow B$ if for all w' such that wRw', $w' \not\models A$ or $w' \models B$.

We use the standard notions of satisfiability and validity. We write $\models_{C+J} A$ to express that A is valid.

First of all, we note that the rule of uniform substitution cannot be valid in $C+J$. E.g. the formula $p \Rightarrow (q \Rightarrow p)$ is valid, whereas ${\sim}p \Rightarrow (q \Rightarrow {\sim}p)$ is not.

In fact, this illustrates a more general problem:

Proposition. The weakening axiom schema $A \Rightarrow (B \Rightarrow A)$ is not valid in $C+J$-models.

Corollary. $C+J$ cannot be axiomatized as a conservative extension of intuitionistic logic J.

In other words, we cannot get an axiomatization of the class of $C+J$-models just by putting together the axiomatics of J and that of C (possibly adding some other axioms to deal with the interaction between classical and intuitionistic implication). [2]

Therefore we have chosen to study $C+J$ in the framework of conditional logics.

Formally, what is studied in conditional logics are implications that are weaker than classical implication, in the sense that the former (classically) imply the latter, but not the converse. This is exactly what happens in our $C+J$-models: We always have $w \models A \to B$ whenever $w \models A \Rightarrow B$, but it may be the case that $w \models A \to B$ without having $w \models A \Rightarrow B$.

More precisely, what people in the field of conditional logics study is how to invalidate several principles that are valid in classical logic and that are generally felt to be too strong for a natural reading of the natural language conditional "if . . . then" (Nute, 1984), (Nute, 1980). These principles are

---

[2]In fact, if we just joined the axiomatics of J and C, both ${\sim} p \to (A \Rightarrow {\sim}p)$ and $p \to (A \Rightarrow p)$ would be instances of the weakening axiom schema (for every atom p and every formula A), and therefore every world of a $C+J$-model would have the same interpretation. Consequently our logic would collapse into classical logic in the sense that it would contain $A \Rightarrow B. \leftrightarrow .A \to B$.

TRANS: $(A \Rightarrow B) \wedge (B \Rightarrow C). \to .A \Rightarrow C$
MON: $A \Rightarrow C. \to .(A \wedge B) \Rightarrow C$
CONTR: $A \Rightarrow C. \to . \sim C \Rightarrow \sim A$

They are closely related: in one of the weakest logics of conditionals W (which nevertheless contains the principle $A \Rightarrow B. \to .A \to B$), to adopt one of them means to adopt them all (Nute, 1980). [3]

From that point of view, $C + J$ will be somewhat degenerated, because it has all of these principles. On the other hand, given Gödel's translation of intuitionistic logic into modal logic $S4$ we should not be surprised that $C + J$ collapses into modal logic.

4. Axiomatics

In order to restrict the intuitionistic weakening axiom we need the notion of a persistent formula.

A formula of $C + J$ is said to be **persistent** iff every occurrence of classical implication, equivalence or negation is in the scope of some intuitionistic implication or negation. [4] Formally, persistent formulas are defined inductively as being the smallest set such that

- A is persistent if $A \in ATOMS$,
- $A \Rightarrow B$ is persistent,
- $\neg A$ is persistent,
- $A \wedge B$ is persistent if A and B are both persistent,
- $A \vee B$ is persistent if A and B are both persistent.

We have the following **axiom and inference rule schemas** for $C + J$:

CL: all theorems of classical logic
CK: $A \Rightarrow (B \to C). \to .(A \Rightarrow B) \to (A \Rightarrow C)$
ID: $A \Rightarrow A$
CMP: $A \Rightarrow B. \to .A \to B$
PER: $A \to .B \Rightarrow A$ if A is persistent
MP: If A and $A \Rightarrow B$ then B
RCN: If A then $B \Rightarrow A$

We use the standard notions of consistency and theoremhood. We write $\vdash_{C+J} A$ to express that A is a theorem of $C + J$.

---

[3] Note that a conditional logic having these principles does not collapse into classical logic, but into modal logic K, in the sense that $A \to B. \leftrightarrow \Box(A \to B)$ (Nute, 1980).

Note also that the picture would be different if our logic was not constructed as an extension of classical logic, i.e. if there was no other implication than the intuionistic one. Then transitivity TRANS would not imply monotony MON. Examples of this type are intuitionistic logic J as well as relevance logics (where transitivity is desirable, whereas monotony is felt to be in contradiction with the intuitions of relevance).

[4] Another possibility is to define persistent formulas to be either atoms or formulas of the form $A \Rightarrow B$ or $\neg A$.

Remark. It is important to have axiom schemas here (instead of axioms). This enables us to avoid the rule of uniform substitution (which would conflict with our axiom PER.)

Remark. All the axiom names except that of the axiom of persistence PER are standard in the conditional logic literature (Chellas, 1975). (We have just changed the name of Chellas's modus ponens axiom for conditionals from MP to CMP to avoid a notational conflict with the classical Modus Ponens inference rule.)

Remark. Now we can also give a syntactic proof that unrestricted acceptation of either the intuitionistic weakening axiom schema $A \Rightarrow (B \Rightarrow A)$ or our weaker variant $A \rightarrow (B \Rightarrow A)$ would make $C+J$ collapse into classical logic. First, we have that $A \Rightarrow B. \rightarrow .A \rightarrow B$ (this is CMP). Now with the unrestricted version of the persistence axiom schema we can prove the opposite direction $A \rightarrow B. \rightarrow .A \Rightarrow B$: such a schema would give us $A \rightarrow B. \rightarrow .A \Rightarrow (A \rightarrow B)$, and from the consequens part we would get $A \Rightarrow B$ with ID and CK. Consequently we would have $A \Rightarrow B. \leftrightarrow .A \rightarrow B$.

Given that all the other schemas are standard in conditional logic, it is clear that it is the axiom schema PER which is most powerful here. In fact, it is this schema which allows to derive most of the other schemas that are studied in conditional logics:

Theorem. The following schemas
>   RCEA: If $A \leftrightarrow B$ then $A \Rightarrow C. \leftrightarrow .B \Rightarrow C$
>   CSO: $(A \Rightarrow B) \wedge (B \Rightarrow A). \rightarrow .(A \Rightarrow C) \leftrightarrow (B \Rightarrow C)$
>   CUT: $(A \Rightarrow B) \wedge ((A \wedge B) \Rightarrow C). \rightarrow .A \Rightarrow C$
>   CA: $(A \Rightarrow C) \wedge (B \Rightarrow C). \rightarrow .(A \vee B) \Rightarrow C$
>   CV: $(A \Rightarrow C) \wedge \neg(A \Rightarrow \neg B). \rightarrow .(A \wedge B) \Rightarrow C$
>   TRANS: $(A \Rightarrow B) \wedge (B \Rightarrow C). \rightarrow .A \Rightarrow C$
>   MON: $A \Rightarrow C. \rightarrow .(A \wedge B) \Rightarrow C$

are derivable in $C+J$.

Proof. CSO, CUT, and CV follow from TRANS and MON. From the rest, we only prove MON.

First, as $A \Rightarrow C$ is persistent we have

$$A \Rightarrow C. \rightarrow .(A \wedge B) \Rightarrow (A \Rightarrow C)$$

by PER. Second, as $A \Rightarrow C. \rightarrow .A \rightarrow C$ with CMP, we get

$$(A \wedge B) \Rightarrow (A \Rightarrow C). \rightarrow .(A \wedge B) \Rightarrow (A \rightarrow C)$$

by RCN and CK. Third,

$$(A \wedge B) \Rightarrow (A \rightarrow C). \rightarrow .(A \wedge B) \Rightarrow C$$

with ID and CK. From these three formulas we get MON by the transitivity of classical implication.

Note that in the field of conditional logics, the principles TRANS and MON are considered to be undesirable properties of a conditional operator.

Theorem. The axiomatics of $C+J$ is sound.

Proof. The validity of all the axiom and rule schemas except PER can be established easily by refering to standard conditional logics as e.g. the systems in (Chellas, 1975). That PER is valid can be proved by showing the following

Lemma. Let A be a persistent formula. Whenever $w \models A$ for some world w in some model $M =< W, R, I >$, then $w' \models A$ for every w' such that wRw'.

Proof. We use induction on the structure of A. W.l.o.g. we suppose that abbreviations have been eliminated. The base follows from the restriction on the interpretation I in the models. In the case of the intuitionistic connective $\Rightarrow$ we use transitivity of the accessibility relation. Finally, the case of the classical connective $\sim$ never occurs, because else A would not be persistent. This ends the proof of the lemma.

Now suppose A is persistent, and $w \models A$. The above lemma warrants that $w' \models A$ for all w' such that wRw'. Hence $w' \models A$ for all w' such that wRw' and $w' \models B$. But this means that $w \models A \Rightarrow B$.

This completes the proof of the soundness theorem.

Theorem. The axiomatics of $C+J$ is complete.

The completeness theorem will be proven in the next section, making use of semantic tableaux.

5. Semantic tableaux and completeness proof

In this section we give a tableau method, which furnishes us at the same time a completeness proof for $C+J$.

$S, S_1, S_2, \ldots$ denote sets of formulas. A **form** is a set of sets of formulas. $\mathcal{S}, \mathcal{S}', \ldots$ denote forms.

Tableau rules rewrite parts of a form $\mathcal{S}$. In this way, a new form is obtained.

A binary tableau rule takes a set of formulas $S \in \mathcal{S}$ and rewrites it into a new set of formulas S_1. A ternary tableau rules takes a set of formulas $S \in \mathcal{S}$ and rewrites it into two new sets of formulas S_1 and S_2. Binary rules are noted $S \rightsquigarrow S_1$, and ternary ones $S \rightsquigarrow S_1, S_2$.

A binary tableau rule rewrites a form $\mathcal{S}$ into $\mathcal{S}' = (\mathcal{S} - S) \cup \{S_1\}$, and a ternary tableau rule rewrites it into $\mathcal{S}' = (\mathcal{S} - S) \cup \{S_1, S_2\}$.

Here come the **tableau rules for** $C+J$.

- $S \cup \{A \wedge B\} \rightsquigarrow S \cup \{A, B\}$
- $S \cup \{\sim(A \wedge B)\} \rightsquigarrow S \cup \{\sim A\}, S \cup \{\sim B\}$
- $S \cup \{A \vee B\} \rightsquigarrow S \cup \{A\}, S \cup \{B\}$
- $S \cup \{\sim(A \vee B)\} \rightsquigarrow S \cup \{\sim A, \sim B\}$
- $S \cup \{\sim \sim A\} \rightsquigarrow S \cup \{A\}$
- $S \cup \{A \Rightarrow B\} \rightsquigarrow S \cup \{\sim A\}, S \cup \{B\}$
- $S \cup \{\sim(A \Rightarrow B)\} \rightsquigarrow S^{\sharp} \cup \{A, \sim B\}$

where $S^{\sharp}$ is the set of persistent formulas of S:

$$S^{\sharp} = \{C : C \in S \text{ and } C \text{ is persistent}\}$$

Remark. Note that we need not use the machinery of signed formulas *à la* Fitting (Fitting, 1983) because we have classical negation at our disposal.

Given a formula A, a **tableau for** A is a sequence of forms $S_1, \ldots, S_n$ such that

- $S_1 = \{\{A\}\}$, and
- S_{i+1} is obtained from S_i via a tableau rule.

A *set of formulas is closed* if it contains some formula A and its classical negation $\sim A$. A *form is closed* if each set of formulas in it is closed. A *tableau is closed* if some form in it is closed.

Note that (in accordance with what we have said about the rule of uniform substitution) we can use tableaux only to prove theorems (as opposed to theorem schemas). This is illustrated by the following two examples.

Example. The formula $p \Rightarrow (q \Rightarrow p)$ has a closed tableau.

Example. The formula $\sim p \Rightarrow (q \Rightarrow \sim p)$ does not have a closed tableau.

The next two theorems connect our tableaux with semantics and axiomatics.

Theorem. If $\models_{C+J} A$ then there exists a closed tableau for $\sim A$.

Proof. We construct a $C+J$-model for A from the non-closed tableau for A in the usual way (v. (Fitting, 1983)).

Theorem. If there exists a closed tableau for $\sim A$ then $\vdash_{C+J} A$.

Proof. We must show that the tableau rules preserve consistency. The only interesting one is

$$S \cup \{\sim(A \Rightarrow B)\} \rightsquigarrow S^{\sharp} \cup \{A, \sim B\}$$

Confusing the set $S^\sharp$ and the conjunction of its elements we can show that

$$(S^\sharp \wedge \sim(A \Rightarrow B)) \to \sim(A \Rightarrow \sim(S^\sharp \wedge A \wedge \sim B))$$

is a theorem of $C+J$. As $S^\sharp \cup \{ \sim(A \Rightarrow B)\}$ is consistent, the above theorem warrants that

$$\sim(A \Rightarrow \sim(S^\sharp \wedge A \wedge \sim B))$$

is consistent as well.

We immediately get a proof of the completeness theorem for $C+J$ by putting together the above two theorems.

6. Decidability

Here, we establish the decidability of $C+J$ by giving a translation $(.)^\tau$ into modal logic $S4$. It is the natural extension of Gödel's translation and is defined by

- $A^\tau = \Box A$ if $A \in ATOMS$
- $(\sim A)^\tau = \sim A^\tau$
- $(A \Rightarrow B)^\tau = \Box(A^\tau \to B^\tau)$
- $(A \wedge B)^\tau = A^\tau \wedge B^\tau$
- $(A \vee B)^\tau = A^\tau \vee B^\tau$

The following two theorems establish soundness and completeness of the translation.

Theorem. If $\vdash_{C+J} A$ then $\vdash_{S4} A^\tau$.

Proof. We transform the $C+J$-proof of A into an $S4$-proof of A^τ by induction on the length of the former.

The cases of all the axioms are immediate, except that of PER. We show the following

Lemma. Let B be a persistent formula. Then $\vdash_{S4} B^\tau \leftrightarrow \Box B^\tau$.

Proof. We use induction on the structure of B. The only non-trivial case is when B is of the form $C \vee D$. Then $B^\tau = C^\tau \vee D^\tau$. By induction hypothesis, $\vdash_{S4} C^\tau \leftrightarrow \Box C^\tau$ and $\vdash_{S4} D^\tau \leftrightarrow \Box D^\tau$. Hence $C^\tau \vee D^\tau \leftrightarrow \Box C^\tau \vee \Box D^\tau$. This establishes the lemma.

By the lemma, we have that the translation of PER: $(C \to (D \Rightarrow C))^\tau = C^\tau \to \Box(D^\tau \to C^\tau)$ is equivalent to $\Box C^\tau \to \Box(\Box D^\tau \to \Box C^\tau)$, which is a theorem of $S4$.

Theorem. If A is $C+J$-satisfiable then A^τ is $S4$-satisfiable.

Proof. Let A be $C+J$-satisfiable. There is a $C+J$-model $M = <W, R, I>$ and a world $w \in W$ such that $w \models A$. M is a $S4$-model as well, and it is sufficient to prove $w \models A^\tau$ in that $S4$-model. We use induction on the structure of A. We only give the cases of atoms and intuitionistic implication.

If a formula B is an atom then we have $w \models \Box B$, because $C+J$-models satisfy the condition $I(w_1) \subseteq I(w_2)$ for all w_1, w_2 such that $w_1 R w_2$.

If B is of the form $C \Rightarrow D$ then $B^\tau = \Box(C^\tau \to D^\tau)$. $w \models C \Rightarrow D$ means that for all w' such that wRw' we have $w \models C \to D$. By induction hypothesis, $w \models C^\tau \to D^\tau$ for all w' such that wRw'. Hence $w \models \Box(C^\tau \to D^\tau)$.

As $S4$ is sound and complete, soundness and completeness of the translation into $S4$ follow from the above two theorems.

Moreover, as $S4$ is decidable, we immediately get the

Corollary. $C+J$ is decidable.

Finally, $S4$ and intuitionistic logic J being both PSPACE-complete ((Ladner, 1977), (Statman, 1979)), we get another result concerning the complexity of the decision problem for $C+J$:

Corollary. The decision problem for $C+J$ is PSPACE-complete.

7. Conclusion

In this paper, we have proved soundness and completeness of our logic $C+J$ combining classical and intuitionistic logic. Completeness has been shown via semantic tableaux, which give us at the same time a proof method for the $C+J$. Another proof method comes with the translation into $S4$ that we have established as well.

We have shown that the Hilbert axiomatization of that semantics *cannot* be attained by simply extending the union of the axiomatizations of C and J by some supplementary interaction axiom. In other words, classical and intuitionistic logic cannot be combined in a conservative way.

Humberstone (Humberstone, 1979) has studied the combination of intuitionistic and classical logic in a similar way, aiming at an interval-based semantics for tense logics. Contrarily to ours, his language is based on negations and not on implications, and $A \Rightarrow B$ is introduced as an abreviation of $\neg(A \wedge \sim B)$.

Humberstone's completeness proof is in terms of canonical models. Contrarily to our tableau-based completeness proof, this does not give a theorem proving procedure.

Humberstone's axiomatization is in natural deduction style. He has a notion of persistent formulas that is just as ours. The rule for negation introduction on the right is

If all formulas in Γ are persistent and $\Gamma, \alpha \vdash \bot$ then $\Gamma \vdash \sim\alpha$

At first glance this rule looks rather different from our Hilbert-style axiom of persistence PER, but it is – as one could expect – quite closely related to it.

8. Acknowledgements

Thanks to two anonymous referees for their helpful comments.

LUIS FARIÑAS DEL CERRO AND ANDREAS HERZIG
IRIT, Université Paul Sabatier
Toulouse, France

References

L. Catach (1988), Normal multimodal logics. *Proc. Nat. Conf. on AI (AAAI-88)*, pp 491-495.

B. Chellas (1975), Basic conditional logics. *Journal of Philosophical Logic* 4, pp 133-153.

M. Fitting (1983), *Proof methods for modal and intuitionistic logics.* Synthese Library, Vol. 169, D. Reidel Publishing Company, Dordrecht.

D. Gabbay (1996), *Labelled Deductive Systems.* Oxford University Press, to appear.

L. Humberstone (1979), Interval semantics for tense logic: some remarks. *Journal of Philosophical Logic* 8, pp 171-196.

M. Kracht (1993), How completeness and correspondence theory got married. In: M. de Rijke (ed.), *Diamonds and Defaults.* Synthese, Kluwer.

S. Kripke (1965), Semantical analysis of intuitionistic logic, I. In: J. N. Crossley, M. A. E. Dummet (eds.), *Formal systems and recursive functions.* Amsterdam, North Holland Publishing Co., pp 92-129.

D. Ladner (1977), The computational complexity of provability in systems of modal propositional logic. *SIAM J. Computing 3, pp. 467-480.*

D. Nute (1980), *Topics in conditional logic.* Philosophical Studies Series, D. Reidel, Dordrecht.

D. Nute (1984), Conditional logic. In Dov M. Gabbay and Franz Guenthner (eds.), *Handbook of Philosophical Logic.* D. Reidel, Dordrecht.

R. Statman (1979), Intuitionistic propositional logic is polynomial-space complete. *Theoretical Computer Science 9, pp 67-72.*

A NEW CORRECTNESS PROOF OF THE NELSON-OPPEN COMBINATION PROCEDURE

Abstract. The Nelson-Oppen combination procedure, which combines satisfiability procedures for a class of first-order theories by propagation of equalities between variables, is one of the most general combination methods in the field of theory combination. We describe a new nondeterministic version of the procedure that has been used to extend the Constraint Logic Programming Scheme to unions of constraint theories. The correctness proof of the procedure that we give in this paper not only constitutes a novel and easier proof of Nelson and Oppen's original results, but also shows that equality sharing between the satisfiability procedures of the component theories, the main idea of the method, can be confined to a restricted set of variables.

While working on the new correctness proof, we also found a new characterization of the consistency of the union of first-order theories. We discuss and give a proof of such characterization as well.

1. Introduction

Nelson and Oppen were among the first to provide a fairly general method to combine logical theories and relative satisfiability procedures (Nelson & Oppen, 1979). Since then, almost all the effort in the field of combination has been concentrated on equational theories and unification algorithms (Baader & Schulz, 1992; Baader & Schulz, 1995a; Baader & Schulz, 1995c; Boudet, 1990; Boudet, 1993; Domenjoud *et al.*, 1994; Herold, 1986; Kirchner & Ringeissen, 1992; Ringeissen, 1992; Schmidt-Schauß, 1988; Schmidt-Schauß, 1989; Tidén, 1986; Yelik, 1987b; Yelik, 1987a). Others have worked on combinations of more general theories as well (see (Kirchner & Ringeissen, 1994; Shostak, 1979; Shostak, 1984), for instance) but to date the Nelson-Oppen method appears to be

F. Baader and K.U. Schulz (eds.), Frontiers of Combining Systems, 103–119.
© 1996 *Kluwer Academic Publishers.*

still one of the most general in the field.

The need of extending the focus from unification to more general satisfiability problems is well felt in the combination literature after the emergence and consolidation of several constraint-based computational paradigms that operate on more general constraint domains than those described by equational theories.

We have shown in (Tinelli, 1995) how the generality of the Nelson-Oppen method allows us to easily incorporate a combination procedure based on that method into a constraint-based computation framework, namely, the CLP scheme of Jaffar and Lassez (Jaffar & Lassez, 1986; Jaffar & Lassez, 1987; Jaffar & Maher, 1994), with few modifications of the scheme itself. In that work, we first describe a non-deterministic version of the combination procedure originally devised by Nelson and Oppen. Then we integrate the procedure into a modified version of the CLP scheme to obtain a new scheme that operates with unions of constraint theories by using constraint solvers for the single component theories.

For space limitations, we are not able to describe here the integration in CLP nor prove that the main properties of CLP lift to the new scheme. For this we refer the reader to (Tinelli, 1995). Instead, in this paper, we describe and discuss the combination procedure used in (Tinelli, 1995) and provide a novel proof of Nelson and Oppen's original results. In addition, we provide a characterization of the consistency of the union of first-order theories.

1.1. NOTATION AND CONVENTIONS

In general, we will adhere to the notation and definitions given in (Shoenfield, 1967). The most notable notational conventions followed are given below with the understanding that other notations that may appear in the paper follow the common conventions of the field.

The letters v, x, y, z denote logical variables, φ, ψ first order formulas, and ϑ a value assignment, or valuation, to a set of variables.

Some of the above symbols may be subscripted or have an over-tilde which will represent a finite sequence. For instance, $\tilde{x}$ stands for a sequence of the form $(x_1, x_2, \ldots, x_n)$ for some natural number n. When convenient, we will use the tilde notation to denote just sets of symbols—as opposed to sequences. The notation $\varphi(\tilde{x})$ is used to indicate that the free variables of φ are *exactly* the ones in $\tilde{x}$. In general, $var(\varphi)$ is the set of all the free variables of φ. The shorthand $\tilde{\exists}\,\varphi$ stands for the existential closure of φ.

Where $\mathcal{M}$ is a structure and φ a sentence, that is, a closed formula, the notation $\mathcal{M} \models \varphi$ means that $\mathcal{M}$ satisfies φ or, equivalently, that φ is true in $\mathcal{M}$. If φ is in general a formula and ϑ a valuation on $\mathcal{M}$ of φ's free variables, if any, the notation $\mathcal{M} \models \varphi\vartheta$ means that ϑ satisfies φ in $\mathcal{M}$. Notice that, in analogy with substitutions, we write valuation applications in postfix form. For convenience, where ϑ is a valuation of $\tilde{x}$, we will also indicate with ϑ the reduction of ϑ to $\tilde{y}$ if $\tilde{y} \subset \tilde{x}$, or an arbitrary expansion of it to $\tilde{y}$ if $\tilde{x} \subset \tilde{y}$.

We will generally identify first-order theories with the set of their theorems. We will also identify union of multi-sets of formulas with logical conjunction.

1.2. ORGANIZATION OF THE PAPER

In Sect. 2, we first recall and briefly discuss the Nelson-Oppen method; then we describe our non-deterministic version of their combination procedure. In Sect. 3, we formalize the combination problem more rigorously and discuss the conditions under which union of theories are consistent. Then we prove the correctness of the combination procedure introduced in Sect. 2. We conclude in Sect. 4 with some remarks on the possible extension of the procedure and give direction for further development.

2. The Nelson and Oppen Combination Method

In (Nelson & Oppen, 1979), Nelson and Oppen show how a satisfaction procedure for a theory built by combining several first-order theories can be derived as a combination of the satisfaction procedures for each of these theories. The main idea is to combine the satisfiability procedures by means of equality sharing. We will clarify this in a moment, but first let us set the problem more rigorously.

We say that a formula is in *simple Conjunctive Normal Form* if it is a conjunction of literals. Consider n first-order theories with equality, $\mathcal{T}_1, \ldots, \mathcal{T}_n$, with respective signatures $\Sigma_1, \ldots, \Sigma_n$. Assume that no two signatures have any non-logical symbols in common.[1] For each theory, let $sCNF(\mathcal{T}_i)$ be the class of simple Conjunctive Normal Form formulas built with the symbols of Σ_i. Where $\Sigma := \bigcup_{i=1 \ldots n} \Sigma_i$, let $\mathcal{T}$ the Σ-theory defined as the (deductive closure of) the union of all the above theories and let $sCNF(\mathcal{T})$ be the class of sCNF Σ-formulas.

If for each $\mathcal{T}_i$ we have a procedure that decides the satisfiability in $\mathcal{T}_i$ of formulas of $sCNF(\mathcal{T}_i)$, we can easily derive a procedure that decides the satisfiability in $\mathcal{T}$ of any formula of $sCNF(\mathcal{T})$. Because of the possible presence of "mixed" terms and predicates (that is, expressions built with symbols from different signatures), φ cannot be processed directly by any of the satisfiability procedures unless it is of the form $\varphi_1 \wedge \varphi_2 \wedge \ldots \wedge \varphi_m$—call it

---

[1]Where we treat the equality symbol "$=$" as a logical constant.

separate form—where each sub-formula φ_i is a formula of some $sCNF(\mathcal{T}_j)$. If that is the case, we know that φ is unsatisfiable in $\mathcal{T}$ if and only if any φ_i is. Now, if $\varphi_i \in sCNF(\mathcal{T}_j)$ say, by construction it is unsatisfiable in $\mathcal{T}$ exactly when it is unsatisfiable in $\mathcal{T}_j$ and so we can use directly the satisfiability procedure for $sCNF(\mathcal{T}_j)$ to verify the unsatisfiability of φ.

If φ is not already in separate form, we can apply a conversion procedure that, given φ, returns an equivalent separate form. The separation procedure (and its correctness proof) is straightforward but to describe it we need some definitions and notation first. We have adapted these from those in (Baader & Schulz, 1992), among others, which appear to be well established in the field.

Consider the theories described above. For $i = 1, \ldots, n$, a member of Σ_i is an *i-symbol*. A Σ-term t is an *i-term* if it is a variable or it has the form $f\tilde{s}$ and f is an i-symbol. An i-predicate is defined analogously. A sub-term of an i-term t is an *alien* sub-term of t if it is a j-term, with $j \neq i$, and all of its super-terms in t are i-terms. An i-term is *pure* (or also, *i-pure*) if it only contains i-symbols. Analogously we can define alien predicate arguments. An i-predicate (including equations) is pure if all of its arguments are i-terms. Pure formulas are defined as obvious. Observe that, given our assumption on the various signatures, a variable is an i-term for any i and so an equation is always pure if one of its members is a variable and the other is a pure term.

The separation procedure consists of the following steps. Consider a formula $\varphi \in sCNF(\mathcal{T})$ and see it as a multi-set of literals.

1. **Variable Abstraction**
   In φ, recursively replace each alien predicate argument or sub-term t with a newly generated variable z and add the equation $z = t$ to φ. Purify each equations of the form $t_1 = t_2$, where neither of t_1 and t_2 is a variable, by replacing it with the equations $z = t_1, z = t_2$ (where z is a new variable) and purifying t_1 and t_2 in turn.

2. **Partition**
   Partition the new multi-set in $m \leq n$ blocks containing only i-pure literals.[2]

The resulting partition can be seen as a sCNF formula of the form $\varphi_1 \wedge \cdots \wedge \varphi_m$ where each φ_i is a j-pure sCNF formula for some $j \in \{1, \ldots, n\}$.

A formula may have many separate forms, but they are all equivalent modulo variable renaming and the standard properties of logical conjunction and equality, hence it is appropriate to speak of *the* separate form of a formula. We indicate the separate form of a formula $\varphi \in sCNF(\mathcal{T})$ with $\ddot{\varphi}$. For notational convenience, we will always think of $\ddot{\varphi}$ as a conjunction

---

[2] Equations between variables are partitioned arbitrarily.

of the form $\varphi_1 \wedge \cdots \wedge \varphi_n$, with n being the number of component theories, where for each $i \in \{1, \ldots, n\}$, φ_i is an i-pure sCNF formula, even if φ may not contain any i-symbol for some i. In that case, φ_i is defined as the identically true formula—which can be thought of as belonging to all $sCNF(\mathcal{T}_i)$'s.

It is immediate that any $\varphi \in sCNF(\mathcal{T})$ is logically equivalent to $\exists \tilde{z} \, \ddot{\varphi}$ where $\tilde{z}$ is the set of new variables introduced by the separation procedure. This entails the following

Proposition 2.1 *A sCNF formula is satisfiable iff its separate form is.*

Clearly, the problem with deciding the satisfiability of a formula φ by analyzing its separate form is that, in general, each sub-formula φ_i could be singly satisfiable without their conjunction being satisfiable. Therefore, to be able to apply distinct satisfiability procedures to each φ_i and correctly decide the satisfiability of φ we need to establish some sort of communication between the various procedures. In the Nelson-Oppen method, such communication is achieved by propagating from one procedure to the others any implied equalities between the variables of $\ddot{\varphi}$.

Actually, the method is a little more complex because, in general, it is possible that at a certain step, not one, but a proper disjunction of variable equalities is implied and so reasoning by cases becomes necessary. Such complication does not arise though if the component theories are *convex*, that is, such that their formulas never imply proper *splits*[3].

The issue of theory convexity is important to assessing the time and space complexity of a combination procedure based on the above method because of the case reasoning required with non-convex theories. Computational complexities issues related to the implementation of the Nelson-Oppen method have been extensively investigated in (Oppen, 1980) and we refer the reader to that work. We will ignore those issues here by considering a non-deterministic combination procedure that we have adapted from those in (Oppen, 1980) itself and (Baader & Schulz, 1992) and which applies to convex as well as non-convex theories. Essentially, instead of propagating variable equalities back and forth and having to reason by cases when splits are implied, the non-deterministic procedure guesses in advance all the equalities that hold between the variables of the input formula.

We describe our version of the procedure in the next section where we consider only the simple case of two component theories, the general case being an easy generalization.

---

[3]A *split* of a formula φ is a disjunction of variable equalities which is implied by φ and is such that none of its proper sub-formulas are implied by φ. A proper split is a split with at least two disjuncts.

2.1. THE COMBINATION PROCEDURE

If P is any partition on a set S of words and R is the corresponding equivalence relation, we call the *arrangement* of S given by P the set

$$ar(S) \quad := \quad \{x = y \mid x, y \in S \text{ and } xRy\} \cup \{x \neq y \mid x, y \in S \text{ and not } xRy\}$$

containing, modulo reflexivity and symmetry of "$=$", all the equations between any two equivalent words and all the disequations between any two non-equivalent words of S. For instance, if $S := \{x_0, x_1, x_2, x_3\}$ and $P := \{\{x_0, x_1, x_2\}, \{x_3\}\}$,

$$ar(S) := \{x_0 = x_1, \ x_0 = x_2, \ x_1 = x_2, \ x_0 \neq x_3, \ x_1 \neq x_3, \ x_2 \neq x_3\}.$$

In the following, we will make use of arrangements over sets of variables. Since, where $\tilde{z}$ is a set of variables, $ar(\tilde{z})$ is a set of formulas, we will also treat it when convenient as the conjunction of all its equations and disequations.

Given two theories $\mathcal{T}_1$ and $\mathcal{T}_2$, for $i = 1, 2$, we assume the availability of a procedures, Sat_i, that decides the satisfiability in $\mathcal{T}_i$ for the formulas in $sCNF(\mathcal{T}_i)$.

The combination procedure given below decides the satisfiability in the union of $\mathcal{T}_1$ and $\mathcal{T}_2$ in two phases. In the first phase, which is non-deterministic, two i-pure formulas $\langle \psi_1, \psi_2 \rangle$ are generated from the input formula[4]. In the second phase, each ψ_i is tested for satisfiability in $\mathcal{T}_i$. The combination procedure succeeds when both ψ_i's are satisfiable and fails otherwise.

More specifically, let L be the (multi)set of literals of the input formula. The procedure is composed of the following steps:

- **Decomposition Phase**

  1. For $i = 1, 2$, let L_i be the i-pure part of L's separate form.
  2. Where $\tilde{x}$ is the set of all variables shared by some literal in L_1 and some in L_2, choose an arrangement $ar(\tilde{x})$.
  3. Pass the pair $\langle L_1 \cup ar(\tilde{x}), L_2 \cup ar(\tilde{x}) \rangle$ to the next phase.

- **Check Phase**

  1. Run Sat_1 on $L_1 \cup ar(\tilde{x})$.
  2. Run Sat_2 on $L_2 \cup ar(\tilde{x})$.

---

[4]This phase corresponds to the decomposition algorithm of Baader and Schulz (see, (Baader & Schulz, 1995a) or (Baader & Schulz, 1995b)).

3. Succeed if both Sat_1 and Sat_2 succeed. Fail if either of Sat_1 or Sat_2 fails.

Observe that, the choice of an arrangement corresponds to the *variable identification* step of most combination methods (Baader & Schulz, 1992; Baader & Schulz, 1995b; Boudet, 1993; Kirchner & Ringeissen, 1992; Ringeissen, 1992) and to the equality propagation mechanism of the original Nelson-Oppen procedure. In our case though, variable identification is only performed on the *shared* variables of the two i-pure halves of the input instead of on *all* of them. That this is enough is also recognized in (Baader & Schulz, 1995a), for a similar combination method over structures, on the basis of model-theoretic justifications analogous to the ones we give later.

This combination procedure is provably sound and complete for a restricted class of first-order theories. In addition, it is allows incremental implementations when incremental satisfiability procedures for the component theories are available. A simple incremental implementation is the following.

Let L be any permutation of the literals in the separate form of the input formula. We feed the two satisfiability procedures by picking one literal at a time from L and passing it to Sat_1 or Sat_2 according to whether it is a 1-literal or a 2-literal. In addition, if the literal shares some variables $\tilde{v}$ with the literals already passed to the other satisfiability procedure, we choose some $ar(\tilde{v})$ and pass it to both procedures.[5]

It is an incremental version of the combination procedure that it is actually used in (Tinelli, 1995) to extend the CLP scheme.

Finally, notice that, although we have restricted our attention to sCNF formulas, the above procedure can also be used to decide the satisfiability of quantifier-free formulas in general, since a quantifier-free formula is satisfiable iff some disjunct—which is in turn a sCNF formula—of its disjunctive normal form is satisfiable.

3. Correctness of the Combination Procedure

All the theories we consider in the following are first-order theories incorporating the theory of equality. For convenience, we will follow the common convention that considers the theory of equality as an integral part of the logical machinery of First Order Predicate Logic and so, from now on, when we say "theory" we will mean an theory in FOPL with equality.

Earlier, we defined the combination, or union, of two or more theories as the deductive closure of the union of the component theories. In the fol-

---

[5]It should be obvious that since implementations are necessarily deterministic, some sort of backtracking mechanism is required in this case to recover from *wrong* choices of arrangements.

lowing, where T_1 and T_2 are two theories, we will denote their combination simply as $T_1 \cup T_2$.

3.1. CONSISTENT UNIONS OF THEORIES

Before proving the correctness of the combination procedure, we make a brief digression on the combination of theories that will later justify some of the restrictions to the application of the procedure.

Combining theories is, in general, a non-trivial operation because most of the model and proof-theoretic properties of theories are not modular, including the most important one: consistency. Some papers in the combination literature do provide a proof of consistency for their combined theories, or structures, but their proofs are often ad hoc. Others (including Nelson and Oppen's) seem to ignore the issue by either assuming consistency or giving it for granted in their case.

Craig and Robinson have identified a while ago a *local* criterion for the consistency of combined theories which justifies the choice of signature-disjoint component theories if some conditions on the cardinality of the theories' models are met.

We formalize this in the following theorem by giving a necessary and sufficient condition for combining signature-disjoint theories meaningfully. The theorem and its following extension to the general case of non-signature-disjoint theories essentially subsume all previous similar results in the literature.

In our proofs, we will use the class of formulas defined below.

Definition 3.1 *A first order formula is called an* equational formula *iff all its atomic sub-formulas are equalities between variables.*[6]

Proposition 3.1 *Let T_1 and T_2 be two theories. Assume that they are consistent and their respective signatures, Σ_1 and Σ_2, are disjoint. Then, their union is consistent iff there is a cardinal κ such that both T_1 and T_2 have a model of cardinality κ.*

Proof. ($\Rightarrow$) Let $T := T_1 \cup T_2$ and consider any $\mathcal{M} \in Mod(T)$—where $Mod(T) \neq \emptyset$ as T is consistent by assumption. By construction of T, the reduct of $\mathcal{M}$ to Σ_i is a model of T_i, for $i = 1, 2$. Obviously, both reducts have the same cardinality.

($\Leftarrow$) Let $\mathcal{M}_1$ and $\mathcal{M}_2$ be models of T_1 and T_2, respectively, and assume they have the same cardinality. By the Craig-Robinson Theorem (Shoenfield, 1967), T is inconsistent iff there is a sentence φ, whose non-logical symbols are in $\Sigma_1 \cap \Sigma_2$, such that $T_1 \models \varphi$ and $T_2 \models \neg\varphi$. If such φ exists, we have

---

[6]Nelson-Oppen call such formulas *simple* formulas.

that

$$\mathcal{M}_1 \models \varphi \quad \text{and} \quad \mathcal{M}_2 \models \neg\varphi. \tag{1}$$

Now, as $\mathcal{T}_1$ and $\mathcal{T}_2$ are signature-disjoint, φ can only be an equational formula. It is a well-known result of Model Theory that the reducts of any two structures to the the language of equational formulas are isomorphic whenever they have the same cardinality[7]. This means that either both $\mathcal{M}_1$ and $\mathcal{M}_2$ model φ or neither of them does, which contradicts (1). $\square$

The above proof immediately suggests an extension of the previous result to the combination of any two consistent theories.[8]

Proposition 3.2 *Let $\mathcal{T}_1$ and $\mathcal{T}_2$ be two consistent theories with respective signatures Σ_1 and Σ_2. Their union is consistent iff there is a model $\mathcal{M}_1$ of $\mathcal{T}_1$ and a model $\mathcal{M}_2$ of $\mathcal{T}_2$ such that their reducts to $\Sigma_1 \cap \Sigma_2$ are isomorphic.*

Proof: Analogous to that of Prop. 3.1. $\square$

A well-known general result on the union of theories is Robinson's Consistency Theorem (see (Chang & Keisler, 1990), for instance). This theorem, however, provides only a sufficient condition for the consistency of the union theory and a somewhat stronger one: $\mathcal{T}_1 \cup \mathcal{T}_2$ is consistent if $\mathcal{T}_1 \cap \mathcal{T}_2$ is a complete theory for the language of $(\Sigma_1 \cap \Sigma_2)$-sentences.

In our terms, this condition is expressed as follows: for every model $\mathcal{M}$ of one theory there is a model $\mathcal{M}'$ of the other such that the reducts of $\mathcal{M}$ and $\mathcal{M}'$ to Σ are elementarily equivalent[9]. Now, elementary equivalence between structures is a strictly more general relation than isomorphism. In some sense, however, our consistency result can be seen as more general than Robinson's for requiring the existence of just one pair of structures, related through the isomorphism of their $(\Sigma_1 \cap \Sigma_2)$-reducts, as opposed to an infinite number of them, related through the elementary equivalence of their $(\Sigma_1 \cap \Sigma_2)$-reducts.

In general, the conditions of Prop. 3.2 on the models of the component theories are not so easy to verify. With signature-disjoint theories, however, some cases are immediate. For instance, a consequence of Prop. 3.1, to which we will appeal later, is that the combination of signature-disjoint theories admitting infinite models is always safe.

Corollary 3.3 *Let $\mathcal{T}_1$ and $\mathcal{T}_2$ be as in Prop. 3.1 but such that they both admit an infinite model. Then, their union is consistent.*

---

[7] In other words, the theory of equational formulas is κ-*categorical* for any cardinal κ (see, (Shoenfield, 1967)).

[8] This extension was also independently obtained by Franz Baader (Baader, 1995) as a generalization of our Prop. 3.1.

[9] Recall that two Σ-structures are *elementarily equivalent* iff they satisfy exactly the same Σ-sentences and that a theory is complete exactly when every two models of it are elementarily equivalent.

Proof. Immediate consequence of the fact that, by the Löwenheim-Skolem-Tarski Theorem (Chang & Keisler, 1990), if each $\mathcal{T}_i$ admits an infinite model then it admits infinite models of any cardinality ($\geq$ the cardinality of its signature). $\square$

In the field of equational theories[10] consistency in the general sense is not an issue since all equational theories admit trivial models. A stronger version of consistency is then used, let us call it *E-consistency* here: an equational theory is *E*-consistent iff it admits models of cardinality greater than 1. Franz Baader has shown in (Baader, 1995) that a well-known result in Unification Theory (see (Schmidt-Schauß, 1988; Schmidt-Schauß, 1989)) is easily derivable as a consequence of Cor. 3.3.

Corollary 3.4 *Let $\mathcal{E}_1$ and $\mathcal{E}_2$ be two signature-disjoint equational theories. If $\mathcal{E}_1$ and $\mathcal{E}_2$ are E-consistent, then their union is E-consistent as well.*

Proof. For $i = 1, 2$, let E_i be a presentation of $\mathcal{E}_i$ and consider the set $F_i := E_i \cup \{\tilde{\exists}(x \neq y)\}$. Observe that, since $\mathcal{E}_i$ is *E*-consistent by assumption, E_i admits infinite models[11] and clearly so does F_i. By Cor. 3.3 then $F_1 \cup F_2$ admits at least one model; moreover, since it entails $\tilde{\exists}(x \neq y)$, it only admits non-trivial models. The claim follows by monotonicity observing that $\mathcal{E}_1 \cup \mathcal{E}_2$ is included in $F_1 \cup F_2$. $\square$

3.2. SOUNDNESS AND COMPLETENESS RESULTS

We are now ready to prove the correctness of the combination procedure. We start with a rigorous definition of the notions that we have been using informally in the previous section.

Definition 3.2 *Consider a Σ-theory $\mathcal{T}$. We say a that a formula φ is satisfiable in $\mathcal{T}$ iff it is satisfiable in some model of $\mathcal{T}$, that is, iff there exists a model $\mathcal{M} \in \mathrm{Mod}(\mathcal{T})$ such that $\mathcal{M} \models \tilde{\exists}\,\varphi$.*

This definition is the dual of the standard definition of unsatisfiability for formulas (as opposed to sentences) when we follow the convention of considering free variables as implicitly universally quantified.

Recall that we apply the combination procedure to signature-disjoint theories. Prop. 3.1 provides a condition on the component theories, namely that there is a model for one theory and a model for the other which have the same cardinality, that guarantees the consistency of their union and, as a consequence, that the satisfiability problem in it is not trivial. Unfortunately, that conditions alone is not sufficient for the correctness of

[10]Recall that an equational theory is a first order theory admitting an axiomatization, or *presentation*, all of whose axioms are universally quantified equations.
[11]For instance, the free algebras in countably many generators.

the combination procedure. Problems might arise with theories that admit only finite models. We explain this point with the help of an example.

Example. Consider a theory T_1 admitting models of cardinality at most 2 and a signature-disjoint theory T_2 admitting models of any cardinality. Assume that f is a functor of T_1, g a functor of T_2 and neither of them is defined as a constant function in its respective theory[12]. The union T of T_1 and T_2 is consistent by the Prop. 3.1, so consider the input formula

$$\varphi \quad := \quad fx \neq fy \wedge gx \neq gz \wedge gy \neq gz$$

The procedure splits φ into

$$L_1 := \{fx \neq fy\} \quad \text{and} \quad L_2 := \{gx \neq gz, gy \neq gz\}.$$

Observe that only possible arrangements of the variables shared between L_1 and L_2 are $\{x = y\}$ and $\{x \neq y\}$.

Now, $L_1 \cup \{x = y\}$ is clearly unsatisfiable and so the procedure fails on that arrangement. With the other arrangement however, both L_i's are satisfiable in their respective theories and so the procedure concludes that φ is satisfiable in T. Unfortunately,

$$T \models \varphi \rightarrow (x \neq y \wedge x \neq z \wedge y \neq z)$$

which means that φ is unsatisfiable in T because, as T_1, T only has models of cardinality less than 3.

It is an easy consequence of the Lówenheim-Skolem-Tarski Theorem that cases like the above do not appear if we only consider formulas that are satisfiable in infinite models. Hence, we will restrict our attention to the class of theories in which formulas are satisfiable if and only if they are satisfiable in an infinite model and prove that for this class the combination procedure is sound and complete.

Definition 3.3 (Stable-infiniteness (Oppen, 1980)) *A   consistent, quantifier-free theory T with signature Σ is called* stably-infinite *iff any quantifier-free Σ-formula is satisfiable in T iff it is satisfiable in an infinite model of T.*

It follows from the above definition that every stably infinite theory admits at least one infinite model.[13]

---

[12]That is, $T_1 \not\models \forall x, y\ fx = fy$ and similarly for g.

[13]We would like to point out that just having infinite models, although necessary, is not sufficient for stable-infiniteness. Consider the theory $\{p(z) \rightarrow x = y\}$; the theory admits infinite models but the quantifier-free formula $p(z)$ is only satisfiable in the trivial models of the theory.

In the following, we will indicate with $\Delta(\tilde{x})$ the quantifier-free formula obtained as the conjunction of all possible disequalities (modulo symmetry) between distinct variables of $\tilde{x}$. For instance, if $\tilde{x}$ is $\{x_1, x_2, x_3\}$, then $\Delta(\tilde{x})$ is $x_1 \neq x_2 \wedge x_1 \neq x_3 \wedge x_2 \neq x_3$. Recalling the definition of arrangement, it is immediate that $\Delta(\tilde{x})$ is the arrangement generated by the discrete partition of $\tilde{x}$.

We start with some lemmas involving equational formulas.

Lemma 3.5 *Consider a theory $\mathcal{E}$ over the language of equational formulas. If a closed equational formula ψ is valid in an infinite model of $\mathcal{E}$, then it is valid in every infinite model of $\mathcal{E}$.*

Proof: Again, immediate consequence of the Upward Löwenheim-Skolem-Tarski theorem (see (Chang & Keisler, 1990)), since, as we saw earlier, all the models of $\mathcal{E}$ with the same cardinality are isomorphic. $\square$

Lemma 3.6 *Consider a theory $\mathcal{E}$ as above. Assume that an equational formula ψ with free variables $\tilde{x}$ is satisfied in a model $\mathcal{M}$ of $\mathcal{E}$ by an assignment ϑ of different individuals to each variable in $\tilde{x}$. Then ψ is satisfied in $\mathcal{M}$ by any assignment of different individuals to each variable in $\tilde{x}$. Equivalently, for all $\mathcal{M} \in \text{Mod}(\mathcal{E})$,*

$$\text{if } \mathcal{M} \models \exists \tilde{x} \, (\Delta(\tilde{x}) \wedge \psi) \text{ then } \mathcal{M} \models \forall \tilde{x} \, (\Delta(\tilde{x}) \rightarrow \psi).$$

Proof: It can be shown[14] that ψ is equivalent to a formula of the form

$$\bigvee_{i \in I} (\psi_i \wedge \varphi_i)$$

where each φ_i is a sentence and ψ_i is an arrangement of $\tilde{x}$. If $\mathcal{M} \models (\Delta(\tilde{x}) \wedge \psi)\vartheta$ for some model $\mathcal{M} \in Mod(\mathcal{E})$ and assignment ϑ, there exists an $i \in I$ such that $\mathcal{M} \models (\psi_i \wedge \varphi_i)\vartheta$. Since ϑ also satisfies $\Delta(\tilde{x})$, ψ_i cannot contain any equality and so must be equal to $\Delta(\tilde{x})$. The claim follows immediately from the fact that φ_i is closed and so satisfied in $\mathcal{M}$ by any assignment. $\square$

The proof of the theorem below is based on the following corollary (see (Shoenfield, 1967)) of the above mentioned consistency result by Craig and Robinson.

Lemma 3.7 (Craig Interpolation Lemma) *If $\mathcal{T}_1 \cup \mathcal{T}_2 \models \varphi_1 \rightarrow \varphi_2$, where for $i = 1, 2$, φ_i is a formula in the language of $\mathcal{T}_i$, there exists a formula ψ, whose free variables are among the free variables shared by φ_1 and φ_2, such that $\mathcal{T}_1 \models \varphi_1 \rightarrow \psi$ and $\mathcal{T}_2 \models \psi \rightarrow \varphi_2$.*

We are now ready for the main result of this section.

[14]See Lemmas 1.5.6 and 1.5.7 of (Chang & Keisler, 1990), for instance.

Proposition 3.8 *Let T_1 and T_2 be two stably-infinite, signature-disjoint theories and let $\varphi_1 \in \text{sCNF}(T_1)$ and $\varphi_2 \in \text{sCNF}(T_2)$. Let $\tilde{v}$ be the set of variables shared by φ_1 and φ_2. If $\varphi_i \wedge \Delta(\tilde{v})$ is satisfiable in T_i for $i = 1, 2$, then $\varphi_1 \wedge \varphi_2$ is satisfiable in $T_1 \cup T_2$.*

Proof. Ad absurdum, assume that $\varphi_1 \wedge \varphi_2$ is unsatisfiable in $T := T_1 \cup T_2$, then $T \models \varphi_1 \to \neg\varphi_2$. By the lemma above, there exists a formula ψ, whose free variables $\tilde{x}$ are in $\tilde{v}$, such that $T_1 \models \varphi_1 \to \psi$ and $T_2 \models \varphi_2 \to \neg\psi$. Again, since T_1 and T_2 are signature disjoint, ψ must be an equational formula. Now, T_1 is stably infinite and $\varphi_1 \wedge \Delta(\tilde{x})$ is satisfiable in T_1, therefore it is satisfiable in an infinite model $\mathcal{M}$ of T_1 and so is $\psi \wedge \Delta(\tilde{x})$.

Observe that since T_1 contains $\mathcal{E}_\emptyset$, the theory of equality, $\mathcal{M}$ is a model of $\mathcal{E}_\emptyset$ as well. By Lemma 3.6, it follows that $\forall \tilde{x}\, (\Delta(\tilde{x}) \to \psi)$ is valid in $\mathcal{M}$. By Lemma 3.5, it follows that $\forall \tilde{x}\, (\Delta(\tilde{x}) \to \psi)$ is valid in *every* infinite model of $\mathcal{E}_\emptyset$.

In the same way, we can show that $\forall \tilde{x}\, (\Delta(\tilde{x}) \to \neg\psi)$ is valid in every infinite model of $\mathcal{E}_\emptyset$, which leads to a contradiction. $\square$

In the above proposition, we assumed that the two φ_i's are satisfied by an assignment of a different individual to each of their shared variables. The following corollary shows that there is no loss of generality in considering only such assignments since we can always *eliminate*, by identification, those shared variables that would be assigned to the same individual.

Corollary 3.9 *Consider φ_1 and φ_2 as above and let $ar(\tilde{v})$ be an arrangement of their shared variables. If $\varphi_i \wedge ar(\tilde{v})$ is satisfiable in T_i for $i = 1, 2$, then $\varphi_1 \wedge \varphi_2$ is satisfiable in $T_1 \cup T_2$.*

Proof. Assume that φ_i has the form $\varphi_i(\tilde{v}, \tilde{z}_i)$ where $\tilde{z}_i$ are the non shared variables of φ_i. Given the equivalence relation that generates $ar(\tilde{v})$, we choose an *identification* of the elements of $\tilde{v}$, that is, a substitution σ from $\tilde{v}$ to $\tilde{v}$ that substitutes each variable in a same equivalence class with a given representative for that class. Now, let $\tilde{u} := \tilde{v}\sigma$

Clearly, $(\varphi_i(\tilde{v}, \tilde{z}_i) \wedge ar(\tilde{v}))\sigma$ is still satisfiable in T_i. Observe however that $\varphi_i\sigma$ has now the form $\varphi_i(\tilde{u}, \tilde{z}_i)$ and $ar(\tilde{v})\sigma$ is actually (equivalent to) $\Delta(\tilde{u})$. By Prop. 3.8 then, $\varphi_1(\tilde{u}, \tilde{z}_1) \wedge \varphi_2(\tilde{u}, \tilde{z}_2)$ is satisfiable in $T_1 \cup T_2$ and hence $\varphi_1(\tilde{v}, \tilde{z}_1) \wedge \varphi_2(\tilde{v}, \tilde{z}_2) \wedge ar(\tilde{v})$ is satisfiable in $T_1 \cup T_2$ as well. $\square$

We can now prove the correctness of the combination procedure for the union T of two signature-disjoint, stably-infinite theories, T_1 and T_2.

Proposition 3.10 (Soundness) *If one of the of the pairs $\langle \psi_1, \psi_2 \rangle$ output by the decomposition phase of the combination procedure is such that ψ_i is satisfiable in T_i for $i = 1, 2$, then the input formula is satisfiable in T.*

Proof. We have already seen that Step 1 of the decomposition phase preserves satisfiability. The claim then is an immediate consequence of Cor. 3.9. $\square$

Proposition 3.11 (Completeness) *If a formula $\varphi \in$ sCNF$(\mathcal{T})$ is satisfiable in $\mathcal{T}$, then there exists an output pair $\langle \psi_1, \psi_2 \rangle$ of the decomposition phase such that ψ_i is satisfiable in $\mathcal{T}_i$ for $i = 1, 2$.*

Proof. Assume φ is satisfiable in a model $\mathcal{M}$ of $\mathcal{T}$, then $\ddot{\varphi} := \varphi_1 \wedge \varphi_2$ is satisfiable in $\mathcal{M}$ for some valuation ϑ. This valuation induces an arrangement $ar(\tilde{v})$ on the set $\tilde{v}$ of shared variables between φ_1 and φ_2, where for any $x, y \in \tilde{v}$, $(x = y) \in ar(\tilde{v})$ if $x\vartheta = y\vartheta$ and $(x \neq y) \in ar(\tilde{v})$ otherwise.

Clearly, we have that $\mathcal{M} \models (ar(\tilde{v}) \wedge \varphi_1 \wedge \varphi_2)\vartheta$, which implies that $\mathcal{M} \models (ar(\tilde{v}) \wedge \varphi_1)\vartheta$ and $\mathcal{M} \models (ar(\tilde{v}) \wedge \varphi_2)\vartheta$. Since, for $i = 1, 2$, the reduct of $\mathcal{M}$ to the signature of $\mathcal{T}_i$ is a model of $\mathcal{T}_i$, we have that $ar(\tilde{v}) \wedge \varphi_i$ is satisfiable in $\mathcal{T}_i$.

The claim follows from the fact that $\langle \varphi_1 \wedge ar(\tilde{v}), \varphi_2 \wedge ar(\tilde{v}) \rangle$ is indeed a possible output pair of the decomposition procedure. $\square$

Notice that for the results above we need neither to postulate that the satisfiability problem is decidable in the component theories nor that the theories are axiomatizable. When satisfiability is in fact decidable, we obtain the stronger correctness result below.

Corollary 3.12 *Assume that for $i = 1, 2$, Sat$_i$ is a decision procedure for the satisfiability in $\mathcal{T}_i$ of formulas of sCNF$(\mathcal{T}_i)$. Then, a formula $\varphi \in$ sCNF$(\mathcal{T})$ is satisfiable in $\mathcal{T}$ if and only if the combination procedure succeeds on φ.*

Proof. Immediate consequence of Propositions 3.10 and 3.11 and the easily proved fact that the procedure halts on every input whenever both Sat_1 and Sat_2 do.

4. Conclusions and Further Developments

In this paper we have described a non-deterministic version of the Nelson-Oppen combination procedure and given a novel proof of Nelson and Oppens' original results.

We believe that our proof is relevant for at least two reasons. First, it avoids the problematic concept of *residue* of a formula, introduced in Nelson and Oppen's proofs, which is only defined for infinite interpretations of the formula itself.[15] Second, it shows that equality propagation between the satisfiability procedures of the component theories, the main idea of the method, can be confined to a restricted set of variables.

---

[15]It looks like the authors had not realized this initially. As a matter of fact, the proofs given in (Nelson & Oppen, 1979) were incorrect. In later papers on the methods, (Oppen, 1980) and (Nelson, 1984), the problem was side-stepped by restricting attention to infinite models only and implicitly claiming that such restriction did not invalidate the generality of the results. This is indeed true but not totally immediate.

Another contribution of the paper was a characterization of the consistency of the union of first-order theories.

In an attempt to extend the combination procedure to more general cases, we are confronted with two issues, among others, that we believe are very significant and should deserve further investigation.

The first issue is the stable-infiniteness requirement on the component theories. There certainly are interesting constraint theories that are not stably-infinite[16]. We have seen in Sect. 3.2 what kind of problems can arise if one of the component theories is not stably infinite, in particular if it only admits finite models. Stable-infiniteness does not seem to be a necessary condition for the correctness of the combination procedure although it is the most general sufficient condition identified so far. We conjecture that there might exist weaker requirements on the component theories which are sufficient for the procedure's correctness in the case of signature-disjoint theories and quantifier-free input formulas.

The second issue is the disjointness requirement on the signatures of the component theories. The procedure can be easily extended to the case of component theories with signatures sharing only constant symbols[17]; unfortunately, the correctness proof reported here does not lift to this extension because—like Nelson and Oppen's— it is based on the exclusive model-theoretic properties of equational formulas. We have found, however, a simple constructive proof of Prop. 3.2 that leads naturally to a constructive proof of an analogous to Prop. 3.8 for the case of component theories whose signatures share a finite number of constants. The case of shared function symbols is, understandably, much harder because of the infinite number of terms that are then shared by the languages of the component theories. We are currently trying to identify further model-theoretic restrictions on the component theories, beside stable infiniteness, that might lead to some controlled form of term sharing and therefore suggest further extensions of the procedure.

5. Acknowledgements

We would like to thank Alan Frisch for initially pointing out Nelson and Oppen's method, Franz Baader for a long and constructive series of discussions on combination methods and their deep implications, Joshua Caplan for a number of illuminating discussions on some model-theoretic issues related to this work, and the anonymous referees for their valuable comments and suggestions.

[16]With theories of finite domains being the most prominent examples, of course.

[17]In essence this can be done by including the shared constants in the computation of arrangements.

This work is partially supported by grant DACA88-94-0014 from the US Army Construction Engineering Laboratories.

CESARE TINELLI AND MEHDI HARANDI
University of Illinois at Urbana-Champaign, USA

References

Franz Baader, July 1995. Personal communication.

Franz Baader and Klaus U. Schulz. Unification in the union of disjoint equational theories: Combining decision procedures. In *Proceedings of the 11th International Conference on Automated Deduction*, volume 607 of *Lecture Notes in Artificial Intelligence*, pages 50–65. Springer-Verlag, 1992.

Franz Baader and Klaus U. Schulz. Combination of constraint solving techniques: An algebraic point of view. In *Proceedings of the 6th International Conference on Rewriting Techniques and Applications*, volume 914 of *Lecture Notes in Computer Science*, pages 50–65. Springer-Verlag, 1995.

Franz Baader and Klaus U. Schulz. Combination techniques and decision problems for disunification. *Theoretical Computer Science*, 142:229–255, 1995.

Franz Baader and Klaus U. Schulz. On the combination of symbolic constraints, solution domains, and constraint solvers. In *Proceedings of the First International Conference on Principles and Practice of Constraint Programming, Cassis (France)*, September 1995.

Alexandre Boudet. Unification in a combination of equational theories: An efficient algorithm. In M. E. Stickel, editor, *Proceedings of the 10th International Conference on Automated Deduction*, volume 449 of *Lecture Notes in Artificial Intelligence*. Springer-Verlag, 1990.

Alexandre Boudet. Combining unification algorithms. *Journal of Symbolic Computation*, 16(6):597–626, December 1993.

C. C. Chang and H. Jerome Keisler. *Model Theory*, volume 73 of *Studies in logic and the foundations of mathematics*. North-Holland, Amsterdam-New York-Oxford-Tokyo, 1990.

E. Domenjoud, F. Klay, and C. Ringeissen. Combination techniques for non-disjoint equational theories. In A. Bundy, editor, *Proceedings 12th International Conference on Automated Deduction, Nancy (France)*, volume 814 of *Lecture Notes in Artificial Intelligence*, pages 267–281. Springer-Verlag, 1994.

A. Herold. Combination of unification algorithms. In J. Siekmann, editor, *Proceedings 8th International Conference on Automated Deduction, Oxford (UK)*, volume 230 of *Lecture Notes in Artificial Intelligence*, pages 450–469. Springer-Verlag, 1986.

Joxan Jaffar and Jean-Louis Lassez. Constraint Logic Programming. Technical Report 86/74, Monash University, Victoria, Australia, June 1986.

Joxan Jaffar and Jean-Louis Lassez. Constraint Logic Programming. In *POPL'87: Proceedings 14th ACM Symposium on Principles of Programming Languages*, pages 111–119, Munich, January 1987. ACM.

Joxan Jaffar and Michael Maher. Constraint Logic Programming: A Survey. *Journal of Logic Programming*, 19/20:503–581, 1994.

Hélène Kirchner and Christophe Ringeissen. A constraint solver in finite algebras and its combination with unification algorithms. In K. Apt, editor, *Proc. Joint International Conference and Symposium on Logic Programming*, pages 225–239. MIT Press, 1992.

Hélène Kirchner and Christophe Ringeissen. Constraint solving by narrowing in combined algebraic domains. In P. Van Hentenryck, editor, *Proc. 11th International Conference on Logic Programming*, pages 617–631. The MIT press, 1994.

Greg Nelson. Combining satisfiability procedures by equality-sharing. *Contemporary Mathematics*, 29:201–211, 1984.

Greg Nelson and Derek C. Oppen. Simplification by cooperating decision procedures. *ACM Trans. on Programming Languages and Systems*, 1(2):245–257, October 1979.

Derek C. Oppen. Complexity, convexity and combinations of theories. *Theoretical Computer Science*, 12, 1980.

Christophe Ringeissen. Unification in a combination of equational theories with shared constants and its application to primal algebras. In *Proceedings of the 1st International Conference on Logic Programming and Automated Reasoning*, volume 624 of *Lecture Notes in Artificial Intelligence*, pages 261–272. Springer-Verlag, 1992.

Manfred Schmidt-Schauß. Unification in a combination of disjoint equational theories. In *Proceedings of the 9th International Conference on Automated Deduction*, volume 310 of *Lecture Notes in Computer Science*, pages 378–396. Springer-Verlag, 1988.

Manfred Schmidt-Schauß. Combination of unification algorithms. *Journal of Symbolic Computation*, 8(1–2):51–100, 1989.

Joseph. R. Shoenfield. *Mathematical Logic*. Addison-Wesley, Reading, MA, 1967.

Robert E. Shostak. A practical decision procedure for arithmetic with function symbols. *Journal of the ACM*, 26(2):351–360, April 1979.

Robert E. Shostak. Deciding combinations of theories. *Journal of the ACM*, 31:1–12, 1984.

E. Tidén. Unification in combinations of collapse-free theories with disjoint sets of function symbols. In J. Siekmann, editor, *Proceedings 8th International Conference on Automated Deduction, Oxford (UK)*, volume 230 of *Lecture Notes in Artificial Intelligence*, pages 431–449. Springer-Verlag, 1986.

Cesare Tinelli. Extending the CLP scheme to unions of constraint theories. Master's thesis, Department of Computer Science, University of Illinois, Urbana-Champaign, Illinois, October 1995.

K. Yelik. Combining unification algorithms for confined equational theories. *Journal of Symbolic Computation*, 3(1), April 1987.

K. Yelik. Unification in combinations of collapse-free regular theories. *Journal of Symbolic Computation*, 3(1–2):153–182, April 1987.

CHRISTOPHE RINGEISSEN

COOPERATION OF DECISION PROCEDURES
FOR THE SATISFIABILITY PROBLEM

Abstract. Constraint programming is strongly based on the use of solvers which are able to check satisfiability of constraints. We show in this paper a rule-based algorithm for solving in a modular way the satisfiability problem w.r.t. a class of theories Th. The case where Th is the union of two disjoint theories Th_1 and Th_2 is known for a long time but we study here different cases where function symbols are shared by Th_1 and Th_2. The chosen approach leads to a highly non-deterministic decomposition algorithm but drastically simplifies the understanding of the combination problem. The obtained decomposition algorithm is illustrated by the combination of non-disjoint equational theories.

1. Introduction

In recent years, the problem of combining decision procedures became of greatest interest in many fields of computer science, especially for constraint programming and automated deduction. The modular construction of decision procedures has been considered for the first time by Shostak (Shostak, 1979; Shostak, 1984) in order to solve heterogeneous formulae involving arithmetic and additional function symbols. Approximately in the same time, Nelson & Oppen (Nelson & Oppen, 1979; Oppen, 1980; Nelson, 1981) proposed an algorithm dedicated to the union of theories axiomatizing reals, arrays, lists and additional function symbols. The aim was to build a validity checker for programming languages in which such formulae frequently appear.

Formally, the *combined* decision problem can be stated as follows: Given two first-order theories Th_1 and Th_2 built respectively on the signatures Σ_1 and Σ_2, how is it possible to build a decision algorithm for the $\Sigma_1 \cup \Sigma_2$-theory $Th_1 \cup Th_2$ thanks to the decision algorithms provided for the Σ_1-theory Th_1 and the Σ_2-theory Th_2?

F. Baader and K.U. Schulz (eds.), Frontiers of Combining Systems, 121–139.
© *1996 Kluwer Academic Publishers.*

Until now, the main assumption is the disjointness of signatures Σ_1 and Σ_2. A further assumption needed in the framework of Nelson & Oppen is that single theories must be universal and *stably infinite* which means that a formula φ is satisfiable if and only if there exists a model of the theory with an infinite domain satisfying φ. So, even for the disjoint case, abstract assumptions on theories are needed in order to obtain a completeness result of the decomposition algorithm.

Another modularity problem has been thoroughly studied during the last decade for unification (Tidén, 1986; Yelick, 1987; Herold, 1987; Schmidt-Schauss, 1989; Boudet *et al.*, 1990; Baader & Schulz, 1992) in the union of two equational theories E_1 and E_2, that is for solving equations in the term algebra $\mathcal{T}(\Sigma_1 \cup \Sigma_2, \mathcal{X})/ =_{E_1 \cup E_2}$. Then, the combination techniques developed in this context have been extended in two directions: first, to allow other constraints (Baader & Schulz, 1995c; Ringeissen, 1993; Kirchner & Ringeissen, 1994) in $\mathcal{T}(\Sigma_1 \cup \Sigma_2, \mathcal{X})/ =_{E_1 \cup E_2}$ and second, to permit shared function symbols (Ringeissen, 1992; Domenjoud *et al.*, 1994).

In this paper, we present combination techniques for solving the original satisfiability problem in a non-deterministic manner (the simplest one) and in the non-disjoint case. Section 2 recalls the basic definitions of the notions used in the following. In Section 3, we present the rule-based decomposition algorithm which is correct and complete for all unions of theories considered along the paper. Section 4 gives the general construction of a model for the union of theories $Th_1 \cup Th_2$. Section 5 introduces sufficient assumptions on disjoint or non-disjoint theories to be combined. We investigate the case where shared function symbols are interpreted in the trivial way (i.e. syntactically) in both models of the theories. Section 6 presents some applications related to equational theories. In Section 7, we conclude with final remarks and future works.

2. Definitions

We first briefly introduce some basic notations. Let $\Sigma = (\mathcal{F}, \mathcal{P})$ be a mono-sorted first-order signature where $\mathcal{F}$ is a finite set of function symbols and $\mathcal{P}$ is a finite set of predicate symbols. The set $\mathcal{P}$ does not contain $=$ which is always interpreted as the identity relation. The subset of function symbols in $\mathcal{F}$ (resp. predicate symbols in $\mathcal{P}$) of arity m is denoted by $\mathcal{F}_m$ (resp. $\mathcal{P}_m$). The arity of a function symbol f (resp. a predicate symbol p) is denoted by $ar(f)$ (resp. $ar(p)$). The set of Σ-terms over a set A and of height n is defined recursively as follows:

1. $\mathcal{T}_0(\Sigma, A) = A$,
2. $\mathcal{T}_n(\Sigma, A) = \{f(\vec{a}) \mid f \in \mathcal{F}_m, \vec{a} \in (\mathcal{T}_{n-1}(\Sigma, A))^m\} \cup \mathcal{T}_{n-1}(\Sigma, A)$.

The set of Σ-terms over A is $\mathcal{T}(\Sigma, A) = \cup_{n \geq 0} \mathcal{T}_n(\Sigma, A)$. An equivalence relation $\equiv$ on A can be extended as usual on A^m:

$$(a_1, \ldots, a_m) \equiv (a'_1, \ldots, a'_m) \text{ if } \forall k \in [1, m], \ a_k \equiv a'_k$$

and defines a congruence relation $\equiv$ on $\mathcal{T}(\Sigma, A)$ as follows: $f(\vec{a_1}) \equiv f(\vec{a_2})$ if $\vec{a_1} \equiv \vec{a_2}$ and $f \in \mathcal{F}$. Note that we use vectors to denote tuples. The equivalence class of $a \in A$ w.r.t. $\equiv$ is denoted by $[a]_{\equiv}$.

Let us consider the set of Σ-terms over $\mathcal{X}$ where $\mathcal{X}$ is an infinite denumerable set of variables. The terms $t_{|\omega}$ and $t[\omega \hookleftarrow s]$ denote respectively the subterm of t at the position ω and the replacement in t of $t_{|\omega}$ by s. The symbol of t occurring at the position ω (resp. the top symbol of t) is written $t(\omega)$ (resp. $t(\epsilon)$). The set of variables occurring in a term t is denoted by $\mathcal{V}(t)$. Let $\mathcal{M}$ be a $\mathcal{F}$-algebra with A as domain. An assignment α is a mapping from $\mathcal{X}$ to A; it uniquely extends to an homomorphism $\underline{\alpha}$ from $\mathcal{T}(\Sigma, \mathcal{X})$ to $\mathcal{M}$. The restriction of α to a set of variables V is denoted by $\alpha_{|V}$. The range of α is denoted by $Ran(\alpha)$. A $\mathcal{M}$-solution of a quantifier-free Σ-formula φ is an assignment α such that $\underline{\alpha}(\varphi)$ holds in $\mathcal{M}$. The formula φ is valid in $\mathcal{M}$, denoted by $\mathcal{M} \models \varphi$, if any assignment α is a $\mathcal{M}$-solution of φ.

A *substitution* is an assignment from $\mathcal{X}$ to $\mathcal{T}(\Sigma, \mathcal{X})$ with only finitely many variables not mapped to themselves. A substitution uniquely extends to an endomorphism of $\mathcal{T}(\Sigma, \mathcal{X})$. We use letters $\sigma, \mu, \gamma, \phi, \ldots$ to denote substitutions and do not distinguish σ and $\underline{\sigma}$. Application of substitutions is written out by postfixed juxtaposition. We call *domain* of the substitution σ the (finite) set of variables $\mathcal{D}om(\sigma) = \{x | x \in \mathcal{X} \text{ and } x\sigma \neq x\}$, *range* of σ the set of terms $\mathcal{R}an(\sigma) = \cup_{x \in \mathcal{D}om(\sigma)} \{x\sigma\}$ and *variable range* of σ the set of variables $\mathcal{V}\mathcal{R}an(\sigma) = \cup_{x \in \mathcal{D}om(\sigma)} \mathcal{V}(x\sigma)$. A substitution σ is *idempotent* if $\sigma = \sigma\sigma$.

Definition 1 A Σ-theory (resp. universal Σ-theory) is a (possibly infinite) set of first-order Σ-sentences (resp. universally quantified first-order sentences), where sentences are formulaes without free variables. The Σ-theory (resp. universal Σ-theory) of a Σ-structure $\mathcal{M}$ is denoted by $\mathcal{T}\mathcal{H}(\mathcal{M})$ (resp. $Th(\mathcal{M})$) and is defined as the set of first-order Σ-sentences (resp. universally quantified first-order Σ-sentences) φ such that $\mathcal{M} \models \varphi$. A Σ-structure $\mathcal{M}$ is a model of a Σ-theory Th, denoted by $\mathcal{M} \models Th$, if $\mathcal{M} \models \varphi$ for any $\varphi \in Th$. A conjunction of atomic Σ-*literals* (i.e. an atomic formula, an equation or its negation) with (possibly) some universally quantified variables is simply called here a Σ-*formula*. A disjunction of Σ-formulae must be viewed as usual as a sequence of Σ-formulae to consider separately. The set of free variables occurring in a formula φ is denoted by $\mathcal{V}(\varphi)$. A formula φ is *satisfiable* w.r.t. Th if there exist a model $\mathcal{M}$ of Th and a $\mathcal{M}$-solution of

φ (or equivalently $\mathcal{M} \models \exists \mathcal{V}(\varphi) : \varphi$). The *satisfiability value* [1] of a formula φ w.r.t. Th is denoted by $(\varphi)_{Th}$ and is equal to $\top$ (true) if φ is satisfiable w.r.t. Th, else $(\varphi)_{Th} = \bot$ (false).

Let Σ' be a signature such that $\Sigma' \subseteq \Sigma$. If $\mathcal{M}$ is a Σ-structure, then $\mathcal{M}^{\Sigma'}$ denotes the Σ'-structure with the same domain and where function and predicate symbols are interpreted as in $\mathcal{M}$. A position ω of a Σ-term t is an alien position w.r.t. Σ' if $t(\omega)$ is not in Σ' and if for any other position ω' on the path from the root to ω, $t(\omega')$ is in Σ'. The set of alien positions of t w.r.t. Σ' is denoted by $AlienPos_{\Sigma'}(t)$.

3. Decomposition algorithm

We present informally in this section a decomposition algorithm for solving the satisfiability problem w.r.t. a $\Sigma_1 \cup \Sigma_2$-theory $Th_1 \cup Th_2$ where Th_1 is a Σ_1-theory, Th_2 a Σ_2-theory and $\Sigma_1 \cap \Sigma_2$ a set of function symbols $\mathcal{SF}$ (there is no shared predicate). This algorithm transforms any heterogeneous $\Sigma_1 \cup \Sigma_2$-formula into either $\bot$ or $\top$ thanks to decision procedures dedicated respectively to pure Σ_1-formulae and pure Σ_2-formulae. The reader is assumed familiar with algorithms based on transformation rules (Jouannaud & Kirchner, 1991). Any rule given in this paper transforms a formula φ into φ' in such a way that φ is satisfiable if and only if φ' is satisfiable. So, transformation rules preserve here satisfiability (but not necessarily the set of solutions). The non-determinism of the decomposition algorithm is due to some steps which transform a formula into a finite disjunction of formulae. Note also that the differents steps are executed sequentially and each of these finitely many steps terminates obviously. Consequently, the whole algorithm terminates and this is one of the greatest advantage of the non-deterministic approach. The normal form of φ with respect to the transformation rules is either $\top$ if φ is satisfiable or $\bot$ if φ is unsatisfiable.

The first step of the decomposition algorithm consists in splitting the input heterogeneous formula into pure equations and formulae. Hence, a heterogeneous $\Sigma_1 \cup \Sigma_2$-formula is transformed into a conjunction $\varphi_1 \wedge \varphi_2$ where φ_i is a Σ_i-formula for $i = 1, 2$. Note that formulae built on the set of shared function symbols $\mathcal{SF} = \mathcal{F}_1 \cap \mathcal{F}_2$ are considered both in φ_1 and in φ_2.

In the second and third (non-deterministic) steps, the same shared equations and disequations are added to each pure formula:

- In Step 2, add equations and disequations of the form $x = t$ and $y \neq t$ where t is a $\mathcal{SF}$-term (over $\mathcal{X}$) of height n such that each variable of t occurs only once and does not appear elsewhere in the current formula.

---

[1] We will use the boolean operators $\vee, \wedge$ for manipulating satisfiability values.

The integer n is somehow a parameter of the decomposition algorithm. We will see later in Section 5 on what relies the choice of n.

- In Step 3, add equations and disequations between variables of the current formula like $x = y$ and $y \neq z$.

In the last step, the solvers related to each theory are called in order to replace each pure formula by true ($\top$) or false ($\bot$).

The decomposition algorithm is described in Figure 1 and looks like one of the most simple we could imagine. It contains for instance only one non-deterministic step if $\mathcal{SF}$ is empty. In the following, we are interested in the assumptions needed on the input theories and formulae for proving the soundness of the decomposition algorithm. We first establish abstract properties which are proved sufficient and necessary.

4. Construction of a combined model

We show how to construct a $\Sigma_1 \cup \Sigma_2$-structure by combining a Σ_1-structure $\mathcal{M}_1$ and a Σ_2-structure $\mathcal{M}_2$. This construction yields a model of a $\Sigma_1 \cup \Sigma_2$-theory $Th_1 \cup Th_2$ when $\mathcal{M}_1$ and $\mathcal{M}_2$ are respectively models of a Σ_1-theory Th_1 and a Σ_2-theory Th_2. We assume that Σ_1 and Σ_2 share only a set of function symbols denoted by $\mathcal{SF}$. In the rest of the paper, φ_i denotes a quantifier-free Σ_i-formula for $i = 1, 2$.

Definition 2 Let $\mathcal{M}_1$ be a Σ_1-structure and $\mathcal{M}_2$ be a Σ_2-structure such that there exists a one-to-one mapping o from the domain A_1 of $\mathcal{M}_1$ to the domain A_2 of $\mathcal{M}_2$ verifying

$$\forall f \in \mathcal{SF}_m, \ \forall \vec{a_1} \in A_1^m, \ \forall \vec{a_2} \in A_2^m, \ \vec{a_1} =_o \vec{a_2} \Rightarrow f_{\mathcal{M}_1}(\vec{a_1}) =_o f_{\mathcal{M}_2}(\vec{a_2})$$

where $=_o$ denotes the equivalence relation on $A_1 \cup A_2$ such that: $a =_o b$ if $a = b$ or $a = o(b)$ or $b = o(a)$. The *combined structure* of $\mathcal{M}_1$ and $\mathcal{M}_2$ is the $\Sigma_1 \cup \Sigma_2$-structure $\mathcal{M}_1 \circledast_o \mathcal{M}_2$ defined as follows:

- its domain is $(A_1 \cup A_2)/ =_o$,
- $\forall g_i \in (\mathcal{F}_i)_m, \ \forall \vec{a_i} \in A_i^m, \ g_{i,\mathcal{M}_1 \circledast_o \mathcal{M}_2}([\vec{a_i}]_{=_o}) = [g_{i,\mathcal{M}_i}(\vec{a_i})]_{=_o}$,
- $\forall p_i \in (\mathcal{P}_i)_m, \ \forall \vec{a_i} \in A_i^m, \ p_{i,\mathcal{M}_1 \circledast_o \mathcal{M}_2}([\vec{a_i}]_{=_o})$ iff $p_{i,\mathcal{M}_i}(\vec{a_i})$.

According to the previous definition, the combined structure $\mathcal{M}_1 \circledast_o \mathcal{M}_2$ exists only if the equivalence relation $=_o$ is preserved after respective application of shared functions. Under this existence assumption, we can choose the combined structure as a model of $Th_1 \cup Th_2$.

Proposition 1 If $\mathcal{M}_1 \circledast_o \mathcal{M}_2$ exists and $\mathcal{M}_1$ is a Σ_1-model of Th_1 and $\mathcal{M}_2$ is a Σ_2-model of Th_2, then $\mathcal{M}_1 \circledast_o \mathcal{M}_2$ is a $\Sigma_1 \cup \Sigma_2$-model of $Th_1 \cup Th_2$.

This combined structure gives us implicitly a satisfiability criterion for the conjunction of a Σ_1-formula and a Σ_2-formula.

1. Purification
   Atom
   $$\frac{\varphi \wedge p(t_1, \ldots, t, \ldots, t_n)}{\varphi \wedge p(t_1, \ldots, t[\omega \leftarrow x], \ldots, t_n) \wedge x = t_{|\omega}}$$
   $$\text{if } \begin{cases} p \in \mathcal{P}_i, \\ \omega \in AlienPos_{\Sigma_i}(t), \\ x \notin \mathcal{V}(\varphi \wedge p(t_1, \ldots, t, \ldots, t_n)). \end{cases}$$

   Equation
   $$\frac{\varphi \wedge s = t}{\varphi \wedge s[\omega \leftarrow x] = t \wedge x = s_{|\omega}} \quad \text{if } \begin{cases} s(\epsilon) \in \Sigma_i, \\ \omega \in AlienPos_{\Sigma_i}(s), \\ x \notin \mathcal{V}(\varphi \wedge s = t). \end{cases}$$

   Disequation
   $$\frac{\varphi \wedge s \neq t}{\varphi \wedge x = s \wedge y = t \wedge x \neq y} \quad \text{if } \begin{cases} s \in \mathcal{T}(\Sigma, \mathcal{X}) \backslash \mathcal{X}, \\ x, y \notin \mathcal{V}(\varphi \wedge s \neq t). \end{cases}$$

   Conflict
   $$\frac{\varphi \wedge t_1 = t_2}{\varphi \wedge x = t_1 \wedge x = t_2} \quad \text{if } \begin{cases} t_i \in \mathcal{T}(\Sigma_i, \mathcal{X}) \backslash \mathcal{T}(\mathcal{SF}, \mathcal{X}) \text{ for } i = 1, 2, \\ x \notin \mathcal{V}(\varphi \wedge t_1 = t_2). \end{cases}$$

2. Instantiation[a]
   $$\frac{\varphi_1 \wedge \varphi_2}{\bigvee_{\rho \in SUBS^{\mathcal{SF}}_{\mathcal{V}(\varphi_1 \wedge \varphi_2), n}} (\varphi_1 \rho \wedge \rho_{\neq}) \wedge (\varphi_2 \rho \wedge \rho_{\neq})}$$

3. Identification[b]
   $$\frac{\varphi_1 \wedge \varphi_2}{\bigvee_{\xi \in ID_{\mathcal{V}(\varphi_1 \wedge \varphi_2)}} (\varphi_1 \xi \wedge \xi_{\neq}) \wedge (\varphi_2 \xi \wedge \xi_{\neq})}$$

4. Decision
   Satisfiability
   $$\frac{(\varphi_i)}{\top} \quad \text{if } \varphi_i \text{ is satisfiable w.r.t. } Th_i.$$

   Unsatisfiability
   $$\frac{(\varphi_i)}{\bot} \quad \text{if } \varphi_i \text{ is unsatisfiable w.r.t. } Th_i.$$

[a]This step is superfluous for disjoint theories. See Subsection 5.2 for the definition of $SUBS$.

[b]See Subsection 5.1 for the definition of ID.

Figure 1. Decomposition algorithm

Proposition 2 A quantifier-free formula $\varphi_1 \wedge \varphi_2$ is satisfiable w.r.t. $Th_1 \cup Th_2$ if and only if there exist

1. a Σ_i-model $\mathcal{M}_i$ of Th_i and a $\mathcal{M}_i$-solution α_i of φ_i for $i = 1, 2$.
2. a one-to-one mapping o from the domain A_1 of $\mathcal{M}_1$ to the domain A_2 of $\mathcal{M}_2$ such that

$$\forall f \in \mathcal{SF}_m, \ \forall \vec{a_1} \in A_1^m, \ \forall \vec{a_2} \in A_2^m, \ \vec{a_1} =_o \vec{a_2} \Rightarrow f_{\mathcal{M}_1}(\vec{a_1}) =_o f_{\mathcal{M}_2}(\vec{a_2})$$

and $\forall x \in \mathcal{V}(\varphi_1 \wedge \varphi_2), \ \alpha_1(x) =_o \alpha_2(x).$

Proof: ($\Leftarrow$) The assignment $\alpha : \mathcal{X} \to (A_1 \cup A_2)/ =_o$ defined as follows:

$$\forall x \in \mathcal{V}(\varphi_1 \wedge \varphi_2), \ \alpha(x) = [\alpha_1(x)]_{=_o}$$

is obviously a $\mathcal{M}_1 \circledast_o \mathcal{M}_2$-solution of $\varphi_1 \wedge \varphi_2$. We could also choose α_2 instead of α_1 in the definition of α since $[\alpha_1(x)]_{=_o} = [\alpha_2(x)]_{=_o}$.
($\Rightarrow$) If $\mathcal{M}$ is a $\Sigma_1 \cup \Sigma_2$-model of $Th_1 \cup Th_2$, then $\mathcal{M}_i = \mathcal{M}^{\Sigma_i}$ is a Σ_i-model of Th_i. We can obviously choose the identity for o and the $\mathcal{M}$-solution α for the $\mathcal{M}_i$-solution α_i. $\square$

This criterion is not directly usable. We will propose in the next section more practical criteria but restricting also the class of considered theories. However, this proposition can be obviously applied to the case where φ_1 and φ_2 are simply the true formula.

Corollary 1 The $\Sigma_1 \cup \Sigma_2$-theory $Th_1 \cup Th_2$ is consistent if and only if there exist a Σ_i-model $\mathcal{M}_i$ of Th_i for $i = 1, 2$ such that $\mathcal{M}_1^{\Sigma_1 \cap \Sigma_2}$ and $\mathcal{M}_2^{\Sigma_1 \cap \Sigma_2}$ are isomorphic.

Proposition 2 is also obviously applicable when Th_1 and Th_2 are identical modulo a renaming of signatures and if φ_1 and φ_2 are identical modulo the same renaming. Then, we are able to solve the satisfiability problem w.r.t. $Th_1 \cup Th_2$ in a modular way and even if Th_1 and Th_2 share some function symbols.

Example 1 Consider $E_i = \{x +_i (-x) = 0\}$ for $i = 1, 2$ and the formula $\varphi = (x +_1 (-x) = x +_2 (-x))$ which is equivalent to $(\varphi_1 = (x +_1 (-x) = y)) \wedge (\varphi_2 = (x +_2 (-x) = y))$. The formula φ is satisfiable w.r.t. $E_1 \cup E_2$ due to the solution $\alpha = \alpha_1 = \alpha_2 = \{x \mapsto 0, y \mapsto 0\}$ of φ_1 and φ_2.

In the example above, one could remark that $(\varphi_1 \wedge \varphi_2)_{Th_1 \cup Th_2} = (\varphi_i)_{Th_i}$ and so there is no need here to combine decision procedures. The reader should note however that the problem of solving a conjunction of two renamed formulae is no more obvious in the context of unification (Domenjoud *et al.*, 1994).

5. Assumptions on theories

We study now three cases of disjoint and non-disjoint unions of theories for which we give sufficient assumptions for satisfying the second point of Proposition 2.

5.1. STABLY INFINITE DISJOINT THEORIES

When the signatures of theories are disjoint, then it is sufficient to assume that each universal theory admits, for each satisfiable quantifier-free formula, a model having an infinite domain and which satisfies the formula. If this assumption holds, then it is easy to prove that there exists also a model with an infinite *denumerable* domain. Moreover, there exists obviously a one-to-one mapping between two infinite denumerable sets.

Definition 3 (Nelson, 1981) A Σ-structure is *infinite* if its domain is infinite. A universal Σ-theory Th is *stably infinite* if for any quantifier-free Σ-formula φ satisfiable w.r.t. Th, there exists an infinite Σ-structure $\mathcal{M}$ verifying $\mathcal{M} \models Th$ and $\mathcal{M} \models \exists \mathcal{V}(\varphi) : \varphi$.

The following proposition provides interesting examples of stably infinite theories.

Proposition 3 If $\mathcal{M}$ is an infinite structure, then $Th(\mathcal{M})$ is stably infinite and any quantifier-free formula is satisfiable w.r.t. $Th(\mathcal{M})$ if and only if it is satisfiable in $\mathcal{M}$.

Proof: If a quantifier-free formula φ is satisfiable in $\mathcal{M}$, then φ is satisfiable w.r.t. $Th(\mathcal{M})$ since $\mathcal{M}$ is obviously a model of $Th(\mathcal{M})$. Conversely, if φ is unsatisfiable in $\mathcal{M}$, then $\mathcal{M} \models \overline{\varphi}$ where $\overline{\varphi}$ is $\forall \mathcal{V}(\varphi) : (\varphi \Rightarrow \bot)$, and so $Th(\mathcal{M}) \models \overline{\varphi}$. Therefore, φ is unsatisfiable w.r.t. $Th(\mathcal{M})$. $\square$

Lemma 1 A quantifier-free formula satisfiable w.r.t. a stably infinite theory Th is satisfiable in a model of Th with an infinite *denumerable* domain.

Proposition 4 If $\mathcal{M}_1$ and $\mathcal{M}_2$ are infinite structures with denumerable domains built on disjoint signatures, then there is a one-to-one mapping o such that $\mathcal{M}_1 \circledast_o \mathcal{M}_2$ exists and is an infinite structure with a denumerable domain.

It remains to verify the assumption which states that solutions α_1 and α_2 of respectively φ_1 and φ_2 are equivalent modulo $=_o$. Hence, two variables must have the same solution (or must have different solutions) simultaneously in both models. This explains why we need to consider all possible identifications of variables occurring in $\varphi_1 \wedge \varphi_2$. Formally, the set of identifications of variables in V is $ID_V = \{\xi \mid \mathcal{D}om(\xi) \subseteq V, \mathcal{R}an(\xi) \subseteq$

V, ξ is idempotent}, where V is assumed to be finite. Given an identification $\xi \in ID_V$, $\xi_{\neq}$ denotes the equational formula

$$\bigwedge_{x,y \in V \setminus \mathcal{D}om(\xi)} x \neq y.$$

Proposition 5 Let Th_1 and Th_2 be two disjoint stably infinite theories. Then $(\varphi_1 \wedge \varphi_2)_{Th_1 \cup Th_2} = \bigvee_{\xi \in ID_{V(\varphi_1 \wedge \varphi_2)}} (\varphi_1 \xi \wedge \xi_{\neq})_{Th_1} \wedge (\varphi_2 \xi \wedge \xi_{\neq})_{Th_2}$ where φ_i is a quantifier-free Σ_i-formula for $i = 1, 2$.

Proof: ($\Rightarrow$) If $(\varphi_1 \wedge \varphi_2)_{Th_1 \cup Th_2}$ is true, then there exists a $\mathcal{M}$-solution α of $\varphi_1 \wedge \varphi_2$, where $\mathcal{M}$ is a model $Th_1 \cup Th_2$. Thus, the disjunct corresponding to an identification ξ such that $\forall x, y \in \mathcal{V}(\varphi_1 \wedge \varphi_2)$, $x\xi = y\xi \Leftrightarrow \alpha(x) = \alpha(y)$ is true.

($\Leftarrow$) If one disjunct is true, then there exist a model $\mathcal{M}_i$ of Th_i (with an infinite denumerable domain) and a $\mathcal{M}_i$-solution α_i of φ_i for $i = 1, 2$ such that $\forall x, y \in \mathcal{V}(\varphi_1 \wedge \varphi_2)$, $\alpha_1(x) = \alpha_1(y) \Leftrightarrow \alpha_2(x) = \alpha_2(y)$. So, we are able to choose the one-to-one mapping o such that

$$\forall x \in \mathcal{V}(\varphi_1 \wedge \varphi_2), \ \alpha_2(x) = o(\alpha_1(x)).$$

Then, the assignment α from $\mathcal{X}$ to $(A_1 \cup A_2)/ =_o$ defined by:

$$\forall x \in \mathcal{V}(\varphi_1 \wedge \varphi_2), \ \alpha(x) = [\alpha_1(x)]_{=_o}$$

is a $\mathcal{M}_1 \circledast_o \mathcal{M}_2$-solution of $\varphi_1 \wedge \varphi_2$. $\square$

Corollary 2 If Th_1 and Th_2 are two disjoint stably infinite theories, then $Th_1 \cup Th_2$ is stably infinite.

Corollary 3 The decomposition algorithm described in Figure 1 (without Step 2) solves the satisfiability problem of quantifier-free formulae w.r.t. the union of two disjoint stably infinite theories.

Theorem 1 *Satisfiability of quantifier-free $\Sigma_1 \cup \Sigma_2$-formulae is decidable w.r.t. $Th_1 \cup Th_2$ if*

- $\Sigma_1 \cap \Sigma_2 = \emptyset$,
- *satisfiability of quantifier-free Σ_i-formulae is decidable w.r.t. Th_i,*
- *Th_i is a stably infinite theory (for $i = 1, 2$).*

We thus get the result due to Nelson & Oppen (Oppen, 1980; Nelson, 1981) with a different proof. It is important to note that even for the disjoint case, we have (like them) additional assumptions concerning the individual theories. This result holds however for some non-universal theories since we can also consider arbitrary theories Th_i, provided that each quantifier-free Σ_i-formula is satisfiable w.r.t. Th_i if and only if it is satisfiable in a model of Th_i with an infinite *denumerable* domain. In the following, we prefer to use this new assumption since more theories are taken into account.

5.2. THEORIES STABLY GENERATED BY SHARED TERMS

When the signatures of theories share function symbols, we have now to ensure that the application of the corresponding interpretations of function symbols preserves the one-to-one mapping between the domains of models of Th_1 and Th_2. The idea followed here is simply to assume that shared functions are interpreted syntactically in both models.

Definition 4 A Σ-structure $\mathcal{M}$ is *generated by $\mathcal{SF}$-terms* (over A) if
- its domain is $\mathcal{T}(\mathcal{SF}, A)$ where A is an infinite denumerable set,
- $\forall f \in \mathcal{SF}_m, \forall \vec{a} \in \mathcal{T}(\mathcal{SF}, A)^m, \; f_{\mathcal{M}}(\vec{a}) = f(\vec{a})$.

Given a set $V \subseteq \mathcal{X}$, the conjunction of disequations

$$\bigwedge_{x \in V, f \in \mathcal{SF}} \forall x_{f,1}, \ldots, x_{f,ar(f)} : x \neq f(x_{f,1}, \ldots, x_{f,ar(f)})$$

is denoted by $basic_V$. A formula ϕ of the form $\varphi \wedge basic_V$, where φ is quantifier-free and $V \subseteq \mathcal{V}(\varphi)$, is said *partly basic*. The set of basic variables in ϕ is $\mathcal{BV}(\phi) = V$. A partly basic formula ϕ is *basic* if $\mathcal{BV}(\phi) = \mathcal{V}(\phi)$. A Σ-theory Th is *stably generated by $\mathcal{SF}$-terms* if for any basic Σ-formula φ satisfiable w.r.t. Th, there exists a Σ-structure $\mathcal{M}$ generated by $\mathcal{SF}$-terms verifying $\mathcal{M} \models Th$ and $\mathcal{M} \models \exists \mathcal{V}(\varphi) : \varphi$.

If $\mathcal{M}$ is a structure generated by $\mathcal{SF}$-terms (over A), then an assignment $\alpha : \mathcal{X} \to \mathcal{T}(\mathcal{SF}, A)$ is a $\mathcal{M}$-solution of $basic_V$ if and only if $Ran(\alpha_{|V}) \subseteq A$. Quantifier-free formulae are partly basic formulae φ with $\mathcal{BV}(\varphi) = \emptyset$. Note also that a stably infinite theory is stably generated by $\mathcal{SF}$-terms if $\mathcal{SF}$ is empty.

Similarly to Proposition 3, we can use a unique structure generated by $\mathcal{SF}$-terms to built a theory stably generated by $\mathcal{SF}$-terms.

Proposition 6 If $\mathcal{M}$ is a structure generated by $\mathcal{SF}$-terms, then $\mathcal{TH}(\mathcal{M})$ is stably generated by $\mathcal{SF}$-terms, and any basic formula is satisfiable w.r.t. $\mathcal{TH}(\mathcal{M})$ if and only if it is satisfiable in $\mathcal{M}$.

Proof: The key point is to prove that a basic formula which is unsatisfiable in $\mathcal{M}$ is also unsatisfiable w.r.t. $\mathcal{TH}(\mathcal{M})$. Let $\phi = \varphi \wedge basic_{\mathcal{V}(\varphi)}$ be a basic formula unsatisfiable in $\mathcal{M}$. The first-order sentence $\overline{\phi}$ defined as follows:

$$\forall \mathcal{V}(\varphi) : (\varphi \Rightarrow \bigvee_{x \in \mathcal{V}(\varphi), f \in \mathcal{SF}} \exists x_{f,1}, \ldots, x_{f,ar(f)} : x = f(x_{f,1}, \ldots, x_{f,ar(f)}))$$

is the negation of ϕ and so $\mathcal{M} \models \overline{\phi}$. Then, $\mathcal{TH}(\mathcal{M}) \models \overline{\phi}$ and ϕ is unsatisfiable w.r.t. $\mathcal{TH}(\mathcal{M})$. $\square$

Proposition 7 If $\mathcal{M}_1$ and $\mathcal{M}_2$ are structures generated by $\mathcal{SF}$-terms, then there exists a one-to-one mapping o such that $\mathcal{M}_1 \circledast_o \mathcal{M}_2$ is a structure generated by $\mathcal{SF}$-terms.

Proof: By assumption, there exists a one-to-one mapping o from A_1 to A_2 which can be used to define a congruence relation on $\mathcal{T}(\mathcal{SF}, A_1 \cup A_2)$:

$$\forall f \in \mathcal{SF}, \quad f(\vec{a_1}) =_o f(\vec{a_2}) \text{ if } \vec{a_1} =_o \vec{a_2}.$$

The $\Sigma_1 \cup \Sigma_2$-structure $\mathcal{M}_1 \circledast_o \mathcal{M}_2$ is as follows:

- its domain is $\mathcal{T}(\mathcal{SF}, (A_1 \cup A_2)/=_o)$
- $\forall f \in \mathcal{SF}_m, \forall \vec{a} \in (\mathcal{T}(\mathcal{SF}, (A_1 \cup A_2)/=_o))^m, \ f_{\mathcal{M}_1 \circledast_o \mathcal{M}_2}(\vec{a}) = f(\vec{a})$

Therefore, $\mathcal{M}_1 \circledast_o \mathcal{M}_2$ is still a structure generated by $\mathcal{SF}$-terms. $\square$

We are now faced with the fact that respective pure solutions must be equivalent modulo $=_o$ in order to construct a combined solution in $\mathcal{M}_1 \circledast_o \mathcal{M}_2$. An idea for dealing with this construction is to apply simultaneously in each theory the same kind of shared instance. Hence, partly basic formulae are transformed into basic ones.

Definition 5 A formula φ is n-satisfiable in a Σ-structure $\mathcal{M}$ generated by $\mathcal{SF}$-terms if there exists a $\mathcal{M}$-solution of φ such that its range is included in $\mathcal{T}_n(\mathcal{SF}, A)$. A formula φ is n-satisfiable w.r.t. a theory Th if φ is n-satisfiable in a model of Th generated by $\mathcal{SF}$-terms. If φ is n-satisfiable w.r.t. Th, then we define $(\varphi)_{Th,n} = \top$, else $(\varphi)_{Th,n} = \bot$.

This restricted form of satisfiability is interesting in practice since we are not looking for any solution but only for a "short" solution (it depends on the choice of the height n). However, our non-deterministic approach leads us to enumerate systematically all possible solutions of height less than n. This explains the use of a **finite** set $SUBS_{V,n}^{\mathcal{SF}}$ of substitutions such that their domains and ranges are respectively included in V and $\mathcal{T}_n(\mathcal{SF}, \mathcal{X})$. The variables which are not substituted or those introduced by a substitution $\rho \in SUBS_{V,n}^{\mathcal{SF}}$ must have basic values as solution (i.e. in A_i). This property can be encoded by the conjunction of disequations $\rho_{\neq}$ defined as follows: $\rho_{\neq} = basic_{(V \setminus \mathcal{D}om(\rho)) \cup V\mathcal{R}an(\rho)}$. We assume also that each variable in $V\mathcal{R}an(\rho)$ occurs once in $\mathcal{R}an(\rho)$ and does not appear elsewhere in the input formula.

Proposition 8 If Th_1 and Th_2 are two theories stably generated by $\mathcal{SF}$-terms, then $(\varphi_1 \wedge \varphi_2)_{Th_1 \cup Th_2, n} =$

$$\bigvee_{\rho \in SUBS_{V(\varphi_1 \wedge \varphi_2),n}^{\mathcal{SF}}} \bigvee_{\xi \in ID_{V((\varphi_1 \wedge \varphi_2)\rho \wedge \rho_{\neq})}} ((\varphi_1\rho \wedge \rho_{\neq})\xi \wedge \xi_{\neq})_{Th_1} \wedge ((\varphi_2\rho \wedge \rho_{\neq})\xi \wedge \xi_{\neq})_{Th_2}$$

Proof: ($\Rightarrow$) If $(\varphi_1 \wedge \varphi_2)_{Th_1 \cup Th_2, n}$ is true, then $\varphi_1 \wedge \varphi_2$ is satisfiable in a model of $Th_1 \cup Th_2$ generated by $\mathcal{SF}$-terms (over A), say $\mathcal{M}$, and there exists a $\mathcal{M}$-solution α of $\varphi_1 \wedge \varphi_2$ such that $Ran(\alpha) \subseteq \mathcal{T}_n(\mathcal{SF}, A)$. The structure $\mathcal{M}^{\Sigma_i}$ is obviously a model of Th_i for $i = 1, 2$. We can always choose $\rho \in SUBS_{\mathcal{V}(\varphi_1 \wedge \varphi_2), n}$ such that there exists an assignment $\alpha_i : \mathcal{X} \to \mathcal{T}(\mathcal{SF}, A)$ verifying $\forall x \in \mathcal{V}(\varphi_1 \wedge \varphi_2)$, $\alpha(x) = \underline{\alpha}_i(x\rho)$ with $Ran(\alpha_i) \subseteq A$. Then, we define ξ as follows: $x\xi = y\xi$ iff $\alpha_i(x) = \alpha_i(y)$. The assignment α_i is a $\mathcal{M}^{\Sigma_i}$-solution α_i of $(\varphi_i\rho \wedge \rho_{\neq})\xi \wedge \xi_{\neq}$ for $i = 1, 2$ and so one disjunct is true.

($\Leftarrow$) Conversely, if some $\bigwedge_{i=1}^{2}((\varphi_i\rho \wedge \rho_{\neq})\xi \wedge \xi_{\neq})_{Th_i}$ is true for an instantiation ρ and an identification ξ, then there exists a model $\mathcal{M}_i$ of Th_i generated by $\mathcal{SF}$-terms (over A_i) for $i = 1, 2$. The combined structure $\mathcal{M}_1 \circledast_o \mathcal{M}_2$ is a model of $Th_1 \cup Th_2$ generated by $\mathcal{SF}$-terms (over $(A_1 \cup A_2)/ =_o$). A $\mathcal{M}_1 \circledast_o \mathcal{M}_2$-solution of $\varphi_1 \wedge \varphi_2$ can be constructed thanks to the $\mathcal{M}_i$-solution of $(\varphi_i\rho \wedge \rho_{\neq})\xi \wedge \xi_{\neq}$ for $i = 1, 2$. First, due to the identification of variables, we can assume without loss of generality that $\forall y \in \mathcal{V}((\varphi_1 \wedge \varphi_2)\rho)$, $\alpha_1(y) =_o \alpha_2(y)$. Then, the assignment $\alpha : \mathcal{X} \to \mathcal{T}_n(\mathcal{SF}, (A_1 \cup A_2)/ =_o)$ such that

$$\forall x \in \mathcal{V}(\varphi_1 \wedge \varphi_2), \ \alpha(x) = ([\underline{\alpha}_1(x\rho)]_{=_o}) \ (\text{or } [\underline{\alpha}_2(x\rho)]_{=_o})$$

is a $\mathcal{M}_1 \circledast_o \mathcal{M}_2$-solution of $\varphi_1 \wedge \varphi_2$ and so $(\varphi_1 \wedge \varphi_2)_{Th_1 \cup Th_2, n}$ is true. $\square$

We are now able to present a new algorithm which looks like the one developed for the disjoint case. The difference is that some shared formulae must be taken into account by both theories thanks to an additional non-deterministic step:

Instantiation

$$\frac{\varphi_1 \wedge \varphi_2}{\bigvee_{\rho \in SUBS^{\mathcal{SF}}_{\mathcal{V}(\varphi_1 \wedge \varphi_2), n}} (\varphi_1\rho \wedge \rho_{\neq}) \wedge (\varphi_2\rho \wedge \rho_{\neq})}$$

This new step must be performed after **Purification** and before **Identification**. The corresponding algorithm can be applied to quantifier-free $\Sigma_1 \cup \Sigma_2$-formulae but also more generally for partly basic $\Sigma_1 \cup \Sigma_2$-formulae.

Corollary 4 The decomposition algorithm described in Figure 1 solves the n-satisfiability problem of partly basic formulae w.r.t. the union of two theories stably generated by $\mathcal{SF}$-terms.

Proposition 9 n-satisfiability of partly basic $\Sigma_1 \cup \Sigma_2$-formulae is decidable w.r.t. $Th_1 \cup Th_2$ if

- $\Sigma_1 \cap \Sigma_2 = \mathcal{SF}$,
- satisfiability of basic Σ_i-formulae is decidable w.r.t. Th_i,

- Th_i is stably generated by $\mathcal{SF}$-terms (for $i = 1, 2$).

Proposition 8 and Proposition 9 are also applicable to conjunctions of basic Σ_1-formulae and basic Σ_2-formulae. In the particular case of basic formulae, 0-satisfiability coincides with satisfiability.

Theorem 2 *Satisfiability of conjunctions of basic Σ_1-formulae and basic Σ_2-formulae is decidable w.r.t. $Th_1 \cup Th_2$ if*

- $\Sigma_1 \cap \Sigma_2 = \mathcal{SF}$,
- *satisfiability of basic Σ_i-formulae is decidable w.r.t. Th_i,*
- Th_i *is stably generated by $\mathcal{SF}$-terms (for $i = 1, 2$).*

¿From the decomposition algorithm, we can easily derive a semi-decision procedure for the satisfiability problem in a model of $Th_1 \cup Th_2$ generated by $\mathcal{SF}$-terms. This procedure works as follows: the conjunctions of pure formulae are enumerated according to a breadth-first strategy from smallest shared instances to larger ones and are solved in parallel. If a quantifier-free formula φ is satisfiable, then there exists an integer n such that φ is n-satisfiable w.r.t. $Th_1 \cup Th_2$ and so there is after a finite time a conjunction of pure formulae for which each elementary solver terminates and returns $\top$. In this case, we can halt the process and return $\top$ as result.

Theorem 3 *Satisfiability of partly basic $\Sigma_1 \cup \Sigma_2$-formulae in a model of $Th_1 \cup Th_2$ generated by $\mathcal{SF}$-terms is semi-decidable if*

- $\Sigma_1 \cap \Sigma_2 = \mathcal{SF}$,
- *satisfiability of basic Σ_i-formulae is semi-decidable w.r.t. Th_i,*
- Th_i *is stably generated by $\mathcal{SF}$-terms (for $i = 1, 2$).*

Surprisingly, this semi-decision procedure provides a decision algorithm when $Th_1 \cup Th_2$ is in addition partial recursively axiomatizable (i.e. axiomatized by a semi-decidable set of axioms). In that case, the validity problem w.r.t. $Th_1 \cup Th_2$ is also semi-decidable. Together with the semi-decision procedure for the satisfiability problem w.r.t. $Th_1 \cup Th_2$, this leads to the following corollary:

Corollary 5 Satisfiability of partly basic $\Sigma_1 \cup \Sigma_2$-formulae is decidable w.r.t. $Th_1 \cup Th_2$ if

- $\Sigma_1 \cap \Sigma_2 = \mathcal{SF}$,
- satisfiability of basic Σ_i-formulae is semi-decidable w.r.t. Th_i,
- Th_i is partial recursively axiomatizable (for $i = 1, 2$),
- any partly basic $\Sigma_1 \cup \Sigma_2$-formula satisfiable w.r.t. $Th_1 \cup Th_2$ is satisfiable in a model of $Th_1 \cup Th_2$ generated by $\mathcal{SF}$-terms.

Note that we still need some assumptions on the models of $Th_1 \cup Th_2$. Moreover, the related decision algorithm is not very practical and cannot be compared to the standard decomposition algorithm.

5.3. THEORIES STABLY GENERATED BY BOUNDED SHARED TERMS

In this section, we study how the previous semi-decision procedure can be turned into a decision algorithm for the satisfiability problem w.r.t. $Th_1 \cup Th_2$. This is possible if the definition of a structure generated by $\mathcal{SF}$-terms can be restricted to a bounded set of shared terms. The most trivial example appears when $\mathcal{SF}$ is simply a set of shared constants. Then, $\mathcal{SF}$-terms are necessarily of height 1.

Example 2 If Th is a stably infinite Σ-theory such that

$$\forall c, c' \in \mathcal{SF}_0, \ c \neq c' \Rightarrow Th \models c \neq c',$$

then Th is stably generated by $\mathcal{SF}_0$-terms (of height 1).

In the particular case of shared constants, 1-satisfiability corresponds to satisfiability and there is no universally quantified variables in the formula $basic_{|V}$ introduced in Definition 4. According to Proposition 9, we obtain the following result:

Theorem 4 *Satisfiability of quantifier-free $\Sigma_1 \cup \Sigma_2$-formulae w.r.t. $Th_1 \cup Th_2$ is decidable if*

- $\Sigma_1 \cap \Sigma_2 = \mathcal{SF}_0$ *(a set of constants)*,
- *satisfiability of quantifier-free Σ_i-formulae is decidable w.r.t. Th_i*,
- Th_i *is a stably infinite theory (for $i = 1, 2$) such that*

$$\forall c, c' \in \mathcal{SF}_0, \ c \neq c' \Rightarrow Th_i \models c \neq c'.$$

We could also introduce more complicated (and more artificial) structures generated by $\mathcal{SF}$-terms where only a set of terms of height n is sufficient to define non-shared functions and predicates. However, the decomposition algorithm cannot directly provide a decision algorithm since the answer is false for the satisfiable equation $x = t$, where x does not occur in t and $t \in \mathcal{T}_{n+1}(\mathcal{SF}, \mathcal{X}) \backslash \mathcal{T}_n(\mathcal{SF}, \mathcal{X})$. So, we would need first to transform such formulae having shared symbols at the top of equations and disequations. In this context, the following transformation rules (used for unification) should be added:

Decompose
$$\frac{\varphi \wedge f(s_1, \ldots, s_m) = f(t_1, \ldots, t_m)}{\varphi \wedge \bigwedge_{k=1}^{m} s_k = t_k} \quad \text{if } f \in \mathcal{SF}_m$$

Fail
$$\frac{\varphi \wedge f(\vec{s}) = f'(\vec{t})}{\perp} \quad \text{if } \{f, f'\} \subseteq \mathcal{SF}, f \neq f'$$

Replace

$$\frac{\varphi \wedge x = f(\vec{s})}{\varphi\{x \mapsto f(\vec{s})\}} \quad \text{if } f \in \mathcal{SF}, x \notin \mathcal{V}(f(\vec{s}))$$

These rules preserve satisfiability since structures of interest are generated by $\mathcal{SF}$-terms.

6. Applications

The decomposition algorithm can be applied for solving the satisfiability problem w.r.t. $Th(\mathcal{A}_1) \cup Th(\mathcal{A}_2)$ where $\mathcal{A}_1$ and $\mathcal{A}_2$ are specific structures like term algebras.

6.1. COMBINING EQUATIONAL THEORIES

Most of the decision algorithms related to a set of equational axioms E are in fact developed for decision problems in the particular term algebra $\mathcal{T}(\Sigma, \mathcal{X})/=_E$ like for instance:

– E-unifiability, the satisfiability of a conjunction of equations in $\mathcal{T}(\Sigma, \mathcal{X})/=_E$,
– E-equality, the validity of an equation in $\mathcal{T}(\Sigma, \mathcal{X})/=_E$.

The well-known result of G. Birkhoff (Birkhoff, 1935) states a connection between E-equality and validity w.r.t. E but this is unfortunately not sufficient for an equivalence between satisfiability in $\mathcal{T}(\Sigma, \mathcal{X})/=_E$ and satisfiability w.r.t. E. According to Proposition 3, the theory Th_E of interest is here the theory $Th(\mathcal{T}(\Sigma, \mathcal{X})/=_E)$. We assume that the cardinality of $\mathcal{T}(\Sigma, \mathcal{X})/=_E$ is strictly greater than 1 or equivalently there is no E-equality between two different variables in $\mathcal{X}$. Since $\mathcal{X}$ is an infinite denumerable set, the domain of $\mathcal{T}(\Sigma, \mathcal{X})/=_E$ is still infinite and so Th_E is stably infinite. Due the equivalence between satisfiability w.r.t. Th_E and satisfiability in $\mathcal{T}(\Sigma, \mathcal{X})/=_E$, a E-unification algorithm can be used for computing solutions of equations which are then propagated to disequations. Finally, we simply have to check that left-hand sides and right-hand sides of these resulting disequations are not E-equal.

Proposition 10 Satisfiability of quantifier-free formulae w.r.t. Th_E is decidable if a finitary E-unification algorithm is known and if E-equality is decidable.

Proof: Let $\varphi = ((\bigwedge_{k \in K} s_k = t_k) \wedge (\bigwedge_{k' \in K'} s'_k \neq t'_k))$. Then

$$(\varphi)_{Th_E} = \bigvee_{\sigma \in CSU_E(\bigwedge_{k \in K} s_k = t_k)} \bigwedge_{k' \in K'} (s'_k \sigma \neq t'_k \sigma)_{Th_E}$$

where $(s \neq t)_{Th_E} = \bot$ if $s =_E t$, else $\top$. CSU_E denotes a complete set of E-unifiers (Jouannaud & Kirchner, 1991). $\square$

The decomposition algorithm provides under the adequate assumptions a decision algorithm for satisfiability w.r.t. $Th_{E_1} \cup Th_{E_2}$. But the reader must be aware that this problem is neither satisfiability w.r.t. $Th_{E_1 \cup E_2}$ nor satisfiability in $\mathcal{T}(\Sigma_1 \cup \Sigma_2, \mathcal{X})/ =_{E_1 \cup E_2}$.

We could also suppose that E is a convergent term rewriting system R such that function symbols in $\mathcal{SF}$ are **constructors**. Then $\mathcal{T}(\Sigma, \mathcal{X})/ =_E$ is isomorphic to the algebra of normalized terms w.r.t. R, $\mathcal{T}(\Sigma, \mathcal{X}) \downarrow_R = \mathcal{T}(\mathcal{SF}, A)$, where A denotes the set of normalized terms with a top-symbol not in $\mathcal{SF}$. So, $\mathcal{TH}(\mathcal{T}(\Sigma, \mathcal{X})/ =_E)$ is a theory stably generated by $\mathcal{SF}$-terms over A. Shared function symbols were already assumed to be constructors in the framework developed in (Domenjoud *et al.*, 1994) and it was of greatest interest for the modular construction of $E_1 \cup E_2$-unification algorithms.

6.2. COMBINING EQUATIONAL THEORIES OF NON-DISJOINT FINITE POST ALGEBRAS

In this section, the only shared function symbols are constants. The problem of combining Post algebras has been already studied in (Ringeissen, 1992), but the corresponding cooperation algorithm is much more complicated since it needs elementary solvers based on unification w.r.t. linear constant restriction (Baader & Schulz, 1992).

Definition 6 Let $B_i = \{b_0^i, \ldots, b_{n_i-1}^i\}$ be a finite totally ordered set such that b_0^i denotes the minimal element of B_i and $b_{n_i-1}^i$ denotes the maximal element of B_i and let $\Sigma_{P_i} = \{+^i, *^i, C_{b_0}^i, \ldots, C_{b_{n_i-1}}^i, b_0^i, \ldots, b_{n_i-1}^i\}$ be a set of function symbols. The *Post* Σ_{P_i}-algebra $\mathcal{B}_i$ is the set B_i together with the following operators:

- $\forall b, b' \in B_i,\ b +_{\mathcal{B}_i}^i b' = \max(b, b')$
- $\forall b, b' \in B_i,\ b *_{\mathcal{B}_i}^i b' = \min(b, b')$
- $\forall k \in B_i,\ \forall b \in B_i,\ C_{k\mathcal{B}_i}^i(b) = b_{n_i-1}^i$ if $k = b$, else b_0^i
- $\forall b \in B_i,\ b_{\mathcal{B}_i} = b$

Proposition 11 Let $\mathcal{B}_i$ be a finite Post Σ_{P_i}-algebra. Satisfiability of quantifier-free formulae w.r.t. $Th_{\mathcal{B}_i} = Th(\mathcal{T}(\Sigma_{P_i}, \mathcal{X})/ =_{\mathcal{B}_i})$ is decidable.

A solver for the satisfiability problem can be designed thanks to a $\mathcal{B}_i$-unification algorithm and a decision algorithm for $\mathcal{B}_i$-equality, as previously.

Proposition 12 Let $\mathcal{B}_1$ and $\mathcal{B}_2$ be respectively a finite Post Σ_{P_1}-algebra and a finite Post Σ_{P_2}-algebra such that $B_1 \cap B_2$ is non-necessarily disjoint. Satisfiability of quantifier-free formulae w.r.t. $Th_{\mathcal{B}_1} \cup Th_{\mathcal{B}_2}$ is decidable.

The next example shows how the computation of solved forms helps to eliminate the non-determinism inherent to the decomposition algorithm.

Example 3 Let $\mathcal{B}_1$ be the boolean $\{|, \&, ^-, 0, 1\}$-algebra (where $|$ stands for (or), $\&$ for (and)) and $\mathcal{B}_2$ be the Post $\{+, *, C_0, C_1, C_2, 0, 1, 2\}$-algebra. The formula $\varphi = (\overline{x} = 0 \ \wedge \ 1 + (x|y) = 1 \ \wedge \ C_1(y) = 2)$ is equivalent to $(\varphi_1 = (\overline{x} = 0 \ \wedge \ X = x|y)) \wedge (\varphi_2 = (C_1(y) = 2 \ \wedge \ 1 + X = 1))$. Since $\overline{x} = 0$ is equivalent to the solved form $x = 1$, φ_1 is equivalent to $x = 1 \ \wedge \ X = 1|y$. In the same way, φ_2 is equivalent to $y = 1 \ \wedge \ 1 + X = 1$. Then, we can propagate $y = 1$ into φ_1. Then, we have that $\varphi_1 \wedge y = 1$ is equivalent to $x = 1 \ \wedge \ y = 1 \ \wedge \ X = 1$. Again, we can propagate $X = 1$ into φ_2. We get that $\varphi_2 \wedge X = 1$ is equivalent to $y = 1 \ \wedge \ X = 1$. Finally, φ is satisfiable and a solution is $\{x \mapsto 1, y \mapsto 1\}$.

7. Conclusion

We have presented a non-deterministic decomposition algorithm for solving in a modular way the satisfiability problem w.r.t. a non-disjoint union of theories. The non-determinism is very helpful for the understanding of the algorithm based only on four steps namely purification, instantiation, identification and decision. In the disjoint case, our decomposition algorithm is identical to the one developed recently in (Tinelli, 1995). But, for the moment, the algorithm cannot be efficiently implemented: it was not our aim in this paper. Further work is still necessary to make it more deterministic. One idea could be to propagate as much as possible in one theory the identifications and instantiations by shared terms deduced in the other theory. In that case, algorithms for solving the satisfiability problem should be replaced by more powerful algorithms devoted to the efficient deduction of the "largest shared conclusion" of a formula. We expect also that this strategy can capture more theories than those considered in this paper. In the same way, it would be helpful for the combination process if pure formulae were transformed to some kind of solved forms according to elementary solvers.

Finally, our combined structure has to be compared with the amalgamation product described in (Baader & Schulz, 1995a; Baader & Schulz, 1995b). Even if the contexts are completely different, these two constructions have some similarities and the respective decomposition algorithms are closed.

Acknowledgements: I would like to thank Hélène Kirchner for many helpful comments, Franz Baader, Klaus Schulz and the anonymous referees for a lot of pertinent and constructive remarks.

CHRISTOPHE RINGEISSEN
INRIA-Lorraine & CRIN-CNRS
Villers-lès-Nancy Cedex, France

References

F. Baader and K. Schulz. Combination of constraint solving techniques: An algebraic point of view. In *Proceedings 6th Conference on Rewriting Techniques and Applications, Kaiserslautern (Germany)*, volume 914 of *Lecture Notes in Computer Science*, pages 352–366. Springer-Verlag, 1995.

F. Baader and K. Schulz. On the combination of symbolic constraints, solution domains, and constraint solvers. In *Proceedings of the first International Conference on Principles and Practice of Constraint Programming - CP'95, Cassis (France)*, volume 976 of *Lecture Notes in Computer Science*, pages 380–397. Springer-Verlag, 1995.

F. Baader and K. Schulz. Unification in the union of disjoint equational theories: Combining decision procedures. In Deepak Kapur, editor, *11th International Conference on Automated Deduction, Saratoga Springs (USA)*, volume 607 of *Lecture Notes in Artificial Intelligence*, pages 50–65. Springer-Verlag, June 15–18, 1992.

F. Baader and K. Schulz. Combination techniques and decision problems for disunification. *Theoretical Computer Science*, 142:229–255, 1995.

G. Birkhoff. On the structure of abstract algebras. *Proceedings Cambridge Phil. Soc.*, 31:433–454, 1935.

A. Boudet, J.-P. Jouannaud, and M. Schmidt-Schauß. Unification in boolean rings and abelian groups. In C. Kirchner, editor, *Unification*, pages 267–296. Academic Press, London, 1990.

E. Domenjoud, F. Klay, and Ch. Ringeissen. Combination techniques for non-disjoint equational theories. In Alan Bundy, editor, *Proceedings 12th International Conference on Automated Deduction, Nancy (France)*, volume 814 of *Lecture Notes in Artificial Intelligence*, pages 267–281. Springer-Verlag, June/July 1994.

A. Herold. *Combination of Unification Algorithms in Equational Theories*. PhD thesis, Universität Kaiserslautern (Germany), 1987.

J.-P. Jouannaud and C. Kirchner. Solving equations in abstract algebras: a rule-based survey of unification. In J.-L. Lassez and G. Plotkin, editors, *Computational Logic. Essays in honor of Alan Robinson*, chapter 8, pages 257–321. The MIT press, Cambridge (MA, USA), 1991.

H. Kirchner and Ch. Ringeissen. Combining symbolic constraint solvers on algebraic domains. *Journal of Symbolic Computation*, 18(2):113–155, 1994.

G. Nelson. Techniques for program verification. Technical Report CS-81-10, Xerox Palo Research Center California USA, 1981.

G. Nelson and D. C. Oppen. Simplification by cooperating decision procedures. *ACM Transactions on Programming Languages and Systems*, 1(2):245–257, October 1979.

D. C. Oppen. Complexity, convexity and combinations of theories. *Theoretical Computer Science*, 12:291–302, 1980.

Ch. Ringeissen. Unification in a combination of equational theories with shared constants and its application to primal algebras. In *Proceedings of the 1st International Conference on Logic Programming and Automated Reasoning, St. Petersburg (Russia)*, volume 624 of *Lecture Notes in Artificial Intelligence*, pages 261–272. Springer-Verlag, 1992.

Ch. Ringeissen. *Combinaison de Résolutions de Contraintes*. Thèse de Doctorat d'Université, Université de Nancy 1, December 1993.

M. Schmidt-Schauß. Combination of unification algorithms. *Journal of Symbolic Computation*, 8(1 & 2):51–100, 1989. Special issue on unification. Part two.

R. Shostak. A practical decision procedure for arithmetic with function symbols. *Journal of the ACM*, 26(2):351–360, 1979.

R. Shostak. Deciding combination of theories. *Journal of the ACM*, 31(1):1–12, 1984.

E. Tidén. *First-order unification in combinations of equational theories*. PhD thesis, The Royal Institute of Technology, Stockholm, 1986.

C. Tinelli. Extending the CLP scheme to unions of constraint theories. Master's thesis, Department of Computer Science, University of Illinois, Urbana-Champaign, Illinois,

October 1995.

K. Yelick. Unification in combinations of collapse-free regular theories. *Journal of Symbolic Computation*, 3(1 & 2):153–182, April 1987.

JOHN SLANEY AND TIMOTHY SURENDONK

COMBINING FINITE MODEL GENERATION WITH THEOREM PROVING

PROBLEMS AND PROSPECTS

Abstract. This paper is about automatic searching for proofs, automatic searching for models and the potentially fruitful ways in which these traditionally separate aspects of reasoning may be made to interact. It takes its starting point in research reported in 1993 (Slaney, *SCOTT: A Semantically Guided Theorem Prover*, Proc. 13th IJCAI) on a system which combines a high performance first order theorem prover with a program generating small models of first order theories. The main theorem is an incompleteness result for a certain range of problems to which this combined system has been successfully applied. While the result may not be unexpected, the proof is worth examining and it is important to reflect on its relationship to the research program in combining methods.

1. Proof search, model search and their interaction

1.1. BACKGROUND

Traditional theorem provers search without much intelligence. They may reason forwards from the axioms or backwards from the desired theorem or both, but in either direction they rapidly find themselves in an exponentially growing search space of possible proof fragments which they explore in a manner at once admirably industrious and remarkably dull. Much good work in automatic proof search has gone into the discovery and refinement of methods for controlling the explosion of the search space or for reducing the amount of duplicated work undertaken in traversing it. At the same time, there has been vigorous research on heuristics for directing the search for proofs, but many of the most successful ideas in that field are either unexciting suggestions such as exploring short formulae first or else mysterious ones such as preferring some operator to be nested to the left rather

141

F. Baader and K.U. Schulz (eds.), Frontiers of Combining Systems, 141–155.
© 1996 *Kluwer Academic Publishers.*

than to the right.[1] The root problem is that most powerful theorem provers work only locally, focussing on the specific formulae being transformed by an inference rather than on global aspects of the situation, and more significantly they are based on pure syntax.[2] They may take into account questions like how many function symbols a formula has, which is the first literal in a clause, whether this unifies with that and the like, but they do not consider what the inference step is supposed to achieve, whether it is establishing a general law for the structures under consideration or an accidental property of the case, whether the conclusion of the inference says the same thing as one already proved or the like. That is, traditional provers do not understand what they are doing.

Still less do they understand the problems they are attempting to solve. The contrast with human theorem provers is striking. When we reason, we appeal constantly to conceptual structures within which particular problems make sense. We are not capable of exploring search spaces of millions of clauses, and it seems that we do not need to. What we are able to use is some sense of when a proof search is getting closer to the goal. What lies behind this capacity is not so much an ability to recognise syntactic patterns in formulae (though we importantly have that too) as an understanding of the problem: we know what the symbols, theories, axioms, goals and subgoals *mean*, and this puts us at an advantage in the investigation.

These remarks may be taken as leading to a recommendation that researchers in automated reasoning direct their efforts towards mechanical emulation of human cognitive processes. This, however, would be a mistake, certainly in the present state of development of the discipline. In the details of what they do—in what they find easy and what hard—computers are quite unlike us. Without becoming totally discouraged from projects in programming naturalistic intelligence, we should recognise that for practical purposes the indicated way ahead is to continue to let the machines be mechanical, doing what they do best at the high speeds of which they are capable.[3] Algorithmic proof search is good for some things anyway; it no

---

[1]Not all research in the area has been like that. The present paper is part of a minority tradition characterised by attempts to direct semantic reasoning for syntactic purposes. Plaisted and Caferra should be mentioned as important recent contributors to this tradition, whose roots go back to very early work by Gelernter and others. The present point is just that this is a *minority* tradition.

[2]This is not to deny a semantic basis for the usual rules and methods, but at the level of the individual inference step everything is most easily characterised in syntactic terms.

[3]And of course to facilitate the right kind of human-machine interactions. In order to secure the insights of a mathematician, it is more effective to plug in a mathematician at some appropriate point than to try to emulate one in an unsuitable medium like that provided by current computer technology.

more needs pseudo-human cognition in order to out-infer a mathematician than a car needs legs to outrun an athlete.

So we do not know how to program understanding, and in any case attempting to do so is likely to lead to inefficient systems. What we may rather hope to do is to secure some of the useful effects of understanding in a way congenial to algorithmic processes. To understand something is to know what it means. While that may be unattainable by means of software of the kind we know how to build, it can rather simply be *approximated*. It remains to be seen whether such approximations really pay their way, though the initial indications are good.

What is computationally possible is to *interpret* a formal language as referring to objects in a particular domain of discourse. An interpretation in this sense is just what we were all told it was when, at our mother's knee, we were introduced to model theory. That is, it consists of a nonempty set called the domain and a function which assigns to each predicate symbol a relation of appropriate arity defined on the domain and to each function symbol a function similarly. Elaborations to allow for possible worlds, impossible worlds, higher order structures, multiple sorts or whatever may be pasted into this framework as desired. With an interpretation is associated a notion of truth, fleshed out by the familiar inductive conditions for evaluating molecular formulae in terms of the values of their parts. Thus an interpretation—any interpretation—divides the language into the true and the false. That is, it marks a difference in meaning between half of what may be asserted and the other half. Even a crude interpretation is rich in semantic information of the kind which human reasoners may glean from their understanding and which may help automatic reasoning systems behave more intelligently.

1.2. USES OF INTERPRETATIONS

The oldest and simplest way of using an interpretation to help guide the search for a proof is to delete unprovable subgoals. In searching backwards from a goal, the prover typically looks at the available rules of inference and asks how they may have been applied to generate this particular goal (theorem) from subgoals (lemmata). Sometimes the rules are invertible, so that if the goal is provable then so are the subgoals, but this is not always the case. For example, if the goal is for example $\exists x A$ then *one* way to derive it would be to infer it from some particular A_t, so one possible subgoal would be to derive A_t. However, it is possible that A_t does not follow from the axioms of the problem even if $\exists x A$ does. If we have an interpretation which is a model of the theory in which the proof is sought (that is, of the axioms of the problem) then we may detect the unprovability of some subgoals by testing them against the model and finding that they are false. Then there is no point in trying to derive them, so the search can immediately backtrack and try another subgoal. Diagrams in geometry have this function among others: we do not waste time trying to prove two triangles congruent if it is obvious from the diagram that they have different shapes.

In the case of backward reasoning, there is no danger that the formulae considered will be irrelevant to the goal, but there is a danger that they will not follow from the assumptions. In the case of forward reasoning, the problem is the opposite one: all of the formulae considered are derivable, but most of them play no part in proofs of the target theorem. Here there is another use for interpretations. Given an interpretation, or a set of them, in which the goal is *false*, it makes sense to explore most vigorously the consequences of axioms and lemmata which are also false. This *false preference strategy* rests on the thought enunciated above, that interpretations reveal something of the semantic character of formulae. In the context of the question "How does this set of formulae entail that conclusion?" it corresponds to saying "Here is a way for the conclusion to be false, and *those* formulae fail with it, so concentrate on them."

Another strategy, most germane to the technical part of this paper, is semantic restriction of rules of inference. Given an interpretation—usually again one in which the target theorem is false—the rules of inference may be barred from applying unless one of the premises of the inference is false in the interpretation. Sometimes, as in the case of resolution as a refutation procedure with the empty clause as goal, this kind of restriction is complete in that if there is a derivation at all there is one obeying the model-based condition. In other cases it is incomplete and so may cause proof searches to fail where they would otherwise have succeeded. In such cases it may be weakened to become a variety of false preference, assigning weights or the like not to prohibit the inferences which violate it but to delay them or render them less likely to be selected.

1.3. DYNAMIC MODEL GENERATION

It may be that some particular interpretation is known to be apposite for a class of problems, in which case it may be given to a theorem prover along with the problem definition or even hard-coded into the prover itself. Ho- wever, in many cases we do not know in advance what interpretation will be useful. One option is then to use an all-purpose interpretation, perhaps selected to be such as to make testing for truth value very fast. For exam- ple, ordinary hyper-resolution uses an interpretation in which all atoms are false. It has long been observed (Slagle, 1967) that such an interpretation is unlikely to be ideal for any particular problem, and that an important project in automated reasoning is to find ways of suiting interpretations automatically to problems. Another option, therefore, is to combine a theorem prover with a program which searches for models of theories. This should be done in such a way that:

1. The models generated are adapted to the proof search in hand.
2. Testing formulae in the models is fast.
3. Searching for models does not occupy more time than is saved by using them to provide semantic information.
4. Replacing models by "better" ones as the proof search progresses does not compromise the soundness or completeness of the prover.

Item 1 means that, for example, in a case of goal deletion the interpretations should falsify a lot of the subgoals encountered in the present proof search while remaining models of the axioms. In the case of forward reasoning strategies, the models should not only falsify the goal but make a high proportion of the derived formulae true so that they focus the search significantly. Items 2 and 3 are obvious given that it is a strategic error to put more resources into any strategy than are covered by the return from it. Item 4 is an integrity condition.

One (generic) system capable of meeting these conditions consists of three modules:

Prover. Some form of theorem prover which takes as input a problem (axioms, rules, goal, ...) and gives as output either a proof or failure. It can use information as to whether individual formulae are true or false in some model, but it knows nothing about the content of the model.

Modeller. A program which takes as input two theories, a background theory about the domain of investigation and a set of formulae to be interpreted, together with conditions which limit its search to force termination, and gives as output a model of the theories if it finds one, or a failure message otherwise. It uses formulae from the prover as axioms, but knows nothing about inference rules or proof strategy.

Tester. A module which mediates between the other two. The prover may send a formula to the tester and get back the formula's truth value. The tester may send a theory to the modeller and get back either a model of that theory or failure if there is no model in the search space.

The entire system is in one of two states: active or passive. In passive mode, the modeller is inactive; the tester has an interpretation in memory and it merely tests formulae in that and returns the results. In active mode the tester maintains not only the current interpretation M but also a current theory T, of which M is a model. During the proof search, a series of

formulae arrive at the tester to be assigned truth values. With each formula
A the tester does:

> If A is false in M then
>> Call the modeller with theory $T \cup \{A\}$
>> If a model N is returned then
>>> $M \leftarrow N$
>> Else return FALSE
> Endif
> $T \leftarrow T \cup \{A\}$
> Return TRUE

A 'theory' here is just a set of formulae. Initially, T is null and the initial
M is obtained by calling the modeller with the null theory, thus getting
a model of the background theory (which may also be null, in which case
some dummy model is returned). As a result of all this, after a while the
tester's theory T consists of formulae which have occurred during the proof
search and which have been evaluated as true, and its interpretation M has
been generated specifically to make true as many as possible of the formulae
from the prover. Thus M is not arbitrary, but is adapted to the particular
problem being addressed by the prover and to the particular proof strategy
used.

The theory T is there in order to comply with item 4 on the list of
desiderata. It ensures that once the prover has been told that a formula is
true, that formula remains true even when the guiding model is changed.

Because searching for models is computationally expensive compared
with checking a formula against a given model, the system should switch
at some point from active to passive mode. This should happen when it is
likely that the current model is as good as any within the search space. Of
course, the ways of recognising when that point is reached are likely to be
fallible, but then heuristics generally are fallible. A simple way would be to
stop generating after a certain pre-defined number of formulae have been
tested, or alternatively after a given number of consecutive failures to find
models.

The generic combined system has been implemented (Slaney, 1993;
Slaney et al., 1994). The program SCOTT combines the theorem prover
Otter (McCune, 1990) with the model generator FINDER (Slaney, 1994).[4]
It worth remarking that FINDER searches for *small* models, so usually
testing formulae against them is fast compared with generating and pro-

---

[4]The program is available from ftp://arp.anu.edu.au/pub/SCOTT/ and comes with
both Otter's and FINDER's sources. Since the system is described in the cited papers,
we do not repeat the account of it here.

cessing consequences in the prover. The techniques so far implemented are semantic restriction of rules and false preference. The latter is effected by assigning weight to each true clause, allowing Otter's normal strategy of choosing lightest clauses first to do the rest. Techniques for automatically determining an appropriate value for this extra weight have not yet been implemented.

SCOTT must be regarded as a preliminary essay in combining semantic and syntactic methods. Nonetheless, it exhibits some of the pleasing features one might expect from such a combination. Otter is powerful and fast, but its search is not closely directed by any sensitivity to the meaning of the problem it is trying to solve. Its technique is to spray out consequences in all directions, remove some dross (such as subsumed clauses) and hope. FINDER is capable of extracting rich information from a few axioms, but it is powerless to deduce consequences. Moreover, it needs the many formulae produced during a proof attempt to tell it what to model. In the combined system, prover and modeller inform each other achieving more than either could alone.

2. Incompleteness

2.1. CONDENSED DETACHMENT

Hilbert systems for propositional logic with an implication connective $\to$ traditionally take as primitive some axiom schemata and the rule of Modus Ponens or detachment:

$$\frac{A \qquad A \to B}{B}$$

An interesting variant takes individual formulae (rather than schemata) as axioms and closes under the more general rule of *condensed* detachment—detachment incorporating the substitution required to unify the minor premise with the antecedent of the major one:

$$\frac{C \qquad A \to B}{B\sigma}$$

where σ is the most general unifier of A and C.

Condensed detachment was introduced by Meredith and received a sustained investigation in the 1980s (Kalman, 1983; Meyer & Bunder, 1988). It is clear that condensed detachment and resolution are closely related rules, so interest attaches to the question of whether proof search strategies and heuristics similar to those usual for resolution can be adapted to the case of condensed detachment.

Condensed detachment is also well known in the theorem proving community as a source of maddeningly hard problems for classical first order systems. It is easy to represent the formulae of a propositional logic as first order terms, the connectives being function symbols, and to add a unary predicate p for '...is provable'. The rule of detachment goes over into a clause

$$\neg p(x) \lor \neg p(i(x,y)) \lor p(y).$$

and axioms of the Hilbert systems may simply be asserted as (positive) unit clauses such as

$$p(i(x, i(y, x))).$$
$$p(i(i(x, i(y, z)), i(i(x, y), i(x, z)))).$$

To prove a theorem in the propositional logic, Skolemise its negation

$$\neg p(i(i(i(i(a, b), a), i(i(a, b), b))).$$

and derive the empty clause in first order logic. Since unification is going to be applied to the first order clauses, the rule of inference of the propositional system is exactly condensed detachment.

Experiments with Otter especially have made the condensed detachment problems into a famous challenge (McCune & Wos, 1992).[5] SCOTT has been applied to them with some success (Slaney, 1993). Its performance on such problems is typically better than that of the unaided Otter by a factor of about 2, whether the measure be the number of clauses kept, the number given or the time taken. On certain problems the false preference strategy enables SCOTT to find proofs two orders of magnitude more efficiently than Otter, and in one case SCOTT solved a (minor) open problem in a few seconds after Otter had failed to find a proof in several hours.

It is obvious that where the rule of inference is ordinary binary resolution refutation completeness is not affected, because the tester's theory T ensures that when the guiding model is changed all clauses marked as true remain true in the new interpretation. Where the rule of inference is some other form of semantic resolution such as hyper-resolution, it is equally obvious that further restriction by arbitrary models in the manner of SCOTT may destroy completeness. However, the case of condensed detachment appears to lie somewhere between these two "obvious" cases. In our previous papers (Slaney, 1993; Slaney et al., 1994) the question of whether semantically constrained condensed detachment is in any reasonable sense complete was left open.

---

[5]The challenge of these problems has stimulated other work, for example by the theorem proving groups of ICOT in Tokyo (Hasegawa et al., 1992) and of the Max Planck Institute in Saarbrücken (Graf, 1995).

2.2. DEFINITIONS

We shall deal with the set of terms over variables $(w, x, y, z, \ldots)$ with a binary function symbol $\rightarrow$ which we write in infix. To save on excessive formalism we also take a term to represent all instances of itself under substitutions which replace distinct variables for distinct variables. We shall denote terms by upper-case Roman letters, $A, B, C, \ldots$, and we shall take Γ to range over the set of terms. Moreover, when dealing with two unconnected terms we assume that they do not share variables.

These terms are analogs of propositional formulae and we can interpret the usual rule of *Modus Ponens* in its most general form as the rule of Condensed Detachment (CD) as stated above, tacitly assuming that $A \rightarrow B$ and C have distinct variables. $\Gamma \vdash^{CD} A$ then has its usual meaning, i.e. that it is possible to deduce A from the terms in Γ by applications of CD.[6]

Since we have only an attenuated first order language, our semantic structures or *CD-algebras* will be those of a generalised propositional implication, namely algebras $\langle \mathcal{A}, \leq, \supset, D \rangle$ where $\mathcal{A}$ is a set on which $\leq$ is a binary relation,[7] D is a subset of $\mathcal{A}$ closed under $\leq$ and $\supset$ is a binary operation on $\mathcal{A}$ satisfying the condition:

$$a \supset b \in D \iff a \leq b$$

The notation $\Gamma \models A$ will then be able to take on its usual meaning and we have a fairly standard (and easy) soundness and completeness result:

$$\Gamma \models B \iff \exists A \, (\exists \sigma (B = A\sigma) \, \wedge \, \Gamma \vdash^{CD} A)$$

It is here that we pay a small price for the generality imposed on the conclusions of our CD inferences. In general there may not be a CD derivation of a given semantic consequence of Γ, but there will be a derivation of some term which subsumes it.

Semantic constraint by a specific CD-algebra M means that the Condensed Detachment inference from $A \rightarrow B$ and C to $B\sigma$ can proceed only when either $M \not\models A \rightarrow B$ or $M \not\models C$. If A follows from Γ by M-constrained steps of this form we write $\Gamma \vdash_M A$.

We say that A follows from Γ by Semantically Constrained Condensed Detatchment (SCCD) and write $\Gamma \vdash^{SCCD} A$ when and only when for every

---

[6]Note that we are not capturing all of conclusions that would normally follow in a propositional system from the axioms in Γ. This is because we are not allowing arbitrary substitution during our derivations. This issue is interesting from the perspective of logic (Meyer & Bunder, 1988) but not our present concern.

[7]Typically, for interesting logics, $\leq$ is a partial order. However, in general it need not be. Our presentation is not maximally efficient, since $\leq$ could be defined rather than primitive. We have opted for familiarity rather than economy.

CD-algebra M, if $M \not\models A$ then $\Gamma \vdash_M A$. Then SCCD is complete iff $\Gamma \vdash^{CD} A$ implies $\Gamma \vdash^{SCCD} A$. Note that when trying to prove a goal term by SCCD we must use an algebra which invalidates that term. If this restriction were not imposed, there would generally be no semantically constrained proof. At the extreme, we could choose an algebra in which all of the axioms were true and so block all CD inferences from them.

2.3. THE COUNTER EXAMPLE

As promised, we shall give a specific counter example to the conjecture that SCCD is complete. To do this, it will be sufficient to find a set Γ of terms, a conclusion A which follows from Γ by usual CD reasoning, and a CD-algebra M relative to which we can show that A does not follow from Γ by SCCD. That is:

1. $\Gamma \vdash^{CD} A$,
2. $M \not\models A$, and
3. $\Gamma \not\vdash_M A$.

For our example, we first adopt the notation

$$\begin{aligned} \mathcal{C}_\star(A) &=_{\mathrm{df}} (A \to v) \to v \\ \mathcal{K}(A) &=_{\mathrm{df}} v \to A \end{aligned}$$

where v is a variable (say, the earliest in a standard enumeration) that does not occur in A. Moreover, we write

$$\begin{aligned} \mathcal{C}_\star^0(A) &=_{\mathrm{df}} A \\ \mathcal{C}_\star^{n+1}(A) &=_{\mathrm{df}} \mathcal{C}_\star(\mathcal{C}_\star^n(A)) \end{aligned}$$

Next we define the specific example:

$$\begin{aligned} \alpha &= x \to \mathcal{C}_\star(x) \\ \beta &= x \to \mathcal{K}(x) \\ \Gamma &= \{\alpha, \beta\} \\ A &= \mathcal{K}(\mathcal{C}_\star(\alpha)) \end{aligned}$$

Let M be the algebra given by the following matrix:

$\to$	0	1
0	1	0
1	0	1

$D = \{1\}$

$a \leq b \iff a = b$

So M interprets $\to$ as the Boolean biconditional, and we can take advantage of the well known fact that a propositional formula containing biconditionals as the only connective is a tautology if and only if the number of occurrences of each propositional variable in that formula is even.

Lemma 1 $M \models \alpha$, $M \not\models \beta$, and $M \not\models A$.

Proof. Count the variable occurrences ($mod\ 2$). □

Lemma 2 $\Gamma \vdash^{CD} A$

Proof.

$$\frac{\dfrac{\alpha \qquad\qquad\qquad \alpha}{C_\star(\alpha)} \qquad\qquad \beta}{A}$$

□

Thus we are only left with the requirement that $\Gamma \not\vdash_M A$, for which we shall define a set T which includes the M-constrained consequences of Γ, show that T contains only terms of a specific form, and show that A is not of this form. More specifically let T be the set of CD consequences of Γ except that we require that α cannot be applied to itself in the generation of elements of T. This condition corresponds exactly to following the set of support strategy with α as axiom and β initially in the set of support. Note that since α is true in M, all M-constrained derivations satisfy the condition, so clearly

Lemma 3 *If $\Gamma \vdash_M B$ then $B \in T$.*

It turns out that the converse is also true, but it is not needed for the proof.[8]

To describe the form of the terms in T we shall construct T' (which we will show to be identical to T) as follows:

Definition 4 T' *is the smallest set satisfying:*

1. $\alpha \in T'$,
2. $C_\star^n(\beta) \in T'$ *for all $n \geq 0$*
3. *If $B \in T'$ then $C_\star^n(\mathcal{K}(B)) \in T'$ for all $n \geq 0$*

Lemma 5 $A \notin T'$.

Proof. Assume for reductio that $A \in T'$. Clearly $A \neq \alpha$ and so it was not condition 1 which forced A into T'. Condition 2 does not place A in T' since A is of the wrong form, having a single variable as its antecedent. So $\mathcal{K}(C_\star(\alpha))$ must be there because of condition 3 with $n = 0$, and therefore $C_\star(\alpha) \in T'$. However, $C_\star(\alpha)$ cannot be there in virtue of condition 1 (because $C_\star(\alpha) \neq \alpha$) or of condition 2 (because $\alpha \neq \beta$) or of condition 3 (because α is not of the form $\mathcal{K}(X)$ since x occurs in $C_\star(x)$). There being no other way for $C_\star(\alpha)$ to get into T', this is a contradiction. □

---

[8] α is the only term in T which is validated by M. This fact is non-trivial but is an easy consequence of Lemma 6 below.

Lemma 6 $T = T'$.

Proof. The inclusion from right to left can be obtained simply by applying α and β to themselves (excluding, of course, α to α) and then repeatedly applying them to the resultant terms.

For the inclusion in the reverse direction we proceed by induction on the length of the CD proof of B from $\{\alpha, \beta\}$. The case when B is an axiom (i.e. α or β) is straightforward and so we are left to consider what happens when B is the result of applying our restricted CD rule to major premise C and minor premise D given our inductive hypothesis that $C, D \in T'$.

Let us say that a term is of form i ($1 \leq i \leq 3$) if it is in T' in virtue of condition i of Definition 4.

There are several simple cases to be disposed of. We deal first with the possibility that C is either α or one of the "degenerate" cases of form 2 or 3 with $n = 0$. Then we deal with the cases in which D is of one of those forms.

1. It is not allowed that both C and D are of form 1.
2. If C is of form 1 (i.e. $C = \alpha$) and D is of form 2 or 3, then B is just $C_\star(D)$ which is also of form 2 or 3, so $B \in T'$.
3. If C is of form 2 with $n = 0$ then $C = \beta$ and B is $\mathcal{K}(D)$ so is of form 3 and again $B \in T'$.
4. If C is of form 3 with $n = 0$ then C is $x \to E$ for some $E \in T'$ and so $B = E$.

For the remaining cases, we may safely assume that C is $C_\star(C')$ and that $C' \in T'$. Under this assumption, there are more simple cases:

5. If D is of form 1 (i.e. $D = \alpha$) then the result of unifying D with $C' \to z$ is $C' \to ((C' \to y) \to y)$ where y is not in C', but in that case B is just C, up to rewriting of variables.
6. If D is of form 2 with $n = 0$ then B is $\mathcal{K}(C')$, which is of form 3.
7. If D is of form 3 with $n = 0$ then D is $\mathcal{K}(E)$ for some $E \in T'$, but then $B = E$.

That concludes the special cases. The remaining case has $C = C_\star^n(\gamma)$ and $D = C_\star^k(\delta)$ where each of γ and δ is either β or $\mathcal{K}(X)$ for some $X \in T'$. Now the proof proceeds by induction on the product kn. In the base case ($kn = 0$) at least one of k or n is zero. These cases have been treated above. Now assume for induction that $k > 0$ and $n > 0$ and that the result holds for all such pairs of formulae $C_\star^m(\gamma)$ and $C_\star^j(\delta)$ with $jm < kn$. The result of the CD inference is that of unifying D with $C_\star^{n-1}(\gamma) \to z_n$ and detaching whatever gets substituted for z_n in this process. Evidently, since D is $(C_\star^{k-1}(\delta) \to z_k) \to z_k$, this unifies z_k with z_n and $C_\star^{n-1}(\gamma)$ with $C_\star^{k-1}(\delta) \to z_k$. Hence B is also the result of CD with D as major premise and $C_\star^{n-1}(\gamma)$ as minor, which is in T' by the induction hypothesis. $\square$

Theorem 7 *SCCD is incomplete.*

Proof. Immediate from Lemmas 1, 2, 3, 5 and 6. ☐

3. Remarks

We have seen that the set of support (SOS) strategy is incomplete for condensed detachment problems even when the subtheory initially excluded from the set of support has a model in which the goal theorem is false. This means that it is possible for a prover such as Otter to go wrong on these problems. This incompleteness is not a severe difficulty for Otter, because there is always a way of using the SOS strategy which avoids any failure: at worst, it is possible to put all assumptions except maybe for the detachment clause itself into the set of support, thus guaranteeing completeness at the expense of some efficiency. For SCOTT, however, the situation is different. Because it uses truth values in a model, in effect it constantly re-computes the boundary between axioms and set of support, and because it chooses models dynamically, it sets this boundary aggressively, disallowing as many potential inferences as it can. We want this behaviour, because we want to extract as much guidance as possible from the chosen model, but it is precisely this feature which destroys completeness. Putting everything into the set of support by hand initially avails us nothing if the first thing SCOTT does is to discover the biconditional model which amounts to removing α from the set of support and thus losing the proof.

The situation regarding condensed detachment is but an illustrative example of a much more general problem facing combined systems. Semantic restrictions imposed on proofs as a result of modelling subsets of the formulae derived may, and perhaps typically will, conflict with other constraining mechanisms imposed by the proof strategy. There is no easy general way to control such conflicts except by confining the prover to the use of virtually undirected methods of inference such as crude binary resolution which are not powerful enough for "real" proof search. There are ways out of incompleteness, of course, such as by diluting the semantically directed strategies. Otter, for example, allows the user to decide that, say, nine clauses out of ten will be subject to the guiding strategy but the tenth will be taken from an unregulated breadth-first search just in case it is a bad idea to be overzealous. These ways out are not really what we want, however. We want to be able to trust our methods, not to adopt an attitude of limiting the damage by employing them less than fully.

A more satisfying way to keep completeness without sacrificing too much of the power of restriction strategies for resolution would be to devise an Otter-like prover using full semantic resolution, rather than just model resolution, with respect to dynamically updated models. Semantic resolution as defined by Slagle is like hyper-resolution in being based on a nucleus and satellites, each satelllite picking up a single literal in the nucleus to form a clash. The semantic constraint is that all satellites and the resolvant are required to be false in the guiding model. Whether the clauses involved are positive, negative or mixed is not significant. This form of inference would remain refutation complete, but it is not easy to see how there could be a very fast test for the truth value of the resolvant in the general case.

Another possibility, so far uninvestigated, is to consider ways of detecting incompleteness by some kind of analysis of the prover's behaviour. The particular cases in which incompleteness strikes may not form a decidable set, so methods for detecting them should be expected to be partial. However, there may be insights as well as performance gains to be had from the pursuit of such methods.

Other open lines of research in model-guided theorem proving of the type considered here include the following:

- Generalise the results of the present paper and understand better the issues of completeness and incompleteness for semantically guided proof search. In particular, characterise classes of problems for which SCCD is complete or for which it is incomplete.
- Implement and investigate other systems in the style of SCOTT. For instance, use other model generation methods such as hill-climbing ones or those based on tableaux or on extensions of resolution. Examine the effects on other kinds of proof search, for instance on the "bottom-up" phase of a prover like SETHEO, and on other types of inference such as equational reasoning.
- Work more on the false preference strategy. This seems to be generally effective with a wide range of inference rules, but there is a lack of firm mathematical results concerning it.
- Explore the possibilities for techniques using multiple models. SCOTT only uses one model at a time, which restricts what it is able to do. False preference in particular offers great possibilities for systems capable of working with many models simultaneously. Competitive parallel proof search in the manner of (Schumann, 1995) is another technique obviously suited to guidance by multiple models.

Finally, it must be stressed that the project of harnessing semantic information and putting it in the service of theorem provers is important. The results of this paper are negative for the research program, certainly, but must not be seen as destructive of it.[9]

[9]Among those to whom we owe thanks for their contribution to our understanding of the matters of this paper we would nominate especially Mark Grundy, Ewing Lusk, Bill McCune, Bob Meyer and Greg Restall. We also wish to thank those who attended presentations of the above proof at the Australian National University, DFKI (Saarbrücken)

JOHN SLANEY AND TIMOTHY SURENDONK
Automated Reasoning Project
The Australian National University
Canberra, Australia

References

P. Graf, Substitution tree indexing, *Proceedings of the 6th Conference on Rewriting Techniques and Applications* (1995).

R. Hasegawa,M. Koshimura & H.Fujita, MGTP: A Parallel Theorem Prover Based on Lazy Model Generation, *Proceedings of 11th International Conference on Automated Deduction* (1992), 776–780.

J. A. Kalman, Condensed detachment as a rule of inference, *Studia Logica* 42 (1983), 443–451.

W. McCune, *OTTER 2.0 Users Guide*, Argonne National Laboratory, 1990.

W. McCune & L. Wos, Experiments in automated deduction with condensed detachment, *Proceedings of 11th International Conference on Automated Deduction* (1992), 209–223.

R. K. Meyer & M. W. Bunder, *Condensed detachment and combinators*, Technical report TR-ARP-8/88, Automated Reasoning Project, Australian National University, 1988.

R. K. Meyer, M. W. Bunder & L. Powers, Implementing the fool's model of combinatory logic, *Journal of Automated Reasoning* 7 (1991), .

J. Schumann, *SiCoTHEO: Simple Competitive parallel Theorem Provers based on SETHEO*, Typescript, Institut für Informatik, Technische Universität München, 1995.

J. R. Slagle, Automatic theorem proving with renamable and semantic resolution, *Journal of the ACM* 14 (1967), 687–697.

J. Slaney, SCOTT: A model-guided theorem prover, *Proceedings of 13th International Joint Conference on Artificial Intelligence* (1993), 109–114.

J. Slaney, FINDER: Finite Domain Enumerator (system description), *Proceedings of 12th International Conference on Automated Deduction* (1994), 764–768.

J. Slaney, E. Lusk & W. McCune, SCOTT: Semantically Constrained OTTER (system description), *Proceedings of 12th International Conference on Automated Deduction* (1994), 764–768.

and the Technical University of Munich, for helpful comments. The first author also owes much to Ricardo Caferra, not only for many discussions of matters germane to this research but also for his invitation to LIFIA (Grenoble) where the present paper was written.

FAUSTO GIUNCHIGLIA, PAOLO PECCHIARI, AND CAROLYN TALCOTT

REASONING THEORIES

Towards an Architecture for
Open Mechanized Reasoning Systems

Abstract. Our ultimate goal is to provide a framework and a methodology which will allow users, and not only system developers, to construct complex systems by composing existing modules, or to add new modules to existing systems, in a "plug and play" manner. These modules and systems might be based on different logics; have different domain models; use different vocabularies and data structures; use different reasoning strategies; and have different interaction capabilities. This paper, which is a first small step towards our goal, makes two main contributions. First, it proposes a general architecture for a class of reasoning modules and systems called *Open Mechanized Reasoning Systems* (*OMRSs*). An OMRS has three components: a *reasoning theory* component which is the counterpart of the logical notion of formal system, a *control* component which consists of a set of inference strategies, and an *interaction* component which provides an OMRS with the capability of interacting with other systems, including OMRSs and human users. Second, it develops the theory underlying the reasoning theory component. This development is illustrated by an analysis of the Boyer-Moore system, NQTHM.

1. Introduction

An important problem in the domain of automated reasoning is the development of mechanisms for the interoperation and integration of disparate provers. The components of an integrated prover may be tightly or loosely coupled, they may be based on different logics, they may have different domain models, they may use different vocabularies, representations of information, and reasoning strategies, and they may have different interaction capabilities. Currently, if you need a prover there are two choices: (1) implement your own; or (2) adapt an existing prover to your needs, or, more likely, adapt your needs to an existing prover. Given the state-of-the-art of technology for building provers, neither option is satisfactory. Serious pro-

F. Baader and K.U. Schulz (eds.), Frontiers of Combining Systems, 157–174.
© *1996 Kluwer Academic Publishers.*

vers are difficult to build and there is little in the way of generic parts or tools to help. Existing systems are difficult to connect – they are packaged as stand alone software with inadequately described interfaces. Furthermore it is difficult to extract usable modules from existing provers, since they typically depend upon internal structures of the host prover.

Our long term goal is to be able to compose complex provers from existing modules and to add new modules to existing provers in a principled and sound manner, with minimal changes to the existing modules. To realize this goal we need to think of provers as *logical services* (Sutherland and Platek, 1992), and to develop a general framework for specifying and structuring provers as logical services. Provers implementing logical services must be described at many levels: traditional consequence relations; data structures used for mechanizing deduction; annotations, control information and inference algorithms; interaction capabilities, and the supporting inference mechanisms; and the sharing and updating used for communication and for efficient implementation. An important gap that needs to be filled is an analysis of the structures and protocols that are needed in order to specify interactions, and to support mechanisms for incremental, restartable, reactive deduction. We take this as our starting point. We introduce the notion of *Open Mechanized Reasoning System (OMRS)* as an architecture for specifying and implementing logical services.

Reasoning Theory = Sequent System + Rules

Reasoning System = Reasoning Theory + Control

OMRS = Reasoning System + Interaction

A *reasoning theory* consists of a *sequent system* and a set of inference *rules*. A sequent system determines a set of *sequents* (assertions) and a set of *constraints* (side conditions). A reasoning theory determines a set of *reasoning structures* (proof fragments), a subset of which is the set of *derivations*. Reasoning theories are the OMRS counterpart of the logical notion of formal system. A *reasoning system* consists of a reasoning theory and a set of strategies used to search the space of possible inference rule applications. Reasoning systems are the formal counterpart of the informal notion of standalone provers. Finally, an OMRS is a reasoning system provided with a set of interaction capabilities. An OMRS is the formal counterpart of the informal notion of logical service.

The remainder of this paper focuses on and develops the theory underlying the reasoning theory component of OMRSs. §2 defines the notions of sequent system, rules, and reasoning theory. In §3 the reasoning theory underlying the Boyer-Moore prover, NQTHM, is briefly sketched. The choice of NQTHM is not by chance. Since, NQTHM is a complex and state-of-the-art system, a better understanding of how it works is by itself a major result. In §4 reasoning structures, derivations and proofs are defined as certain labelled graph structures. In §5 a set of primitive operations for constructing reasoning structures is defined and shown to be complete. Finally, §6 contains a discussion of related work.

2. Sequent Systems, Rules, and Reasoning Theories

We start by introducing our notation conventions. Let Y, Y_0, Y_1 be sets. We specify meta-variable conventions in the form: let y range over Y, which should be read as: the meta-variable y and decorated variants such as y', y_0, ..., range over the set Y. $Y_0 \times Y_1$ is the set of pairs with first component from Y_0 and second component from Y_1. Y^* is the set of finite sequences of elements of Y. We write $[y_1, \ldots, y_n]$ for the sequence of length $Len(\bar{y}) = n$ with ith element y_i. (Thus $[\,]$ is the empty sequence.) $u \diamond v$ denotes the concatenation of the sequences u and v. $P_\omega(Y)$ is the set of finite subsets of Y. The empty set is denoted by $\emptyset$. We use the convention that if y ranges over Y, then $\bar{y}$ ranges over Y^* and $\tilde{y}$ ranges over $P_\omega(Y)$. $[Y_0 \xrightarrow{f} Y_1]$ is the set of finite maps from Y_0 to Y_1. We use $\vec{\emptyset}$ to denote the (unique) finite map with empty domain. $[Y_0 \to Y_1]$ is the set of total functions, f, with domain, $Dom(f)$, Y_0 and range, $Rng(f)$, contained in Y_1. If $f \in [Y_0 \to Y_1]$ and $g \in [Y_1 \to Y_2]$, then $g \circ f \in [Y_0 \to Y_2]$ is the composition of f and g: $(g \circ f) = \lambda y.g(f(y))$. For any function f, $f\{y \mapsto y'\}$ is the function f' such that $Dom(f') = Dom(f) \cup \{y\}$, $f'(y) = y'$, and $f'(z) = f(z)$ for $z \neq y, z \in Dom(f)$.

2.1. SEQUENT SYSTEMS

The assertions manipulated by actual provers typically have more structure and procedural information than assertions of traditional logical systems. For instance NQTHM contains special purpose inference modules and special data structures used to represent local context and goals. Many provers make use of some form of schematic assertions — assertions with place holders that can be further instantiated. For example, axiom schemas are often used in first-order languages, and place holders provide the ability to postpone choices. The inference rules underlying an inference procedure often have side conditions that must be met for the rule to be applicable. These conditions may be needed to insure soundness, or may represent heuristic decisions.

We represent assertions by data structures called sequents, side conditions by data structures called constraints, and schematic aspects by instantiation maps. Only certain general features of sequents, constraints, and instantiations are needed to describe the notions of reasoning structure and derivation, and the operations for constructing them. These are abstracted in the notion of *sequent system*. This allows us to decouple the definitions of reasoning structure and derivation from the details of any specific sequent system. A sequent system is a structure:

$$Ssys = \langle S, C, \models, I, \_[\_]\rangle$$

S is the set of sequents and C is the set of constraints. $\models \subseteq (\mathrm{P}_\omega(C) \times C)$, is a consequence relation on constraints, which abstractly represents a constraint checking mechanism. I is the set of instantiation maps (or instantiations). $\_[\_]$ is the operation for application of instantiations to sequents and to constraints, that is $\_[\_] : [S \times I \to S]$ and $\_[\_] : [C \times I \to C]$. Thus, both sequents and constraints can be schematic and instantiation provides a means for filling in schemata. We let s range over S, c range over C, and ι range over I. In the remainder of this section we describe the requirements that such a structure must meet in order to qualify as a sequent system. We also introduce some auxiliary definitions.

Satisfaction must obey the basic laws for a (classical) consequence relation, i.e. reflexivity, monotonicity, and cut (cf. (Avron, 1987; Meseguer, 1989)). We extend satisfaction to a relation between sets of constraints by defining $\tilde{c} \models \tilde{c}_1$ iff $(\forall c \in \tilde{c}_1)(\tilde{c} \models c)$.

We call the collection of entities that I acts on, producing entities of the same sort, *schematic entities*. This collection includes sequents and constraints, and is closed under formation of finite sets or sequences, and finite maps whose range is a set of schematic entities. Instantiation is extended pointwise. Thus, for X any set of schematic entities we have:

$$\tilde{x}[\iota] = \{x[\iota] \mid x \in \tilde{x}\} \quad \text{for} \quad \tilde{x} \in \mathrm{P}_\omega(X)$$

$$\bar{x}[\iota] = [x_i[\iota] \mid i < n] \quad \text{for} \quad \bar{x} = [x_i \mid i < n] \in X^*$$

$$f[\iota] = \lambda y.(f(y)[\iota]) = (\lambda x.x[\iota]) \circ f \quad \text{for} \quad f \in Y \to X$$

idi is the identity instantiation: $x[\text{idi}] = x$ for any schematic entity x. Instantiations are closed under composition, $\iota_0 \circ \iota_1$, and instantiation preserves constraint satisfaction. One of the most basic kinds of constraint is syntactic equality between schematic entities, used for example in matching and unification. We assume that equations $s \sim s'$ between sequents are among the constraints, and that instantiation propagates to the sequent terms, that is $(s \sim s')[\iota] = s[\iota] \sim s'[\iota]$. $\models$ obeys, w.r.t. equations between sequents, the usual laws for equality, i.e. reflexivity, transitivity, and symmetry. Typically schematic entities are obtained by including meta variables of various syntactic sorts among the basic syntactic entities from which others are generated. Then instantiations are just (finite) maps from meta variables to syntactic entities. The axioms above are intended to capture this intuition without forcing a particular model.

Notice that the notion of sequent system meets the desiderata informally stated at the beginning of the subsection, i.e. it allows for the representation of a large variety of concrete sequents; for the use of schematic entities; and for the manipulation of constraints.

2.2. RULES

Inference rules of standard logical systems, for instance natural deduction and resolution systems, are relations on assertions giving the possible conclusions from a collection of premises, possibly with some side conditions which establish their applicability. Most provers have schematic rules, i.e. rules which manipulate sequents containing schematic entities; and many use rules with a variable number of premises. Some provers support provisional reasoning (structures that are valid deductions when certain constraints are met). In these situations, checking for the applicability of an inference rule is sometimes postponed until after the rule is applied. The inference procedures of a prover may apply rules backwards, forwards, or mixed mode.

As for sequent systems, only certain general features of rules are needed to describe the notions of reasoning structure and deduction, and the operations for constructing reasoning structures. In order to allow for provisional reasoning, we treat applicability constraints as rule constituents. Thus a rule is a set of tuples, each consisting of a sequence of sequents, a single sequent, and a finite set of constraints. We require that rules be closed under instantiation. Typically such relations consist of tuples of sequents of the same general structure, but this is not required. A rule set is a finite set of rules each associated with a unique identifier. Such sets are conveniently thought of as finite maps from identifiers to rules. Mathematically, the use of rule sets is just a way of partitioning one rule into several parts and giving each part a name.

Let $Ssys = \langle S, C, \models, I, \_[\_] \rangle$ be a sequent system, and let Id be a set of identifiers. Then the set of rules $R \in \mathbf{Rule}[Ssys]$ over $Ssys$, and the set of rule sets, $\tilde{r} \in \mathbf{Rset}[Ssys, Id]$ over $(Ssys, Id)$ are defined by

$$\mathbf{Rule}[Ssys] = \{R \subseteq (S^* \times S \times \mathrm{P}_\omega(C))$$
$$\mid (\forall \langle \bar{s}, s, \tilde{c} \rangle \in R)(\forall \iota \in I)(\langle \bar{s}, s, \tilde{c} \rangle [\iota] \in R)\}$$

$$\mathbf{Rset}[Ssys, Id] = [Id \xrightarrow{f} \mathbf{Rule}[Ssys]]$$

If $\tilde{r} \in \mathbf{Rset}[Ssys, Id]$ and $id \in Id$ we say that $\langle \bar{s}, s, \tilde{c} \rangle \in \tilde{r}(id)$ is an *instance* of id with premises, $\bar{s}$, conclusion, s, and applicability conditions, $\tilde{c}$. A *rule generator* is any subset rg of $S^* \times S \times \mathrm{P}_\omega(C)$. The rule generated by rg is the set $rg[I]$. An *n-ary rule* is a rule contained in $S^n \times S \times \mathrm{P}_\omega(C)$, i.e. a rule such that its instances have all the form $\langle \bar{s}, s, \tilde{c} \rangle$ where $\bar{s}$ is a list of n sequents.

Notice that the notion of rule satisfies all the desiderata informally stated above. Classical rules with a fixed number n of premises presented by a schema correspond in our framework to n-ary rules whose generator is a singleton set. Our framework allows for complex rules including rules with schematic constituents, rules with a variable number of premises, and rules with variance in the structure of the premises and conclusion. Making constraints an explicit component of rules allows for provisional reasoning. In particular, the checking of constraints can be separated from rule application and hence can be postponed. Finally, since rules are relations, they are adirectional and naturally allow for all possible modes of application.

2.3. REASONING THEORIES

We think of a reasoning theory as presenting a formal system or theory. It specifies a set of sequents and constraints, and a set of rules. A reasoning theory, Rth, is a structure

$$Rth = \langle Ssys, Id, \tilde{r} \rangle$$

such that $Ssys$ is a sequent system, Id is a set of identifiers, and $\tilde{r} \in$ **Rset**$[Ssys, Id]$ is a rule set.

Many systems make use of multiple representations of information, possibly in the context of multiple logics, and may even represent heterogeneous proofs that combine reasoning in the different logics or different mechanizations of the same logic. The natural way to structure such provers using our framework is as the gluing together of separate reasoning theories using additional inference rules. To illustrate this idea we define a simple operation for gluing together a family of reasoning theories. Let $Rth_1 = \langle Ssys_1, Id_1, \tilde{r}_1 \rangle$, ..., $Rth_n = \langle Ssys_n, Id_n, \tilde{r}_n \rangle$ be disjoint reasoning theories, with $Ssys_i = \langle S_i, C_i, \models_i, I_i, \_[\_]_i \rangle$ and $\tilde{r}_i \in$ **Rset**$[Ssys_i, Id_i]$ for $1 \leq i \leq n$. By disjointness we mean that the families of sets S_i, C_i, I_i, and Id_i for $1 \leq i \leq n$ are each pairwise disjoint. Thus $S_i \cap S_j = \emptyset$, for $1 \leq i \neq j \leq n$, etc.

The (disjoint) union, $Ssys$, of the sequent systems $Ssys_i$ for $1 \leq i \leq n$ is defined by

$$Ssys = \bigcup_{1 \leq i \leq n} Ssys_i = \langle S, C, \models, I, \_[\_] \rangle$$

where $S = \bigcup_{1 \leq i \leq n} S_i$, $C = \bigcup_{1 \leq i \leq n} C_i \cup \bigcup_{1 \leq i \neq j \leq n} \{ s \sim s' \mid s \in S_i, s' \in S_j \}$, $I = I_1 \times \ldots \times I_n$. $\models$ and $\_[\_]$, the identity instantiation, and composition are defined as follows:

$$\tilde{c} \models c \quad \text{iff} \quad \tilde{c} \cap C_i \models_i c \quad \text{if} \quad c \in C_i$$

$$\tilde{c} \not\models s \sim s' \quad \text{if} \quad s \in S_i \wedge s' \in S_j \wedge i \neq j$$

$$x[\iota] = x[\iota \downarrow i]_i \quad \text{if} \quad x \in S_i \cup C_i$$

$$\text{idi} = \langle \text{idi}_1, \dots, \text{idi}_n \rangle$$

$$\iota \circ \iota' = \langle \iota \downarrow 1 \circ \iota' \downarrow 1, \dots, \iota \downarrow n \circ \iota' \downarrow n \rangle$$

where $\iota \downarrow j$ is the j-th element of the tuple ι. It is easy to check that all the requirements for a sequent system are satisfied.

Let $Id = \bigcup_{1 \leq i \leq n} Id_i$, and let Id_B be a set of identifiers disjoint from Id. Let $\tilde{r} = \bigcup_{1 \leq i \leq n} \tilde{r}_i$ ($\tilde{r}(id) = \tilde{r}_i(id)$ if $id \in Id_i$), and let $\tilde{r}_B \in \mathbf{Rset}[Ssys, Id_B]$ be a set of inference rules over the joined sequent system. The gluing of the Rth_i via $\tilde{r}_B$ is defined by

$$Rth = glueRth([Rth_1, \dots, Rth_n], Id_B, \tilde{r}_B) = \langle Ssys, Id \cup Id_B, \tilde{r} \cup \tilde{r}_B \rangle$$

We say that Rth is a *composite reasoning theory*, with *components* Rth_i, and *glue* $Id_B, \tilde{r}_B$. The elements of the rule sets $\tilde{r}_i$ are called the *internal rules* (briefly *i-rules*) of Rth_i. The elements of $\tilde{r}_B$ are called *bridge rules*. In the examples of which we are aware bridge rules always have premises and conclusions in different component sequent systems, although this is not a requirement. With a little more technical machinery the disjointness requirements can be relaxed, giving a very general composition mechanism.

3. An Example Reasoning Theory: Rth_{NQTHM}

To illustrate the use of reasoning theories to specify interfaces and interactions between reasoning modules, we give some fragments of a specification of the NQTHM prover (Boyer and Moore, 1979; Boyer and Moore, 1988). We start by giving the reasoning theory underlying the NQTHM simplification process (simplifier), and then we show how to combine this reasoning theory with those specifying other NQTHM processes to obtain the reasoning theory underlying the entire NQTHM prover. This illustrates how our framework may be used to specify a complex system like NQTHM in a modular way. The simplifier maps a formula in clausal form to a logically equivalent conjunction of clauses. The simplifier is itself a combination of modules implementing various techniques: the linear arithmetic specialist, the typeset specialist, the rewriter, and the sweeper. The linear arithmetic specialist reasons about linear inequalities over the natural numbers. The typeset specialist computes the type information associated to a term, e.g. that a term is a number or a list, under some type assumptions on terms (hereinafter we refer to such assumptions as typeset information). The linear arithmetic specialist represents linear inequalities using integer polynomials, and uses simple polynomial arithmetic and linear rewrite rules to deduce new polynomial facts. The rewriter rewrites an input term by applying rewrite rules obtained from definitions, axioms, and lemmas contained in the current theory. The rewriter performs these tasks in a context containing many kinds of information, the most important for our analysis being typeset information and polynomial information (used by the linear arithmetic specialist). The sweeper invokes the rewriter to rewrite (sweep) all the literals of the clause given as input to the simplifier.

In order to make the module structure explicit, we present the reasoning theory underlying the simplifier $Rth_S = \langle Ssys_S, Id_S, \widetilde{r}_S \rangle$ as a composite reasoning theory constructed out of four component reasoning theories

$$Rth_S = glueRth([Rth_P, Rth_{Sw}, Rth_R, Rth_L], Id_B, \widetilde{r}_B)$$

where each component reasoning theory corresponds to a reasoning module of the simplifier, and $\widetilde{r}_B$ is a set of bridge rules describing the interactions among these modules. Rth_P corresponds to the master module of the simplifier. Rth_{Sw}, Rth_R and Rth_L correspond respectively to the sweeper, the rewriter and the linear arithmetic modules.

The reasoning theory description of a complex module like the simplifier can be given in many ways, and at many levels of detail. Here, we have chosen to treat typeset information as constraints, and to use the typeset specialist for constraint checking. An alternative would be to represent the typeset logic as a reasoning theory and replace such constraints by explicit premises of rules. A complete description of the NQTHM reasoning theory includes a rigorous mathematic specification of the constraints and constraint satisfaction. Here we give only informal indications of the constraint semantics. Moreover, the assertions manipulated by the simplifier contain additional (non-logical) information used to control the heuristic proof strategies. We have omitted this information for the present, as we are not treating issues of control here.

According to the definition of $glueRth$, the Rth_S sequent system is given by

$$Ssys_S = \bigcup_{i \in \{P,R,Sw,L\}} Ssys_i$$

where each component sequent system has its own sequents, and constraints. There are meta variables for the various sequent sorts and other syntactic sorts, such as terms and clauses. Instantiations are finite maps (substitutions) from meta variables to syntactic entities of the appropriate sort. As an example we briefly introduce three sequent sorts of $Ssys_S$. All

the reasoning in NQTHM is carried out within the context of an NQTHM
theory, which is a sequence of events. Events include function and shell
definitions, axiom declarations, prove-lemma requests, and instructions for
using lemmas. The simplifier manipulates clauses and sets of clauses. In
$Ssys_P$ this is represented using assertions of the form $h \vdash_P cl \to \widetilde{cl}$, where
h represents an NQTHM theory, cl is a clause (corresponding to the input
to the process), $\widetilde{cl}$ is a set of clauses (corresponding to the output of the
process), and the sign P has been added to distinguish this sort of sequent
from other sorts of sequent used inside $Ssys_S$. A P-sequent asserts that cl is
a consequence of $\widetilde{cl}$ (in the theory h). Reasoning performed by the rewriter
and the linear arithmetic specialist is carried out within a local context
obtained by assuming false all literals of a clause except the one currently
being rewritten. This information is stored both in typeset form and in
polynomial form. A polynomial contains a linear inequation of the form
$z_0 + z_1 \cdot t_1 \ldots z_n \cdot t_n \leq 0$ (where the z_i are integers and the t_i are terms), hypotheses needed for the inequation to be derivable, and a history giving the
facts from which the inequation was derived, e.g. used lemmas. A polynomial information structure, pi, contains a set of polynomials (a polynomial
database) and possibly literals yet to be incorporated into the database, by
linearization. Rewriting assertions are represented in $Ssys_R$ by sequents of
the form $h \vdash_R pi, ti; t \to_m \langle t', \widetilde{l}_1, \widetilde{l}_2 \rangle$ where ti is the typeset representation
of the local context, pi is the polynomial representation of the local context,
m records the mode of rewriting (it can be I or B depending on whether the
corresponding rewriting step in the simplifier preserves equality or proposi-
tional equivalence), and $\widetilde{l}_1, \widetilde{l}_2$ are sets of literals that represent information
propagated (but not used by) the rewriter. An R-sequent asserts that t is
equivalent (relative to the mode) to t' in the theory h (under the addi-
tional assumptions in pi, ti, $\widetilde{l}_1$ and $\widetilde{l}_2$). The linear arithmetic specialist
gets as input typeset information and a polynomial information structure
and returns a new polynomial information structure. In $Ssys_L$, assertions
are represented as sequents of the form $h \vdash_L ti; pi \to pi'$, where ti is the
typeset representation of the local context, and pi and pi' are polynomial
information structures. An L-sequent asserts that pi' is equivalent to pi
under the additional assumptions in ti.

The sequent sorts introduced above allow us to give some examples of
rules of Rth_S. The R-rule NE expresses an equality reasoning step performed
by the rewriter.

$$\text{NE} \quad \frac{}{h \vdash_R ti, pi; (\text{EQUAL } t_1\ t_2) \to_I \langle \text{F}, \emptyset, \emptyset \rangle} \quad \text{if} \quad \text{Ts}(ti, t_1) \cap \text{Ts}(ti, t_2) \sim \emptyset$$

where h is a meta variable standing for NQTHM theories; t_1 and t_2 are
meta variables standing for terms; ti and pi are meta variables standing

for typeset and polynomial information structures, respectively. According to this rule, an equality is rewritten to F (meaning false) if the typesets of the two terms t_1 and t_2, are disjoint in the context ti. The mode I, subscripting the arrow in the conclusion of the rule, indicates that this rewriting step preserves term identity. The figure above should be thought of as describing the following rule generator for the rule NE:

$$\{\langle[\emptyset], h \vdash_R ti, pi; (\text{EQUAL } t_1\ t_2) \rightarrow_I \langle F, \emptyset, \emptyset\rangle, \{\text{Ts}(ti, t_1) \cap \text{Ts}(ti, t_2) \sim \emptyset\}\rangle\}.$$

The top level part of the simplifier invokes the linear arithmetic specialist to derive an impossible polynomial (i.e. a polynomial containing an impossible inequation, e.g. $2 \leq 0$) from the negation of the clause cl given in input to the process. When this succeeds, it means that cl has been proved by linear arithmetic reasoning under additional hypotheses (accumulated during the derivation of the impossible polynomial). The bridge rule buildPI represents this interaction between the two modules.

$$\text{buildPI} \quad \frac{h \vdash_L initTI(cl);\ initPI(cl) \rightarrow pi}{h \vdash_P cl \rightarrow split_c(cl, \tilde{l}_1)} \quad \text{if } impos(pi, \langle \tilde{l}_1, \tilde{l}_2\rangle)$$

where $initTI(cl)$ and $initPI(cl)$ are respectively the typeset and polynomial information structures corresponding to assuming the negation of each literal in cl. The constraint $impos(pi, \langle \tilde{l}_1, \tilde{l}_2\rangle)$ means there is a polynomial in pi, deduced from the literals in $\tilde{l}_2$, that is impossible given additional assumptions $\tilde{l}_1$. These can be considered as a case split and $split_c(cl, \tilde{l}_1)$ yields the clauses that must be proved for the remaining cases.

The reasoning theory underlying NQTHM can be constructed as follows:

$$Rth_{\text{NQTHM}} = glueRth([Rth_U, Rth_W, Rth_S, ...], Id'_B, \tilde{r}'_B)$$

where, Rth_U corresponds to (the reasoning theory part of) the module in charge of interaction with the user, Rth_W corresponds to the module that controls the interactions among the various high-level inference processes of NQTHM, one of which is the simplifier, and the dots stand for the reasoning theories specifying the remaining high-level inference modules (for example the induction process). A modular specification of a prover like the one sketched above allows modifications of a module to be expressed as modifications of the sequent system and rules of the component reasoning theory corresponding to this module. In particular when a theory is modified by adding additional information to sequents and modifying rules to appropriately propagate this information, then inference procedures defined in terms of abstract data types for sequents and rule application operations need not be modified if they make no use of the added information. This suggests a methodology for the integration of a new module into a prover: (1) specification of the module to be added as a reasoning theory; (2) refinement of the specification of the original theories to incorporate any additional information passed to and from the new module; (3) gluing the new module and the modified system, possibly with the addition of new bridge rules.

4. Reasoning Structures and Derivations

Reasoning structures represent proof fragments that occur during the construction of a proof. They provide two independent forms of flexibility: horizontal and vertical. Horizontal flexibility provides flexibility in mode of proof construction, abstraction and reuse of derivations, and schematic reasoning. It comes from being able to stitch together fragments rather like a patchwork quilt and to incrementally refine schematic information by instantiation. Vertical flexibility provides control over the level of immediately visible detail. It comes from nesting of reasoning structures and the ability to encapsulate a substructure into a nesting link, or open up a nesting link. Vertical flexibility is motivated by the need to organize large complex structures hierarchically, to be able to examine them at different levels of depth and detail, or to focus on meaningful substructures.

We let $Rth = \langle Ssys, Id, \tilde{r} \rangle$ be an arbitrary but fixed reasoning theory. We let SN (sequent nodes) and LN (rule nodes) be two disjoint countable sets, used to construct reasoning structures. We give the definition of reasoning structures in two steps. First we define basic reasoning structures. They provide the horizontal dimension of flexibility. Next we add the vertical dimension of flexibility.

A reasoning structure, rs, is a directed labelled graph. The nodes of rs are partitioned into two sets: sequent nodes and link nodes. The edges of rs go from link nodes to sequent nodes or from sequent nodes to link nodes. Each link node has a unique incoming node (the conclusion or goal) and the outgoing nodes (the premises or subgoals) are ordered as a sequence. Sequent nodes are labelled by sequents and link nodes are labelled by *justifications*. One kind of justification is a rule application – represented by a rule identifier and a set of constraints. We call link nodes with such justifications, *rule links*. Another kind of justification is a 4-tuple consisting of a set of constraints, an instantiation map, a sequence of sequent nodes, and a reasoning structure. The instantiation map relates schematic entities of the nested structure to those of its containing structure. The sequent nodes are the nodes in the nested reasoning structure which correspond to the premiss and conclusion nodes of the labelled link node. We call link nodes with such justifications, *nesting links*.

Basic reasoning structures over a reasoning theory Rth and nodes SN,

LN are those with no nesting links.

Definition (Basic Reasoning Structures, $Rs_0[Rth, SN, LN]$):
$Rs_0[Rth, SN, LN]$ is the set of structures

$$rs = \langle Sn, Ln, g, sg, sL, \imath L \rangle$$

such that

(1) $Sn \in P_\omega(SN)$ is the set of sequent nodes of rs, and $Ln \in P_\omega(LN)$ is the set of link nodes of rs;

(2) $g : [Ln \rightarrow Sn]$ maps each link node to its associated goal sequent node;

(3) $sg : [Ln \rightarrow Sn^*]$ maps each link node to its (possibly empty) associated sequence of subgoal sequent nodes;

(4) $sL : [Sn \rightarrow S]$ is the sequent node labelling map;

(5) $\imath L : [Ln \rightarrow [Id \times P_\omega(C)]]$ is the link node labelling map. This map must be such that for $ln \in Ln$ if $\imath L(ln) = \langle id, \tilde{c} \rangle$, $\bar{s} = sL(sg(ln))$, and $s = sL(g(ln))$, then $\langle \bar{s}, s, \tilde{c}' \rangle \in \tilde{r}(id)$ for some $\tilde{c}'$ such that $\tilde{c} \models \tilde{c}'$.

Condition (5) guarantees, informally speaking, that rule links represent only "correct" applications of rules of *Rth*.

We define general reasoning structures by allowing successively deeper levels of nesting, starting with basic reasoning structures at level 0.

Definition (Reasoning Structures, $Rs[Rth, SN, LN]$):   The set of reasoning structures, $Rs[Rth, SN, LN]$, is defined as follows.

$$Rs[Rth, SN, LN] = \bigcup_{n \in N} Rs_n[Rth, SN, LN]$$

where $Rs_n[Rth, SN, LN]$ is the set of reasoning structures of level n. The reasoning structures of level 0 are the basic reasoning structures defined above. The reasoning structures of level $n + 1$ are the structures

$$rs = \langle Sn, Ln, g, sg, sL, \imath L \rangle$$

such that conditions (1-4) in the definition of basic reasoning structures hold, and

(5_{n+1}) $\imath L : [Ln \rightarrow [Id \times P_\omega(C)] + [P_\omega(C) \times I \times [Sn^*, Sn] \times Rs_n[Rth, SN, LN]]]$ such that if ln is a rule link, then condition (5) for basic reasoning structures holds and if ln is a nesting link with $\imath L(ln) = \langle \tilde{c}, \imath, [\overline{sn}, sn], rs' \rangle$ and $rs' = \langle Sn', Ln', g', sg', sL', \imath L' \rangle$ then $[\overline{sn}, sn] \in (Sn')^*$, and $sL'([\overline{sn}, sn])[\imath] = [sL(sg(ln)), sL(g(ln))]$.

Condition (5_{n+1}) for nesting links relates nodes of a nested structure to nodes of its containing structure. This relation makes explicit the idea that a nested structure contains a proof fragment whose conclusion and premises correspond to conclusion and premises of its nesting link.

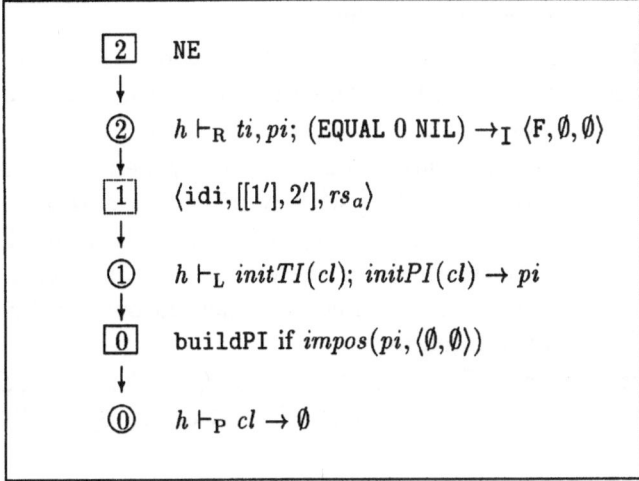

Figure 1. An NQTHM reasoning structure

Figure 1 shows an example of a simple reasoning structure for the NQTHM reasoning theory. There are three sequent nodes (circles) two rule links (boxes 0 and 2) and a nesting link (box 1). rs_a is a reasoning structure with two sequent nodes $1'$, $2'$, appropriately labelled, and no links. Here we can think of the nesting link as suggesting use of the sequent labelling $2'$ to derive the sequent labelling $1'$. Examples of more complex reasoning structures can be found in (Giunchiglia *et al.*, 1994).

In order to define the notion of derivation we need to define how instantiations are applied to reasoning structures.

Definition (instantiation, $rs[\iota]$):   If $rs \in Rs[Rth, SN, LN]$, $rs = \langle Sn, Ln, g, sg, sL, lL \rangle$, and $\iota \in I$, then

$$rs[\iota] = \langle Sn, Ln, g, sg, sL[\iota], lL' \rangle$$

where lL' is defined as follows: for $ln \in Ln$,
(ra)   if $lL(ln) = \langle id, \widetilde{c} \rangle$, then $lL'(ln) = \langle id, \widetilde{c}\,[\iota] \rangle$,
(nest)   if $lL(ln) = \langle \widetilde{c}, \iota_1, [\overline{sn}, sn], rs_1 \rangle$,
        then $lL'(ln) = \langle \widetilde{c}\,[\iota], \iota \circ \iota_1, [\overline{sn}, sn], rs_1 \rangle$.

A reasoning structure is a *derivation* of a conclusion sequent from a set of assumption sequents, if it represents a traditional proof figure. That is, if it satisfies conditions 1-5 below.
(1)   Each rule link has no unsolved constraints.
(2)   Each sequent node is the conclusion of at most one link node.
(3)   There is a unique sequent node that does not occur as the premiss of any inference. The sequent labelling this node is the *conclusion* of the derivation. The sequents labelling occurrences which are not the conclusion of any inference are the *open assumptions*.

(4)   The underlying graph is acyclic.

(5)   For each nesting link, the associated tuple $\langle \tilde{c}, \iota, [\overline{sn}, sn], rs \rangle$ is such that $\tilde{c}$ is the empty set and the reasoning structure $rs\,[\iota]$ is a derivation with conclusion node sn and open assumption nodes $\overline{sn}$.

A sequent s is Rth-derivable from a set of sequents $\tilde{s}$ if there exists a derivation $rs \in Rs[Rth, SN, LN]$ with conclusion s and open assumptions $\tilde{s}$. A reasoning structure is a *proof* if it is a derivation with no open assumptions. The next two lemmas show that, for consideration only of derivability, nesting links and sharing of subderivations can be eliminated (vertical and horizontal unfolding).

Lemma (elimination of nesting):   If rs is a $Rs[Rth, SN, LN]$ derivation of s from $\tilde{s}$ then we can find a level 0 derivation $rs_0 \in Rs_0[Rth, SN, LN]$ of s from $\tilde{s}$.

Lemma (derivation trees):   Let rs be a level 0 derivation of s from $\tilde{s}$ then there exists a level 0 derivation rs' of s from $\tilde{s}$ such that the directed graph underlying rs' is a tree.

5. Operations on Reasoning Structures

There are (at least) two useful algebras based on reasoning structures: an algebra of derivations and an algebra of proof fragments. The algebra of derivations restricts attention to derivations and employs operations such as rule instance, and parallel and sequential composition. It is useful for defining inference procedures and structuring completed derivations. The algebra of proof fragments acts on all reasoning structures, not just derivations, and is based on highly local, incremental operations for constructing reasoning structures. It is useful for interactive reasoning. Here we focus on the algebra of fragments.

We start by giving a set of primitive operations for construction of basic reasoning structures and top level manipulation of nested structures. These are then extended uniformly to build nested reasoning structures. To show how larger interaction steps can be defined in terms of the primitive operations, we define backward inference rule application. To construct basic reasoning structures we define one constant, mtrs, the empty reasoning structure, and four operations: add a sequent, addS; add a rule application link, linkR; solve constraints, solveC. To construct nested reasoning structures we add an operation, linkN, that adds a nesting link.

Definition (empty reasoning structure, mtrs):   The empty reasoning structure mtrs is the structure $\langle Sn, Ln, g, sg, sL, lL \rangle$ where Sn and Ln are the empty set and g, sg, sL, lL are the function with empty domain.

$$\mathtt{mtrs} = \langle \emptyset, \emptyset, \vec{\emptyset}, \vec{\emptyset}, \vec{\emptyset}, \vec{\emptyset} \rangle$$

In the following, each operation, O, has two arguments: a reasoning structure occurrence and a tuple of additional parameters. The reasoning structure occurrence is either a reasoning structure (top level occurrence) or a reasoning structure together with a path consisting of a sequence of nesting links that selects a nested structure. In the following we let rs stand for a reasoning structure $\langle Sn, Ln, g, sg, sL, lL\rangle$. We begin by describing the tuple of arguments, A, appropriate for each operation, O, and the top level action $O(rs, A)$. We then show how this is uniformly lifted to nested substructures.

Definition (adding a sequent, addS(rs, s, sn)):   If $s \in S$ and $sn \in SN - Sn$, then

$$\mathtt{addS}(rs, s, sn) = \langle Sn \cup \{sn\}, Ln, g, sg, sL\{sn \mapsto s\}, lL\rangle$$

Definition (rule linking, linkR($rs, \overline{sn}, sn, r, ln$)):   If $sn \in Sn$, $\overline{sn} \in Sn^*$, $r = \langle id, \overline{s}, s, \widetilde{c}\rangle \in \widetilde{r}$, $\models \{\overline{s} \sim sL(\overline{sn}), s \sim sL(sn)\}$, and $ln \in LN - Ln$, then $\mathtt{linkR}(rs, \overline{sn}, sn, r, ln) = rs'$ where

$$rs' = \langle Sn, Ln \cup \{ln\}, g\{ln \mapsto sn\}, sg\{ln \mapsto \overline{sn}\}, sL, lL\{ln \mapsto \langle id, \widetilde{c}\rangle\}\rangle$$

Definition (constraint solving, solveC($rs, ln, \widetilde{c}'$)):
If $ln \in Ln$, $lL(ln) = \langle id, \widetilde{c}\rangle$, and $\widetilde{c}' \models \widetilde{c}$, then

$$\mathtt{solveC}(rs, ln, \widetilde{c}') = \langle Sn, Ln, g, sg, sL, lL\{ln \mapsto \langle id, \widetilde{c}'\rangle\}\rangle$$

and if $ln \in Ln$, $lL(ln) = \langle \widetilde{c}, \iota, [\overline{sn}, sn], rs_1\rangle$, and $\widetilde{c}' \models \widetilde{c}$, then

$$\mathtt{solveC}(rs, ln, \widetilde{c}') = \langle Sn, Ln, g, sg, sL, lL\{ln \mapsto \langle \widetilde{c}', \iota, [\overline{sn}, sn], rs_1\rangle\}\rangle$$

Note that we can use solveC to add constraints to a link node as well as to eliminate solved constraints.

Nesting links are introduced with minimal structure justifications, just the required sequent nodes, their corresponding sequent labels and the instantiation to put in the link. These can then be extended by applying operations to nested substructures.

Definition (nesting link, linkN($rs, \overline{sn}, sn_0, \iota, \overline{s}, s_0, ln$)):   If $sn_0 \in Sn$, $\overline{sn} = [sn_1, \dots sn_n] \in Sn^*$, $s_0 \in S$, $\overline{s} = [s_1, \dots s_n] \in S^*$, $s_j[\iota] = sL(sn_j)$ for $0 \le j \le n$, and $ln \in LN - Ln$, then $\mathtt{linkN}(rs, \overline{sn}, sn_0, \iota, \overline{s}, s_0, ln) = rs'$ where

$$rs' = \langle Sn, Ln \cup \{ln\}, g\{ln \mapsto sn_0\}, sg\{ln \mapsto \overline{sn}\}, sL, lL'\rangle$$

where $lL' = lL\{ln \mapsto \langle \emptyset, \iota, [\overline{sn}', sn_0'], rs_0\rangle\}$, and $rs_0 = \langle Sn_0, \emptyset, \emptyset, \emptyset, sL_0, \emptyset\rangle$ with $Sn_0 = \{sn_0', sn_1', \dots sn_n'\} \subset SN - Sn$ (distinct fresh sequent nodes), $\overline{sn}' = [sn_1', \dots sn_n']$, and $sL_0(sn_j') = s_j$ for $0 \le j \le n$.

We can construct the reasoning structure of figure 1 as follows. Let $s_0 = h \vdash_P cl \to \emptyset$, $s_1 = h \vdash_L initTI(cl); initPI(cl) \to pi$, $s_2 = h \vdash_R ti, pi$; (EQUAL 0 NIL) $\to_I \langle F, \emptyset, \emptyset \rangle$. Starting with the empty structure, we introduce the goal sequent and the first subgoal and link them with the buildPI rule:

$$rs_1 = \texttt{linkR}(\texttt{addS}(\texttt{addS}(\texttt{mtrs}, s_0, 0), s_1, 1), [1], 0, r_0, 0) \quad \text{where}$$

$$r_0 = \langle \texttt{buildPI}, [s_1], s_0, \{impos(pi, \langle \emptyset, \emptyset \rangle)\} \rangle$$

Next we derive s_2 using addS, linkR, and constraint solving.

$$rs_2 = \texttt{solveC}(\texttt{linkR}(\texttt{addS}(rs_1, s_2, 2), [\ ], 2, r_1, 1), 2, \emptyset) \quad \text{where}$$

$$r_1 = \langle \texttt{NE}, [\ ], s_2, \{\text{Ts}(ti, 0) \cap \text{Ts}(ti, \texttt{NIL}) \sim \emptyset\} \rangle$$

Now we introduce the nesting link: $rs_3 = \texttt{linkN}(rs_2, [2], 1, \texttt{idi}, [s_2], s_1, 1)$.

To define the general application of operations on reasoning structures we first define the set of paths of a reasoning structure. Each path selects a nested substructure. We represent paths as sequences of nesting link nodes. **Definition (path in reasoning structure):** $\overline{ln} \in LN^*$ is a path in a reasoning structure $rs = \langle Sn, Ln, g, sg, sL, lL \rangle$ selecting rs' iff one of the following conditions is satisfied:
(1) $\overline{ln} = [\ ]$, and $rs' = rs$; or
(2) $\overline{ln} = [ln] \diamond \overline{ln}'$, where $ln \in Ln$, $lL(ln) = \langle \widetilde{c}, \iota, [\overline{sn}, sn], rs_1 \rangle$, and $\overline{ln}'$ is a path in rs_1 selecting rs'.
Note that $[\ ]$ is the only path in a basic reasoning structure.
Definition (general application of operations): If O is one of the primitive operations, $rs = \langle Sn, Ln, g, sg, sL, lL \rangle$ is a reasoning structure, $\overline{ln}$ is a path in rs, and A is a tuple of arguments appropriate for O and the nested structure selected by $\overline{ln}$ in rs, then we define $O(\langle rs, \overline{ln} \rangle, A)$ by induction on $\overline{ln}$ as follows:
(mt)   if $\overline{ln} = [\ ]$, then $O(\langle rs, \overline{ln} \rangle, A) = O(rs, A)$
(nmt)   if $\overline{ln} = [ln] \diamond \overline{ln}_1$, $lL(ln) = \langle \widetilde{c}, \iota, [\overline{sn}, sn], rs_1 \rangle$, then $O(\langle rs, \overline{ln} \rangle, A) = rs'$ where

$$rs' = \langle Sn, Ln, g, sg, sL, lL\{ln \mapsto \langle \widetilde{c}, \iota, [\overline{sn}, sn], O(\langle rs_1, \overline{ln}_1 \rangle, A) \rangle\} \rangle$$

The reasoning structure operations presented above are sound and complete in the sense made precise by the following two theorems. The proofs are straightforward and can be found in (Giunchiglia *et al.*, 1994).
Theorem (soundness):
(1) mtrs is a basic reasoning structure.
(2) The operations addS, linkR, solveC, and linkN all map reasoning structures to reasoning structures.

(3) The operations addS, linkR, and solveC all map basic reasoning structures to basic reasoning structures.

Theorem (completeness): If $rs \in Rs[Rth, SN, LN]$, then rs can be constructed from the empty reasoning structure using only the operations addS, linkR, solveC, and linkN. The basic reasoning structures are generated by excluding linkN.

Inference rule applications can be defined by simple composition of the operations defined above. As an example let us consider backwards inference rule applications. We define this operation for basic reasoning structures. It can be lifted uniformly to nested reasoning structures in the same manner as primitive operations are lifted.

Definition (backward application of r): Let $\langle \overline{s}, s_0, \widetilde{c} \rangle \in r(id)$, with $\overline{s} = [s_1, \ldots, s_n]$ such that $\models \widetilde{c}$. Let $rs_0 = \langle Sn, Ln, g, sg, sLlL \rangle$ be a reasoning structure, $sn_0 \in Sn$, $sL(sn_0) = s_0$. Assume $\overline{sn} = [sn_1, \ldots, sn_n]$ with $sn_j \notin Sn$ for $1 \leq j \leq n$, and $ln_0 \notin Ln$. Then $\mathtt{bkwdR}(rs_0, r, sn_0) = rs'$, where rs' is obtained by the following sequence of primitive operations:

(add sequents) $rs_{i+1} = \mathtt{addS}(rs_i, s_{i+1}, sn_{i+1})$, for $0 \leq i < n$;

(add link) $rs_{n+1} = \mathtt{linkR}(rs_n, \overline{sn}, sn_0, r, ln_0)$; and

(solve constraints) $rs' = \mathtt{solveC}(rs_{n+1}, ln_0, \emptyset)$.

For example, in the above reasoning structure construction, the step producing rs_2 could be replaced by a sequent introduction and backwards rule application: $rs_2 = \mathtt{bkwdR}(\mathtt{addS}(rs_1, s_2, 2), r_1, 2)$.

6. Related work

Composite reasoning theories, are similar in spirit to multilanguage systems (Giunchiglia and Serafini, 1994). In multilanguage systems, sequents may have hypotheses and conclusion from different languages (sequent systems), and bridge rules may discharge hypotheses across languages. The notion of n-ary rule generalizes the notion of rule introduced by Scott (Scott, 1974). Reasoning structures generalize the deduction graphs used in IMPS (Monk, 1988). The work presented in this paper is an attempt at an axiomatic presentation of a wide class of deductive systems in the spirit of the work on general logics (Meseguer, 1989). Other meta-logical frameworks have focused on notational systems for presenting logics (more precisely, deductive or entailment systems using the classification of (Meseguer, 1989), as they treat only syntax and deduction, not models). An important example is LF (Harper *et al.*, 1987) – a meta-logical system for describing and prototyping logics; LF and our work share the main goal of characterizing formal systems and provability at an abstract level. However there are important differences, all consequences of the different targets: LF focuses on formalizing (the deductive aspects of) logics while the work on reasoning theories focuses on formalizing and specifying provers, and on analyzing the data structures that support their design and integration.

Acknowledgements.   The authors would like to thank Jussi Ketonen, for many stimulating discussions during the course of the work; and Bob Boyer, Maura Cerioli, José Meseguer, and Toby Walsh for careful reading of earlier drafts. This research was partially supported by ARPA grants NAG2-703, NAVY N00014-94-1-0775, ONR grant N00014-94-1-0857, and CNR grant CN 92.03006.CT26.

FAUSTO GIUNCHIGLIA
IRST and Università di Trento
Italia

PAOLO PECCHIARI
IRST and Università di Genova
Italia

AND

CAROLYN TALCOTT
Stanford University
USA

References

A. Avron. Simple consequence relations. LFCS Report, Laboratory for the Foundations of Computer Science, University of Edinburgh, 1987.

R. S. Boyer and J. S. Moore. *A Computational Logic*. Academic Press, 1979.

R. S. Boyer and J. S. Moore. Integrating decision procedures into heuristic theorem provers: A case study with linear arithmetic. In *Machine Intelligence 11*. Oxford University Press, 1988.

F. Giunchiglia and L. Serafini. Multilanguage hierarchical logics (or: how we can do without modal logics). *Artificial Intelligence*, 65:29–70, 1994.

F. Giunchiglia, P. Pecchiari, and C. L. Talcott. Reasoning theories: Towards an architecture for open mechanized reasoning systems. Technical Report 9409-15, IRST, November 1994. Also appears as Stanford University Computer Science Department Technical Note STAN-CS-94-TN-15.

R. Harper, H. Honsell, and G. Plotkin. A framework for defining logics. In *Second Annual Symposium on Logic in Computer Science*. IEEE, 1987.

J. Meseguer. General logics. In H.-D. Ebbinghaus et al., editor, *Logic Colloquium '87*, pages 275–329. North-Holland, 1989.

L. G. Monk. Inference rules using local contexts. *Journal of Automated Reasoning*, 4:445–462, 1988.

D. S. Scott. Rules and derived rules. In S. Stenlund, editor, *Logical Theory and Semantic Analysis*, pages 147–161. D. Reidel, 1974.

I. Sutherland and R. Platek. A plea for logical infrastructure. In *TTCP XTP-1 Workshop on Effective Use of Automated Reasoning Technology in System Development*, pages 1–3, 1992.

BERND INGO DAHN AND ANDREAS WOLF

NATURAL LANGUAGE PRESENTATION AND COMBINATION OF AUTOMATICALLY GENERATED PROOFS

Abstract. The paper describes a generic tool that generates automatically natural language presentations of proofs from various automated and interactive deductive systems. Proofs from different sources are translated into a unified format and equipped with a block structure. These proofs can be easily combined. Several proof transformation procedures, based only on the analysis of structural aspects of the proofs, are available. The tool is part of the *ILF* system and of the *ILF* mail server.

1. Introduction

Cooperating systems face the user with a complex interface, consisting of a formal union of all the frequently incompatible interfaces of the singular systems that cooperate. Especially in the field of automated and interactive theorem proving – where many systems are in an experimental stage – the user has to learn not only the different input syntax but he also has to understand the output produced by the cooperating provers. These outputs are expressed in different syntax and they describe proofs in different calculi, assuming that the user is acquainted with the specific rules of inference used.

This situation makes it hard to evaluate the results of the provers in order to solve complex tasks by optimal cooperation. But even if only a single system is involved, the user might prefer to have its proofs presented in a format that can be understood by people without special knowledge on the underlying logical calculus and technical arrangements. The user who wants to understand a proof would prefer a calculus that has linear and well-structured proofs formulated in a language which is easy to understand.

In this paper, we put the problem of unified input aside and concentrate on the production of a unified natural language proof presentation for automated theorem provers. We shall describe the principles of proof

175

F. Baader and K.U. Schulz (eds.), Frontiers of Combining Systems, 175–192.
© *1996 Kluwer Academic Publishers.*

presentation that have been implemented in the system *ILF* (Dahn et al., 1994). The *ILF* system is developed at the Institute of Mathematics at the Humboldt University Berlin, supported by the Deutsche Forschungs-gemeinschaft. Within *ILF*, the automated theorem provers *DISCOUNT*[1] (Denzinger & Pitz, 1992), *SETHEO*[2] (Letz et al., 1992), *KoMeT*[3] (Bibel et al., 1994) and *OTTER*[4] (McCune, 1990) cooperate with domain specific provers. *ILF* itself is an interactive system.

The organization of the cooperation of the provers within *ILF* is aside the focus of that paper and will be treated separately.

In early 1993 we intended to present the possibilities achieved by the cooperation of these provers to an audience without going into technical details. Therefore, we presented at a workshop in Darmstadt in March 1993 a natural language proof of Busulinis Theorem from the theory of lattice ordered groups, produced with the help of *ILF*. In fact, only the proof details entered by the user had been presented, while the technical details filled in by the cooperating automated provers had been reduced to mere footnotes which indicated the authorship.

It was only legitimate to demand also a natural language proof presentation for the subproofs produced by the automated provers. Of course, these should be easy to read and fit smoothly into the general proof plan. Therefore, these proofs had to be transformed into a unified logical calculus which has well structured proofs.

Fortunately, it was possible to implement a proof presentation procedure that relied totally on the structure of the proofs and was independent of the specific syntax and set of inference rules used by a particular calculus. This structural approach is in clear contrast with the approach taken by Huang (Huang, 1994) which analyzes pattern of natural deduction inference rules in order to determine the proof presentation. It enabled a unified presentation of proofs from different provers. Moreover, it facilitates the adaptation of the proof presentation to new systems considerably.

Since the proof presentation procedures described below are to a large extent independent of the syntactic format of the formulas in the proof, they are suited equally well for sorted and unsorted first order logic and for higher order logics. Also proofs in other extensions of classical logics – like modal logics – can be presented. However, in these cases some of the tools – especially those dealing with the conversion of indirect proofs – may not be applicable.

[1]University of Kaiserslautern
[2]Munich University of Technology
[3]University of Darmstadt
[4]Argonne National Laboratory

Subsequently, we shall give an outline of the proof presentation process. Then some of the tools which are available to modify the structure or the typesetting of proofs will be presented. We shall also describe the algorithms that are used to transform native output of the provers mentioned above.

This paper is not intended to be a manual. Therefore, we shall present technical details only, if they are absolutely necessary to understand the power and the limits of the concepts involved. The reader, who wants to see more details or to use the available tools for his own prover can receive a manual sending an e-mail to `ilf-serv@mathematik.hu-berlin.de` with the text *help*.

2. An Overview of the Process of Proof Presentation

In this Section we shall give a general survey on the way from the native output of a theorem prover to a natural language LaTeXfile. We shall consider only the most general case. System specific details can be found in later Sections.

The process of proof presentation can be roughly divided into the following phases.

1. Syntactic proof unification,
2. Semantic proof unification,
3. Structural proof transformation,
4. Generation of proof presentation.

It is necessary, that the steps can be performed in this sequence; however, not every prover will demand all steps in order to have its proofs presented.

Lack of space does not permit us, to illustrate all these steps with an example. However, the interested reader can obtain such an example by sending mails to `ilf-serv@mathematik.hu-berlin.de`.

- `help example_setheo` or `help example_komet` will return an example in the native output language of the prover *SETHEO* or *KoMeT* respectively.
- `help example_me` will return the example as it would have been obtained after step 1.
- `help example_block` will return the example as it looks after step 2.
- If in any of the returned mails SEND `.tex` `.out` is added as a first line and the mail is sent back to the *ILF* server, the server will return an answer that contains a section FILE `.out` with the same example after step 3 and a section FILE `.tex` with the final proof presentation obtained after step 4.

When a proof presentation has been obtained, it may be reasonable to repeat steps 3 and 4 in order to obtain a better readable output. This will demand that the user changes certain system settings. However, the steps described above are performed completely automatically.

In order to present the output of a prover as a human readable proof, it is necessary that the prover has produced a complete proof. Especially, it is necessary that the output contains all axioms used in the proof and references to the places where they are used. If axioms are converted to clauses and clauses are used in the proof, the clauses must be linked with the input axioms as well as with the places where they are used. For some provers, this requires specific settings. Provers that are not able to give these links restrict the power of the presentation tool.

2.1. SYNTACTIC PROOF UNIFICATION

The first step mentioned above – syntactic proof unification – has the aim to transform the proof syntactically from the native output of a special prover into a general format - the *ILF* standard proof format. This step is prover-specific and requires acquaintance with the prover output language. Since (almost) no prover has a documentation of its output language, this may be a matter of trial and error. Fortunately, only the axiom system, the formulas in the proof and the derivability relation between the formulas have to be preserved. Additional information (e. g. on search strategies or system parameters) can be easily preserved and used in the final proof presentation, but this is not necessary.

The proof is considered as a finite directed acyclic graph *(DAG)*.

The nodes of the DAG are labeled with formulas. Moreover they can carry status and control information that specify how the formula attached to a node has been proved and which other formulas and rules have been involved in its proof. The proof will be presented as a set of PROLOG facts. The graph is described by attaching at each node the list of its predecessors. There are no restrictions on the syntax of formulas or rules except that they have to be in a PROLOG readable format. Within *ILF* it is possible already at this stage to apply tools to visualize the structure of the proof. This makes use of the *TreeViewer* (Dahn et al., 1994) – a powerful general tool to deal with trees and DAG's.

In the *ILF* standard proof format each node of the encoded DAG is referenced by a pair, the first component of which is unique for the proof. Each proof can contain references to lines in other proofs. This gives a simple possibility to combine proofs from different provers. Since the sequence of facts in the proof description is irrelevant, they can be produced and transmitted in any order. By assigning the status *unproved* to some nodes in the proof, it is possible to handle incomplete proofs. This is of special importance, if a subsystem delivers an incomplete proof that is to be completed by another subsystem.

2.2. SEMANTIC PROOF UNIFICATION

The semantic proof unification transforms the proof into a block structure. This step is crucial for a unified presentation of proofs from different systems. A calculus with block structured proofs which is sound and complete with respect to the standard semantics of first order logic is described in (Dahn & Wolf, 1994). The *Block Calculus* is a special kind of *Natural Deduction* (Gentzen, 1935), enlarged by the block structures.

Definition 1 A *block structured proof* is a structure consisting of the *header* and of the *body* of the proof.

The *header* specifies

– the formula to be proved,
– the axioms to be used and
– system specific control information.

The *body* of the proof consists of a linear sequence of lines.

– Each line has a formula as it's contents.
– Each line has a *status*.
– Each line can have a *control information*.

The *status* of a line can be one of the following: *proved, unproved, assumption* or *subproof(Pf)*. In the latter case, the parameter Pf points to another block structured proof which serves to justify the contents of the actual line.

The *control information* consists of a list of rules and a list of references used to justify the given line.

There are no restrictions on the kind of rules to be used. References are hints to axioms or to lines which are usable for a given line.

Definition 2 The relation A *is usable for* B is the least transitive relation with the following properties:

1. If line A precedes line B in the same proof, then A is usable for B.
2. If line A precedes line B in the same proof and B has the status $subproof(Pf)$ then A is usable for each line in the subproof Pf points to.

Note, however, that if A and B are lines in the same proof with status $subproof(PA)$, $subproof(PB)$ respectively, then lines of subproof PA are not usable for lines of subproof PB.

The general assumption underlying the generation of LaTeX output is, that the contents of each line is a logical consequence of the axioms and the contents of the usable lines.

All proof restructuring tools used in the following steps preserve this discipline of reference.

The process of semantic proof unification is calculus specific. More precisely, it depends on the semantics of the directed graph produced in the first step. An algorithm for the transformation of model elimination proofs into block structured proofs is sketched below. A complete description can be found in (Wolf, to appear).

For the aforementioned systems integrated within *ILF*, the *ILF* mail server has algorithms to do the necessary transformations. For other systems, a default algorithm can be sketched as follows.

We suppose, that there is a relation $\models$ of semantic logical consequence relating sets of sentences with single sentences. For each set S of sentences let $S^\models = \{\varphi : S \models \varphi\}$. We assume that $\models$ is a closure operator, i. e. $S \subseteq S^\models$, $S_1 \subseteq S_2$ implies $S_1^\models \subseteq S_2^\models$ and $S^{\models\models} = S^\models$.

Moreover, we have to assume that a native proof yields a set $\mathcal{P}$ of sentences such that each sentence $\varphi \in \mathcal{P}$ is obtained from sentences of the kind $\varphi_1, \ldots, \varphi_n \in \mathcal{P}$ by some rule of inference which is correct, in the sense that $\{\varphi_1, \ldots, \varphi_n\} \models \varphi$.

Then the algorithm can transform the proof as follows.

1. Build a DAG which has $\mathcal{P}$ as the set of nodes and an edge from φ_i to φ if and only if φ has been obtained from $\varphi_1, \ldots, \varphi_n$ by a single application of a rule of inference.
2. Apply a standard topological sorting algorithm to construct a linear order $\leq$ on $\mathcal{P}$ such that $\varphi_i \leq \varphi$ if there is an edge from φ_i to φ.

This will yield a block structured proof which is correct in the sense, that each sentence in the proof is a semantic logical consequence of the preceding sentences. This proof can be presented by the presentation tools of the *ILF* mail server. However, it cannot be expected, that the resulting presentation is well-readable.

Much better results can be obtained, when structural information can be carried over from the original proof. The algorithm described in (Wolf, to appear) can serve as a guide for the transformation of tree-structured proofs – possibly with cross references – like those used in tableaux or problem reduction calculi. More useful information for structuring proofs in various calculi can be found in (Gabbay, 1994) and (Lingenfelder, 1990).

Though the semantics of block structured proofs is fixed, there are still no restrictions on the syntax of formulas or rules of inference. Hence the designer of a block structuring algorithm for a specific calculus is free to introduce his own set of rules of inference that explain why a given line is a logical consequence of certain preceding lines.

As mentioned in (Dahn & Wolf, 1994) it is a general property of block structured calculi that their combination by joining their sets of rules yields a correct logical calculus provided that the combined calculi were correct. Therefore, at this level block structured subproofs generated from proofs of different provers can be easily combined into a larger block structured proof.

At this stage – having obtained a block structured proof – a natural language proof presentation could be generated. Though this will yield output which is easier to read than the provers native output, it is frequently still hard to understand. The proof may be overloaded with (sometimes even superfluous) details. It may appear as a very complex proof consisting of many fairly small and obvious steps. Formulas may not occur in the place where they are needed. Subproofs may be unnecessarily complex – e. g. indirect subproofs where direct subproofs would do as well.

2.3. STRUCTURAL PROOF TRANSFORMATION

Therefore this is a good place to go to the third step mentioned above, i. e. to apply proof restructuring tools to improve the readability. However, 'readability' is a fairly subjective matter. Some reader may want to see just a rough sketch of the proof where another prefers a detailed explanation. A third reader may be only interested in the role that a certain concept plays in the proof etc. Therefore, ideally, proof restructuring tools should be applied by the reader to produce LaTeX output for his personal demands. This is possible within the *ILF* system as will be explained below.

Nevertheless, a reasonable default sequence of proof restructuring tools can lead to a considerable improvement of readability. This default sequence is in general calculus dependent. Especially it will make use of the control information that the block structured proof has inherited from the primary logical calculus.

2.4. GENERATION OF PROOF PRESENTATION

The final step generates the natural language proof presentation as a LaTeX source file. The *title and author* can be succeeded by a user defined abstract. The rest of the output can be generated completely automatic, based on some user defined typesetting declarations as described below.

When the proof is formulated in a sorted or higher order logic, it can be convenient to use certain *typing conventions* for variables of specific sorts - e. g. n, m for natural numbers. If this is the case, the corresponding explanations are given at the beginning of the proof.

The second Section of the output proof lists the *axioms* that have been indeed used in the proof. Automated theorem provers may demand certain technical axioms – e. g. to axiomatize equality, arithmetic or other domains which are well known to the user. These can be suppressed in the proof presentation on demand.

A complex block structured proof can be presented as a series of lemmas followed by a theorem. In this case, the outermost block consists of the sequence of lemmas and the theorem, each having the status pointing to its particular subproof.

After the statement of the lemma or theorem, the proof is given. By default, the prover that found the proof is mentioned in an automatically generated footnote.

The sentences presenting the proof have by default a fairly simple structure. Each sentence corresponds to a single line in the block structured proof. Such a sentence consists of an introductory phrase like *Hence, Therefore* or *By* followed by a sequence of references, the formula to be displayed and by a sequence of arguments to justify the validity of the formula.

Hence indicates, that only the formula immediately preceding the actual formula has been used. If *Therefore* is used, other formulas, at most R lines before the given formula are used. Formulas which are more distant are explicitly referenced as described below. The value of R can be set by the user. The default setting is 3.

References to distant formulas are normally given by an appropriate numbering. The user can demand that simple formulas are repeated where they are needed instead of giving an explicit reference when their distance from the actual line is greater than twice the value R mentioned above. Simple formulas are formulas of term depth 1.

It is also possible to suppress the justification of lines by rules of inferences. If the proof is presented in the appropriate granularity, it is usually possible to rely on the users ability to derive a logical consequence without being pointed to a rule of inference. Nevertheless, rules maybe output in an appropriate way to provide additional explanations if necessary. As will be explained below, the user can specify how rules should be typeset. This – in connection with the freedom to introduce new rules – considerably enhances the possibilities to display proofs in different styles.

It has been a matter of debate, whether the simple structure of the output sentences should be replaced by a more complex one. In fact, the restriction to a small set of sentence templates can make the output look monotonic. Notably, this may happen when proofs from automated provers with a very small set of rules of inference are converted. E. g. if an equation is transformed step by step, the default output would be a sequence of sentences all starting with *Hence*.

However, it turns out, that this simple structure, that expresses the philosophy to *say the same things with the same words*, is an important help in order to understand a complex proof step by step.

Nevertheless, the experienced user has many possibilities to introduce complex sentences. E. g. formulas from several succeeding lines can be contracted into a single formula which is output into a user defined sentence. Thus, a sequence of equations $a = c1, a = c2, \ldots, a = cn, a = b$ may be transformed into a single formula $= ([a, c1, c2, \ldots, cn, b])$ which is displayed as

$$a = c_1$$
$$= c_2$$
$$\vdots$$
$$= c_n$$
$$a = b.$$

Also by introducing new rules of inference and declaring their typesetting as explained below, the experienced user can gain complete control on the sentences that are finally printed.

3. Adapting the Output of Specific Provers

Since each prover uses its specific output format, the first step described above translates the native proofs into the *ILF* standard proof format. Subsequently we describe some of the problems that occur in this phase and the methods by which they have been solved. We shall not presume acquaintance with the particular systems.

3.1. MODEL ELIMINATION PROOFS FROM SETHEO AND KOMET

Model Elimination (Stickel, 1984) can be represented as a special kind of Tableau Calculus (Beth, 1969), (Smullyan, 1968), that omits the formulas generated while reducing clauses (quantifiers, skolemization, disjunctions, implications). Furthermore, the duplication of partial proofs used multiply in the tableau can be prevented using some kind of *factorization* (lemma generation), if a correct closure of all branches in all factorized partial tableaus is still possible. In general, Model Elimination is a goal oriented top down procedure.

In Tableau Calculus, an extension can be interpreted as a reduction of a clause of the axiom system followed by the closure of one of the generated branches. A reduction is the closure of a leaf against a (not necessarily direct) predecessor in the tree. Factorizations are normally not included in the standard calculus, a description can be found in (Letz *et al.*, 1994).

In the *ILF* tool, there are integrated two different Model Elimination style automated theorem prover: *SETHEO* (Letz *et al.*, 1992) and *KoMeT*

(Bibel *et al.*, 1994). *KoMeT* can additionally deal with the equality predicate in a special way, and *SETHEO* cannot do this (yet). *KoMeT* can be adapted to use an input and output language similar to that of *SETHEO*, so only some small changes in the code have been necessary.

The output of *SETHEO* is a PROLOG list describing the proof tree with references on the contrapositives involved in the inferences. The axioms belonging to these contrapositives can be extracted from the output of *SETHEO's* parser *inwasm*.

From Model Elimination Proofs there are extracted all structures that can be interpreted as linear chains of inferences. These chains are represented by blocks in the Block Calculus. Factorization steps are presented as subproofs and reduction steps lead to indirect subproofs. The structure of the whole proof is preserved in the order of these blocks. Note, however, that the proof transformation steps described in the following sections may change this structure considerably. The transformation algorithm is described in detail in (Wolf, to appear).

3.2. DISCOUNT PROOFS

DISCOUNT (Denzinger & Pitz, 1992) is an prover for equation theories based on cooperating provers performing Knuth-Bendix completion procedures. Special subsystems can also use conditional equations. However, if these subsystems are used, *DISCOUNT* is currently not able to output a proof; it just states the provability of the assertion. If no conditional equations are involved and *DISCOUNT* has established the provability of the goal, a second run is required to generate a proof. *DISCOUNT* has sophisticated proof restructuring tools. These analyze the role of formulas in the proof and can break the proof into a series of lemmas. Each lemma has a proof which derives an equation by continuously transforming its right side until its left side is obtained. DISCOUNT can present the proof by LaTeXbut the user has no possibilities to influence the typesetting.

The presentation of *DISCOUNT* proofs within *ILF* is based on the proof restructuring performed by *DISCOUNT* itself. The sequence of lemmas produced by *DISCOUNT* has already a block structure modulo syntactic details. Within the proofs of the lemmas, each line is a consequence of the line immediately preceding it and an axiom or another lemma. Therefore the sequence of lines cannot be changed without violating the correctness of the proof and there are no possibilities to further restructure these proofs.

The main advantage of the use of the *ILF* procedures to present *DISCOUNT* proofs is a better quality of typesetting, e. g. by using infix operators where *DISCOUNT* demands a prefix notation. Moreover it is possible to suppress simple inferences like applications of associativities. With respect to cooperation, the presentation of block structured *DISCOUNT* proofs in the *ILF* standard proof format makes it easy to combine these proofs with those produced interactively or by other integrated systems.

3.3. OTTER PROOFS

OTTER (McCune, 1990) is a Resolution based automated theorem prover (Robinson, 1965) with a large number of refinements. It can perform the *Set of Support* strategy and Hyper-resolution as well as Paramodulation and other techniques. *OTTER* follows the concept of Bottom Up search, modified by the goal oriented features of its strategies. Therefore *OTTER* has a large number of flags and parameters to influence the search. There is a possibility to let *Otter* determine these settings in accordance with the structure of the proof problem.

The output of *OTTER* consists of a sequence of labeled formulas together with the rules and labels of formulas they were derived from.

Similarly to *SETHEO*, the proofs contain only poor structures, but steps performing Paramodulation or Hyper-resolution can be used to be candidates to become a lemma. *OTTER* itself does not bring along any tools for the visualization of its proofs, the natural output is readable only for users well acquainted with the system.

OTTER's Hyper-resolution steps can be quite complex, especially if clauses with several positive literals are involved. In order to obtain a more detailed explanation, *OTTER* can print a step by step resolution proof, called a *proof object*. In order to link this proof object with the Hyper-resolution proof, *ILF* has to search for corresponding formulas. Unfortunately, the current version 3.0.3.i of *OTTER* strips the axioms of the input theory from the proofs, if these contain quantifiers and proof objects are constructed. Therefore, a second run of *OTTER* using the setting of the parameter *max_given* to 0 is necessary. This will only produce the set of clauses but not the proof and will retain the initial axioms. Then a combination of both output files yields a complete and detailed proof.

4. Improving the Proof Style

In a block structured proof each line can be used at any place below it in the same proof or its subproofs. Therefore, there is no need to duplicate formulas, except when they are proved within different subproofs from different assumptions. Hence it is a good first step to remove unnecessary duplicates. This will also remove some trivial inferences, e. g. when a clause is calculated from a formula which was already a clause. In some cases it is also necessary to remove unused formulas from the proof.

More trivial inferences can be removed by eliminating formulas which are of a specific form (e. g. $T = T$) or which were obtained by specific rules.

Removing proof lines can render subproofs trivial. Such subproofs – especially direct subproofs with one or two lines or indirect subproofs with two or three lines can be eliminated automatically.

Sometimes the transformation of automatically generated proofs yields indirect subproofs which can be converted automatically into direct subproofs. This is the case if the assumption of the indirect proof has in fact been proved earlier or if the assumption is only used in the last step of the subproof to obtain a contradiction. However, in order to realize these possibilities, it must be known which rules of inference generate indirect subproofs and which symbols stand for negation and contradiction.

There are also tools to move lines within a proof to places where they are needed. We have been experimenting with different procedures for this purpose. For small proofs best results are obtained by moving the assumptions needed for a line L successively as close to L as possible. This is done for each L from the last line of the proof to the first. This procedure requires in worst case a time which is quadratic in the length of the proof. It has the advantage, that the sequence of lines in a proof is preserved as far as possible. This maybe useful when the proof has been structured already by the user in an interactive process. For large proofs it may be more useful to save time by building a completely new sequence of proof lines by a topological sorting algorithm which works in a time linear in the proof size. There are also tools to move a particular proof line as far towards another line as possible without violating the correctness of the proof.

It may happen, that a formula is proved within a proof but is used only within one subproof. This formula can be moved inside the subproof.

Deeply nested proofs are hard to read. Therefore, it can be useful to extract subproofs as lemmas in order to reduce the proof depth. The algorithm used in *ILF* determines the deepest chain of nested subproofs. This chain is split in the middle and the subproof starting there is converted into a lemma. External references of the subproof lead to additional assumptions in the lemma.

Besides these tools that change the proof structure, *ILF* provides also facilities that influence the display of the proof. Subproofs can be hidden or the display of the proof can be confined to those parts of the proof that are needed to justify a particular line or to those inferences that use a particular axiom or rule. Thus the role of the concepts involved in the proof can be investigated interactively.

ILF provides procedures to calculate the size, the depth or the list of external references of a subproof. These can be used easily to develop further tools to to transform a given proof for better readability.

5. Determining the Typesetting

The proof presentation we discuss in this paper has been designed for a quick review of complex proofs. Therefore, ease of use was much more important than linguistic elegance. It was not assumed that the user would spend much time to describe the output format. Hence the tools to describe the typesetting had to be simple on the one hand but on the other hand flexible enough to adapt to the varying situations occurring in proofs of theorems from various fields and by various provers. We confine the following description to the most general of these tools.

5.1. TYPESETTING FORMULAS

In principle, it is sufficient to assign an output string to each relation and function symbol in the proof. However, much better results can be obtained when the following structural concept is used.

A structural declaration for displaying a term consists of a *priority number* and a *script* attached to a *term*. The priority number determines (as in operator declarations in PROLOG) whether arguments of the formula are enclosed in parentheses. Arguments with main operator of priority greater than or equal to this number are printed in parentheses. However, the user is not forced to understand the concepts of priority control. Identifiers which do not have a priority declared are treated as operators of maximal priority, i. e. their arguments – except atoms and variables – will be printed in parentheses by default.

A script is a list that contains expressions like

$$x(V), y(V), z(V), x(V, VS), y(V, VS), z(V, VS)$$

where V and VS are terms, and strings. x, y, z determine, in which cases the argument is printed in parentheses. In most cases V will be a variable of the term to be displayed. The second argument VS is only needed if a special typesetting for variables is required.

In order to display a term it is analyzed from the outermost operator down to atoms and variables. If a script for a matching term is found, it is evaluated. Its strings are directly output and they may also contain LaTeX-commands. Then the process continues on the arguments.

The following simple example shows, how declarations can be used to vary the presentation of negation in different contexts.

Example 3 Assume we are given the following declarations for **not**.

```
not(in(X,Y))        :x(X)," \\not\\in ",x(Y)
not(X)              :"\\neg",x(X)
t_all(X,(X:nat),H):"\\mbox{for all naturals }",x(X),\
```

```
"\\ ",y(H)
```

Then

```
t_all(X,(X:nat),(t_all(Y,(Y:nat),not(not(in(X,a)))))))
```

would be displayed as

for all naturals A **for all naturals** B $\neg(A \notin B)$.

By default, variables in output formulas are renamed to $A, B, \ldots$. The user can change this easily. E. g. replacing `x(X)` in the last declaration by `x(X,"n")` would have the effect, that variables bound by the quantifier `t_all(X,(X : nat),...)` will be denoted by $n_1, n_2, \ldots$.

Introducing additional technical terms can enhance the flexibility as in the following example.

Example 4 Consider the following declarations.

```
forall(X,Y) : "\\forall\\ ",x(ilf_varlist(X)),"\\ ",y(Y)

ilf_varlist([]) :- ""
ilf_varlist([X]) :- x(X)
ilf_varlist([X|V]) :- x(X),",",y(ilf_varlist(V))
```

These declarations introduce the technical term `ilf_varlist`. Therefore, instead of displaying the first argument of `forall(X,Y)` the auxiliary term `ilf_varlist(X)` will be typeset. This has the effect, that the formula `forall([X,Y,Z],r([X,Y,Z]))` will be typeset as

$$\forall\ A, B, C\ (r([A, B, C]).$$

Note the different display of the list after the quantifier and as an argument of r.

5.2. TYPESETTING RULES OF INFERENCE

Since the proof presentation formalism should work for any calculus and since the user is free to introduce new ways of justification that explain specific ideas, there is a need for a flexible way to express justifications.

Recall the organization of the control information contained in block structured proofs. That knowledge is encoded by facts which attach a control information of the form *rule(RuleList,ArgList))* to the lines of a proof. The presentation of the argument behind such a control information is governed by rule declarations of the form *rule(RuleList,ArgList) :- RuleScript*. When a control information is found, the first matching rule declaration is determined and the *RuleScript* is evaluated.

RuleScript is a list containing strings and expressions of the following form:

formula the formula being the contents of the current line is printed,

intro an introductory phrase like *Hence* .., *Therefore* ... etc, together with references to the lines in *ArgList* is printed.

proof if the current line has a status *subproof(Pf)*, the subproof with proof handle *Pf* is printed in this place,

ref(H) prints a reference to the contents of the proof line *H*. *H* can occur in *ArgList* or as a sub-term of a member of *RuleList*.

formula(H) prints the contents of the proof line *H*. *H* can occur in *ArgList* or as a sub-term of a member of *RuleList*.

math(X) prints the value of X as a formula.

rules prints *RuleList*.

formula_ref prints a reference to the contents of the actual line.

Example 5 The following shows this concept in some examples.

```
rule([intro(Op)],[H1,H2]) : formula,"\
is obtained by introducing the operator ", math(Op),\
" into the formula ",formula(H1)," This is justified by",\
ref(H2),"."

rule([incomplete],_) : intro," -- as we hope to show --",\
formula," is true.\n\n We only sketch the proof idea.",\
proof,"{\em Check this!}"

rule([ref(R)],_) : formula," has been proved by ",\
math(author(R))," in \\ref{",math(ref(R)),"}."
```

The last example must be augmented by structural declarations which decode the author and the reference from the term substituted for R.

6. An Example

In the following we present an example of a proof presentation as it can be performed using the tools of *ILF*. The example can be found in the *TPTP* (Suttner & Sutcliffe, 1994) as *MSC006-1*.

Example 6 *Suppose there are two relations, p and q. p is transitive, and q is both transitive and symmetric. Suppose further the squareness of p and q: any two things are related either in the p manner or the q manner. Prove that either p is total or q is total.*(Pelletier & Rudnicki, 1986*)*

This task was proved by *SETHEO*, the output will be transformed using the tools of *ILF*. The result is the presentation shown in the figure.

A Proof from the ILF Server[*]

Setheo

November 29, 1995

Axiom 0.1 (The transitivity of the relation ρ) $A \rho B \wedge B \rho C \rightarrow A \rho C$.

Axiom 0.2 (The fact, that the relation ρ is not total) $\neg a \rho b$.

Axiom 0.3 (The symmetry of the relation σ) $A \sigma B \rightarrow B \sigma A$.

Axiom 0.4 (The fact, that all elements are related) $A \rho B \vee A \sigma B$.

Axiom 0.5 (The transitivity of the relation σ) $A \sigma B \wedge B \sigma C \rightarrow A \sigma C$.

Theorem 0.1 $c \sigma d$.

Proof[1].

We show indirectly that

$$c \sigma d. \tag{1}$$

Let us assume that

$$\neg c \sigma d. \tag{2}$$

We show indirectly that

$$c \rho b. \tag{3}$$

Let us assume that

$$\neg c \rho b. \tag{4}$$

Hence by the fact, that all elements are related $c \sigma b$. Hence by the transitivity of the relation σ and by (2) $\neg b \sigma d$. Hence by the symmetry of the relation $\sigma \neg d \sigma b$. Hence by the fact, that all elements are related $d \rho b$. By (2) and by the fact, that all elements are related $c \rho d$. Therefore by the transitivity of the relation ρ $c \rho b$. This contradicts (4). Thus we have completed the proof of (3).

By (2) and by the symmetry of the relation σ

$$\neg d \sigma c. \tag{5}$$

We show indirectly that

$$a \rho c. \tag{6}$$

Let us assume that

$$\neg a \rho c. \tag{7}$$

By (5) and by the fact, that all elements are related

$$d \rho c. \tag{8}$$

Therefore by the fact, that all elements are related $a \sigma c$. Hence by the transitivity of the relation σ and by (5) $\neg d \sigma a$. Hence by the symmetry of the relation $\sigma \neg a \sigma d$. Hence by the fact, that all elements are related $a \rho d$. Hence by the transitivity of the relation ρ and by (8) $a \rho c$. This contradicts (7). Thus we have completed the proof of (6).

Hence by the transitivity of the relation ρ and since $c \rho b a \rho b$. Hence by the fact, that the relation ρ is not total we have a contradiction.

Thus we have completed the proof of (1).

q.e.d.

---

[*]This manuscript was generated by ILF. For information on ILF contact gehne@mathematik.hu-berlin.de.
[1] Setheo and Ilf

1

7. Conclusion

We have described the proof presentation tools which have been developed and implemented for the *ILF* system by Lars Allner, Thomas Honigmann and the authors. They have been used to present interactive proofs constructed within *ILF* as well as proofs from the automated provers integrated into *ILF*. The System is designed to be adapted to the output of further systems. Our experiences showed, that the involvement of new systems could be done with relatively low efforts.

These tools are currently also available through the *ILF mail server* (`ilf-serv@mathematik.hu-berlin.de`) which receives proofs and returns natural language proof presentations as LaTeX files. More information on the *ILF server* can be obtained via email (`ilf-serv-request@mathematik.hu-berlin.de`). A stand-alone version of the proof presentation procedures is in preparation.

A frequently asked question concerns modifications of the tools described above for multi-lingual – especially German – output. This has been accomplished in principle, but the results have been unsatisfactory. The reason is the more complex grammar needed to form even simple German sentences. E. g. consider the sentence *Since A is a group, B is a group*. It can be composed from the introductory phrase *Since* and instances of the phrase *X is a group* which displays a term `group(X)`. the corresponding German sentence reads *Wenn A eine Gruppe ist, so ist B eine Gruppe*. In this sentence, `group(X)` has to be displayed either as *A eine Gruppe ist* or *B ist eine Gruppe* depending on the position of the phrase in the sentence. A similar problem, and not the last one, occurs with the grammatically correct use of the different forms of the definite article in German.

Therefore, the generation of German output would require either

- to restrict the freedom of the user to chose phrases or
- to force the user to provide the correct declarations for all grammatically possible situations or
- a more profound lexical analysis of the user declaration

We found the first two possibilities unsatisfactory, while the third is beyond our present scope of research. Therefore, we decided to leave it as a project for future times.

BERND INGO DAHN
Institute of Mathematics
Humboldt University Berlin,
Germany

ANDREAS WOLF
Department of Computer Science
Munich University of Technology,
Germany

References

Beth, E. W., The Foundations Of Mathematics, North-Holland, 1969

Bibel, W.; Brüning, S.; Egly, U.; Rath, T.: KoMeT (system description), Proceedings of the 12th CADE, LNAI 814, pp. 783-787, Springer, 1994

Dahn, B. I.; Gehne, J.; Honigmann, Th.; Walther, L.; Wolf, A.: Integrating Logical Functions with *ILF*; Preprint 94-10, Humboldt University Berlin, Department of Mathematics

Dahn, B. I.; Wolf, A.: A Calculus Supporting Structured Proofs; Journal for Information Processing and Cybernetics (EIK), 5-6/1994

Denzinger, J., Pitz, W.: Das DISCOUNT-System: Benutzerhandbuch; University of Kaiserslautern, SEKI Working Paper SWP-92-16

Gabbay, D. M.: Labeled Deductive Systems, Vol. 1 - Foundations; MPI-I-94-223, Max-Planck-Institute Saarbrücken (1994)

Gentzen, G.: Untersuchungen über das logische Schließen, Mathematische Zeitschrift 39, Berlin, 1935

Huang, X.: Human Oriented Proof Presentation: A Reconstructive Approach, University of the Saarland, SEKI Report SR-94-07

Letz, R.; Schumann, J.; Bayerl, S.; Bibel, W.: SETHEO: A High-Performance Theorem Prover, Journal of Automated Reasoning, 8 (1992)

Letz, R.; Mayr, K.; Goller, Ch.: Controlled Integration of the Cut Rule into Connection Tableau Calculi, Journal of Automated Reasoning, 4 (1994)

Lingenfelder, Ch.: Transformation and Structuring of Computer Generated Proofs, PhD Thesis, Department of Computer Science, University of Kaiserslautern

McCune, W.: Otter 2.0, in: Stickel, M.E. (ed.): Proceedings of the 10th CADE, pp. 663-664, Springer, Berlin, 1990

Pelletier F.J., Rudnicki, P.: Non-Obviousness, In: Wos L. (Ed.), AAR Newsletter (6/1986), Association for Automated Reasoning, Argonne.

Robinson, J. A., A Machine–Oriented Logic Based On The Resolution Principle, Journal ACM 12/1, 1965

Smullyan, R. M., First-Order Logic, Springer, 1968

Stickel, M. E.: A PROLOG Technology Theorem Prover, in: New generation computing 2 (1984)

Suttner, Ch. B.; Sutcliffe, G.: The TPTP Problem Library, Munich University of Technology, Department of Computer Science, Technical Report AR-94-03

Wolf, A.: A Translation of Model Elimination Proofs into an Structured Natural Deduction, to appear

SYMBOLIC COMPUTATION:
COMPUTER ALGEBRA AND LOGIC

Abstract. In this paper we present our personal view of what should be the next step in the development of symbolic computation systems. The main point is that future systems should integrate the power of algebra and logic. We identify four gaps between the future ideal and the systems available at present: the logic, the syntax, the mathematics, and the prover gap, respectively. We discuss higher order logic without extensionality and with set theory as a subtheory as a logic frame for future systems and we propose to start from existing computer algebra systems and proceed by adding new facilities for closing the syntax, mathematics, and the prover gaps. Mathematica seems to be a particularly suitable candidate for such an approach. As the main technique for structuring mathematical knowledge, mathematical methods (including algorithms), and also mathematical proofs, we underline the practical importance of functors and show how they can be naturally embedded into Mathematica.

1. The Next Goal for Symbolic Computation

By the work of researchers in various areas, we now have powerful tools available that support various aspects of problem solving by computer:

- numerical libraries
- special purpose systems for simulation, CAD, robotics, neural network design, ...
- "symbolic computation" systems for computer algebra, computer analysis, ...
- rewrite labs, logic programming systems, constraint solvers, ...
- special and general theorem provers, theorem generators, theorem checkers, ...
- program verification, transformation, and synthesis systems, ...
- advanced software technology tools in all theses systems,

F. Baader and K.U. Schulz (eds.), Frontiers of Combining Systems, 193–219.
© *1996 Kluwer Academic Publishers.*

- graphics, animation, sound, typesetting, notebook, and hyperlink tools in these systems,
- links between these systems, access through the web, electronic user communities, ...

The availability of these systems has drastically enhanced our problem solving potential. However, we want more. Who is "we"? In this paper, I address people who

- explore,
- invent,
- apply,
- teach,
- study, and
- publish

mathematics or, in other words, people who are "doing" mathematics. In this paper, I do not address people who are casual users of mathematics as a "black box". I think that the next natural goal "we" should go for is to do all of mathematics in one system. Here, by a "system", I mean both

- a *logical* system, which should be a sound and uniform frame for all of our mathematical activities and
- a *software* system, which supports these activities.

Although there are systems on the market that advertise explicitly that they are systems for "doing" mathematics, see (Wolfram 1988), I think that there are still significant gaps between what we have and what we want. Basically, I see four gaps:

- the logic gap,
- the syntax gap,
- the mathematics gap, and
- the prover gap.

I will analyze these gaps in Section 2. Then I will argue, in Section 3, why Mathematica seems to be a good starting point for developing a rapid prototype system that may overcome these gaps. (I will not argue, however, that Mathematica is "the" ideal future symbolic computation system. In my view, this would only be possible if the designers of Mathematica made some basic changes in the design of their system.) In Sections 4 to 7, I will then sketch my proposal how one may overcome the logic, the syntax, the mathematics, and the prover gap, respectively.

2. The Present Gaps

2.1. THE LOGIC GAP

At present, officially, there is one uniform logic system in use as one uniform frame of mathematics (although, inofficially, this frame is not at all applied uniformly in the every-day practice of mathematicians): The system of "Bourbakism", which is first order predicate logic plus set theory formulated as a first order theory (which is called "Zermelo Fraenkel set theory"). The problem with the Bourbakistic system as a frame for all of mathematics, including "algorithmic" mathematics, is that in this system functions are sets. Hence, by the extensionality axiom for sets, functions defined by different algorithms but having identical input/output behavior are identical. This is not appropriate for the needs of algorithmic mathematics where we do not only want to discuss the input/output behavior of functions but also properties of the definitions of functions (the "algorithms") and properties of their computational behavior, for example their complexity. Of course, we could circumvent this problem by defining a specific programming language in the frame of ZF (Zermelo-Fraenkel set theory). For this, what we had to do is basically to define a binary function I, the "interpreter" of the programming language, within ZF. (The object $I(p, d)$ is the "result of applying the program p of the given programming language to the input data d".) Then, of course, it is well possible that we have distinct programs $p_1 \neq p_2$ with identical input/output behavior, i.e. such that $\forall d(I(p_1, d) = I(p_2, d))$.

However, this solution is not very natural because, for example in a textbook on algorithmic polynomial ideal theory, we would not like to spend an additional chapter on introducing an extra programming language. Rather, we would like to use certain predicate logic formulae, for example equalities, directly for defining certain algorithmic functions and predicates.

As an alternative we could use higher order predicate logic. However, the problem remains as soon as we introduce an extensionality axiom in such a logic as this is done in most of the usual textbooks on higher order predicate logic, see for example (Andrews 1986). Therefore, in recent years, some authors proposed to use higher order predicate logic without extensionality as a frame for mathematics, see for example CIC (Huet 1995) and NuPRL (Constable 1995). This is a very promising approach. The problem with the present implementations of this idea is that it seems to be quite hard to carry over the results from Bourbakistic mathematics into these systems. This is, however, important because these results are not only of aesthetical value but are indispensable for specifying problems, and describing a hierarchy of increasingly powerful solutions to problems, even in algorithmic mathematics. For example, when describing the membership decision problem for polynomial ideals, it is necessary first to define the non-algorithmic notion of "ideal" and "residue class domain". Also, when developing an algorithmic solution for this problem, for example by "Groebner bases" (Buchberger 1985), it is absolutely necessary to carry out proofs for the correctness of the solution that involve non-algorithmic concepts of set theory. Also, there is still a practical problem with the present implementations of higher order predicate logic without extensionality: Their practical potential for algebraic computation is not very high.

2.2. THE SYNTAX GAP

Notation used in "mathematical" software systems is still far from the usual mathematical notation. For example, in Mathematica, the function definition

```
f[ { }] := { }

f[ { x1_, x2___}] := Prepend[ f[ { x2}], { x1, x1}]
```

and the function call

```
Integrate[ Sin[ Sqrt[ y + a^2]], y]
```

look quite different from the corresponding formulae

$$
\begin{aligned}
f(\langle\rangle) &:= \langle\rangle \\
f(\langle x_1, x_2, \ldots\rangle) &:= \langle x_1, x_1\rangle \smile f(\langle x_2\rangle)
\end{aligned}
$$

and

$$
\int \sin(\sqrt{y + a^2})\, dy
$$

as they may appear in an ordinary mathematical text. Syntax, however, *is* quite important for practical problem solving.

In fact, the next version of Mathematica will provide "ordinary" syntax for mathematical input and output of amazing sophistication, see (Soiffer 1995). However, some unpleasant gaps will still remain. For example, still, brackets will be used for function application, braces will denote tuples instead of sets, and underscores will be used in order to declare variables and "sequence variables".

2.3. THE MATHEMATICS GAP

For algorithmic mathematics, the traditional algebraic notions are too coarse. For example, the following three algebraic structures are "equal" from the point of view of algebra:

Residue Domain Modulo 3:
Carrier:

$$\{\{0 + 3x \mid x \in \mathbf{I}\}, \{1 + 3x \mid x \in \mathbf{I}\}, \{2 + 3x \mid x \in \mathbf{I}\}\}.$$

Operation:

$$\{y + 3x \mid x \in \mathbf{I}\} \oplus_3 \{z + 3x \mid x \in \mathbf{I}\} := \{y + z + 3x \mid x \in \mathbf{I}\}.$$

Simplified Residue Domain Modulo 3:
Carrier:
$$\{0, 1, 2\}.$$

Operation:
$$y +_3 z := \text{remainder}\,(y + z, 3).$$

"Smallest" Simplified Residue Domain Modulo 3:
Carrier:
$$\{-1, 0, 1\}.$$

Operation:
$$y \,\overline{+_3}\, z := \text{smallest remainder}\,(y + z, 3).$$

From the algorithmic point of view the three structures are significantly different. The first structure is non-algorithmic. It results from the integers with addition by application of the non-algorithmic "functor" "residue class formation modulo a congruence relation". The resulting carrier contains elements that are infinite sets and the operation operates on infinite sets and produces infinite sets. The second and the third structures are both algorithmic, i.e. the objects of the carrier are finitary and can be stored in a computer and the operations are computer-realizable. Both structures result from the integers with addition by applying the functor "simplification modulo a canonical simplifier". However, the two structures are not identical and, in fact, the complexity of the two operations is (slightly) different.

When we "do" mathematics, we want to live in both worlds:

— the world of (non-algorithmic) theorems and proofs,
— the world of algorithms and computation.

The World of Theorems and Proofs: For example, when doing Groebner bases theory, we will start with the *definition*

$$G \text{ is a Groebner basis} :\Longleftrightarrow \forall f \in \text{Ideal}(G)\ (f \longrightarrow_G 0)$$

and will want to prove the following *theorem:*

G is a Groebner basis $\iff \forall f, g \in G$ (S–polynomial$(f, g) \longrightarrow_G 0$).

The proof will involve various non-algorithmic proof techniques of predicate logic, in particular those for handling quantifiers, and non-algorithmic concepts from set theory.

The World of Algorithms and Computation: In the example of Groebner bases theory, the above theorem can be used unchanged as an *algorithm*

G is a Groebner basis $:\iff \forall f \in \text{Ideal}(G)$ $(f \longrightarrow_G 0)$

The following proposition can now be "evaluated" by using the above theorem/algorithm and a subset of the predicate logic proof techniques, namely equational logic, as "evaluation machine": When we enter

$$\{x^2y - 3x, xy^2 - 2xy + 4\} \text{ is a Groebner basis.}$$

The result of the evaluation will be "no".

In the future, we would like to live in both worlds within one uniform logic and software system!

2.4. THE PROVER GAP

The available universal theorem provers (e.g. those based on resolution) are general and, therefore, often too inefficient for supporting practical theorem proving in a wide range of mathematical areas. Also, most times, they are available only as stand-alone systems that are not well connected with computer algebra/analysis systems.

The same is true also for special provers (proof checkers, proof generators) that may be quite efficient for particular mathematical theories. However, again, they are rarely integrated with current algebra/analysis software systems.

What we need is the integration of both universal and special computer-supported theorem proving with the current computer algebra/analysis systems so that, when "doing" mathematics, one can switch between computer-supported theorem/algorithm development and algorithm application.

3. Mathematica is a Good Starting Point

For filling the gaps analyzed in the preceding section, i.e. for supporting all aspects of "doing" mathematics or, in other words, for reaching a new level of sophistication in "symbolic computation" by combining computer algebra and logic, one can adopt one of the following strategies:

1. One can start from successful proving systems like CIC, NuPRL, etc. and add the potential of current computer algebra systems like Maple, Mathematica, Macsyma etc.
2. One can start from one of the practically powerful computer algebra systems and add the potential of current provers.
3. One could design a completely new system.

I do not suggest the third possibility, at least not at this moment, because tremendous work would go into repeating the effort for implementing the wonderful man-machine interfaces current algebra and proof systems already have. The construction of completely new systems or, at least, completely new systems kernels may become reasonable and necessary in the future, after we will have experimented with various early prototypes of combined algebra/logic systems. Rather, I suggest that, at the moment, these experiments should be based on either the first or the second possibility.

In this paper, I argue for basing such an experiment for constructing an early prototype for a combined algebra/logic system on Mathematica. The main reasons can be structured according to the analysis of gaps described in Section 2:

1. Although apparently this was not the intention of the Mathematica designers, a close inspection of the Mathematica language reveals that, in fact, the innermost part of *Mathematica can be viewed as nothing else than directed higher order equational logic without extensionality* and, thus, Mathematica can be viewed as a programming language inside logic. This feature is unique among all existing computer algebra systems.
2. Mathematica has an amazing man/machine interface. Its next version will also have wonderful typesetting facilities and, what is more, there will be facilities that *allow the user to define his own syntax* so that higher order logic with quantifiers and the programming part of this language can be presented as executable code to the system.
3. Using Currying and the module construct, *"functors" can be programmed within Mathematica.* This fact is little known but is essential for structuring non-algorithmic and algorithmic mathematics within a uniform system frame.
4. The directed equational logic facilities of Mathematica are available "twice" in the system: First, at the basic level, they constitute the programming language. Second, at the metalevel, these facilities can also be applied to programs (sets of equalities) at the first level. This allows one, within Mathematica, to program "provers" for proving properties of the programs defined on the first level. Together with the functor

principle, well structured provers that combine special provers tailored to the various functors can be built up.

Of course, it may turn out that basing a uniform logic frame for all of mathematics on top of Mathematica may result in intolerable loss of speed because some of the more advanced language features must be simulated by the available features in Mathematica. If this is the case, in a later stage, a new and specially designed kernel should replace the present Mathematica kernel. Thus, we repeat, the present proposal is only meant as a proposal for implementing a rapid prototype of an algebra/logic system as quickly as possible in order to be able to study the logical, mathematical, algorithmic, and practical implications and gain the necessary experience in using such a combined system for inventing and presenting mathematics.

In the next sections we describe in more detail how we want to overcome the logic, syntax, mathematics, and prover gap by adding features to higher order logic and implementing them in the frame of Mathematica.

4. Overcoming the Logic Gap

We start from higher order logic without extensionality. The fundamental concept of this logic is "application" of an object (a "function") f to some other object x, denoted by

$$f(x).$$

Since we do not have extensionality, this (logic instead of set-theoretic) notion of a function remains "fine grain" (i.e. functions with the same input/output behavior are not necessarily identical), which is indispensable for the algorithmic aspects of mathematics.

In this paper, we do not discuss the subtle question of using "types" in such a logic. Of course this question is very important for obtaining sound proof rules in such a logic. This question is deemed to be important also for the practical work within such a system when building up the "tower of mathematical domains". However, in Section 6, we show how functors can be built up in such a way that we do not need types for the practical mathematical work within the system. Rather, we work with one uniform equality all over the system and, in compensation, we give explicit descriptions of the objects in the various domains.

We observe that (the innermost kernel of) Mathematica can be viewed as an (efficient) implementation of the (directed) equational part of higher order logic without extensionality. For example, the induction definition (in the syntax of Version 2.2 of Mathematica)

```
apply[ f_, { x1_}]    :=  x1
apply[ f_, { x1_, x2___}] := f[ x1, apply[ f, { x2}]]
```

can be viewed as two higher-order logic equalities and, at the same time, as an algorithm whose computations for variable-free input terms are nothing else than proofs using substitution and replacement as inference rules and using the equalities in the direction from left to right. (Do not bother about the strange syntax of the above Mathematica equalities. In the next version (3.0) of Mathematica, the user will be able to define his own syntax and may choose to use the usual mathematical syntax. Variables with three underscores are "sequence variables", i.e. variables for which arbitrary finite sequences of terms may be substituted. In principle, sequence variables are dispensable. However, we think that they are practically attractive and useful. Thus, we propose to have them in our system. Appropriate changes must then be made to the usual proof rules of higher order logic.)

Thus, higher order logic without extensionality contains a practical programming language (efficiently implemented as Mathematica) as a sublanguage. The question is: How can we retain Bourbakistic mathematics within this system?

I think that, in addition to the fundamental "$\in$" predicate, it suffices to add one more predicate "is set" so that we can restrict all set-theoretic axioms to those objects that satisfy "is set". For example, extensionality can be stipulated for objects that are sets:

$$A \text{ is a set} \land B \text{ is a set} \implies (A = B \iff \forall x (x \in A \iff x \in B)).$$

Similarly, the other axioms of set theory guaranteeing the existence of various sets constructed from given sets can be formulated. In the same way, we can also introduce the set braces as special quantifier. From there on, we can develop (carry over) all of Bourbakistic mathematics including the set-theoretic (as opposed to the above logic) notion of "function" and "function application". Of course, for set-theoretic function application, we have to introduce a new notation. For example, we can introduce the binary function constant "'" and may write "$f`x$" for "the set-theoretic function $f$ applied to x". One may then prove for example, within the system, that if $f$ is a set-theoretic function and $x$ is in the domain of $f$ then the pair $(x, f`x)$ is in f. Note again that extensionality will remain to be restricted for the set-theoretic notion of function application whereas, by intention, it is not available for the original, logical, notion of function application. For example, if

$$
\begin{aligned}
f(x) &:= x + 1 \\
g(x) &:= x + 2 - 1
\end{aligned}
$$

then, of course,

$$\forall x \; f(x) = g(x)$$

and also, for example,

$$\{(x, f(x)) \mid x \in \mathbf{N}\} = \{(x, g(x)) \mid x \in \mathbf{N}\}$$

can be proved. However,

$$f = g$$

cannot be proved in the system.

5. Overcoming the Syntax Gap

The syntax used in the present version of Mathematica (Version 2.2) is quite different from the usual mathematical notation. The next version of Mathematica (Version 3.0) will have an amazingly powerful syntax that comes quite close to ordinary mathematical notation. It will have TEX quality, it is wysiwyg *and* at the same time is formal, i.e. the formulae allowed in Mathematica have (formal although not formally defined) semantics as executable code and new formulae can be created whose semantics can be defined. For example, if one does not like that Mathematica uses braces for denoting tuples rather than sets, the following instruction will introduce angle brackets for denoting tuples:

```
MakeExpression[ RowBox[ { "<", x___, ">"}], StandardForm] :=
       MakeExpression[ RowBox[ { "{" x, "}"}], StandardForm]
```

A similar instruction will also let Mathematica print angle brackets instead of braces. Similarly, we can teach the system to take the set braces for denoting set construction instead tuple construction, to use parentheses instead of brackets for denoting function application, and to replace the use of underscores for identifying variables and sequence variables by some other convention, see the examples in Section 6.

6. Overcoming the Mathematics Gap

6.1. BUILDING UP MATHEMATICS BY FUNCTORS

Both from a structural and a problem solving point of view, I think it is important to build up computer-supported mathematics by "functors" that construct new domains from given ones. Structuring mathematics by functors has also an immediate implication for structuring provers as will be made explicit in Section 7.

The functor view of mathematics can equally well be applied to both non-algorithmic and algorithmic mathematics and, in fact, it is the unifying view for both worlds. As explained above non-algorithmic mathematics and algorithmic mathematics are strongly intermingled when "doing" mathematics and, often, the transition from the non-algorithmic to the algorithmic context is the actual challenge of mathematics and of course is the main challenge of what is called "symbolic computation".

6.2. REPRESENTING DOMAINS

We view a domain D to be a function that defines functions for certain "operators". For example, the following object Z is a simple domain:

$$Z(o) := Plus,$$
$$Z(e) := 0,$$

where "Plus" and "0" are the "built-in" addition and the built-in integer zero of our system, which as we said above contains the innermost kernel of Mathematica. For syntactic simplicity, we introduce the convention that "o_D" etc. stands for "$D(o)$" etc. Thus, we could write, for example,

$$2 \; o_Z \; 2$$

which, with the above definitions, evaluates to

$$4.$$

Note that such a notation will be perfectly possible in a user-modified syntax of Mathematica 3.0 and the above evaluation can, hence, be done automatically in Mathematica.

Also we will add the following unary operator "ϵ" to any domain with the convention that, for any domain D, $D(\epsilon)$ is a decision function which yields "True" for exactly the objects we consider to be in the "carrier" of D. For example, we could define

$$Z(\epsilon) := IntegerQ$$

where "IntegerQ" is the built-in decision algorithm for integers. Note that the symbol "ϵ" is different from the symbol "$\in$" that denotes the element predicate of set theory.

6.3. REPRESENTING FUNCTORS

Now, in our terminology (which is basically the terminology used, for example, also in ML), a functor is just a function that maps domains into domains. Thus, a functor is an object F that takes D as an argument and produces $F(D)$ with the view that $F(D)$ can now be applied to any operation symbol o yielding an operation in the domain $F(D)$. In particular, $F(D)(\epsilon)$ is a decision function for the objects which we want to be in the carrier of $F(D)$. Also, normally, one will think of any operation $F(D)(o)$ to be reasonably applied only to objects in the carrier of $F(D)$, i.e. objects for which $F(D)(\epsilon)$ yields "True".

For clarifying notation, let us first give a trivial example: We define a functor F that, for any given domain D with any operations o defines the Cartesian product. Staying within higher order predicate logic, this functor could be defined as follows:

$$F(D) := \text{the } N \text{ such that}$$
$$\forall d_1, d_2 \ \langle d_1, d_2 \rangle \epsilon_N :\Longleftrightarrow d_1 \epsilon_D \wedge d_2 \epsilon_D,$$
$$\forall o, d_1, d_2, e_1, e_2 \ \langle d_1, d_2 \rangle o_N \langle e_1, e_2 \rangle := \langle d_1 o_D e_1, d_2 o_D e_2 \rangle.$$

Instead, we introduce a new quantifier "Functor" that binds N and also the other quantified variables that appear in the above definition so that the following definition can be seen as nothing else than an abbreviation of the above definition:

$$F(D.) :=$$
Functor $(\langle N, d_1, d_2, e_1, e_2, o \rangle,$

$$\langle d_1., d_2. \rangle \epsilon_N :\Longleftrightarrow d_1 \epsilon_D \wedge d_2 \epsilon_D;$$
$$\langle d_1., d_2. \rangle o._N \langle e_1., e_2. \rangle := \langle d_1 o_D e_1, d_2 o_D e_2 \rangle;$$

$$N$$
$$).$$

Note that, at certain places, we use a dot after a symbol as a notation for declaring the symbol to be a variable and not a constant. The respective convention is an adaptation to the syntax of Mathematica and has no deeper relevance. Note, in particular, that "o" is a variable in this definition. Thus, the definition applies to any operator o in the "signature" of D. We could define the signature of a domain explicitly. We omit this in order not to overload the presentation in this expository paper.

Now, as a matter of fact, the construct "Module" of Mathematica, if restricted to algorithmic operations, has exactly the semantics of the "Functor" construct as defined above. Thus, when restricted to algorithmic functors, application of this construct yields executable code in Mathematica.

6.4. AN EXAMPLE OF A NON-ALGORITHMIC FUNCTOR

We define a functor that "constructs" the residue class domain of a domain D that is equipped with a binary operator "$\sim$" (which normally will be a congruence relation):

Residue-Domain(D) :=
Functor($\langle R, r, s, d, e, o, C \rangle$,

$$r. \; \epsilon_R :\Longleftrightarrow \exists d \; \epsilon_D \; (r = C(d));$$
$$r. \; o_R \; s. := C(\text{Choose}(r) \; o_D \; \text{Choose}(s));$$

$$C(d.) := \{e \mid e \; \epsilon_D \; \wedge \; e \sim_D d\};$$

R
).

This functor is non-algorithmic in many respects. First, for defining whether or not an object r is in the residue domain R, we need the existential quantifier: r is in the carrier of R iff there exists a d in the given domain D such that r is the residue class of d. Second, for the definition of the residue class $C(d)$ of a given d we need the set quantifier. And, finally, for defining the result of applying the operation o in R to the two residue classes r and s we need the "Choose" function that chooses one element from the argument set.

In many areas of mathematics, non-algorithmic functors are the best one can do for solving very general problems. For the above functor, one can at least prove that if $D(\sim)$ is a congruence the residue domain of D has "similar" properties as D and, in addition, may have some new desirable properties, namely that certain objects exist that solve some problem at hand.

In a concrete situation it may be a trivial, an easy, a non-trivial, a very difficult, or an impossible task to come up with an *algorithmic* functor that constructs a domain which is isomorphic to the domain constructed by the above *non-algorithmic* functor.

6.5. AN EXAMPLE OF AN ALGORITHMIC FUNCTOR

In contrast to the previous functor, the following functor is algorithmic and, under certain conditions, constructs a domain which is isomorphic to a residue domain. For this functor we suppose that D is equipped with a unary operator σ (which normally will be a "canonical simplifier" for the congruence $D(\sim)$):

Simplified-Domain(D) :=

Functor($\langle S, o, s, t \rangle$,

$$s.\ \epsilon_S :\Longleftrightarrow s\epsilon_D\ \wedge\ \sigma_D(s) = s;$$
$$s.\ o._S\ t. := \sigma_D(s\ o_D\ t);$$

S
).

This functor is algorithmic in the sense that, if all the operations of D are algorithmic, then also all the operations in S are algorithmic, i.e. the functor "preserves" computability. First, in order to determine whether a given s is in the carrier of S, we only have to check whether s is in D and whether $D(\sigma)$, the function associated with the operator σ in D, applied to s is identical to s. Hence, under the assumption that $D(\epsilon)$ and $D(\sigma)$ are algorithmic functions, this is an algorithmic process. Similarly, under the assumption that σ_D and o_D are algorithmic, also o_S is algorithmic.

The main theorem which one can prove about the functor "Simplified-Domain" is the fact that, if σ_D is a "canonical simplifier" w.r.t. $\sim_D$ then the Simplified-Domain(D) is isomorphic to Residue-Domain(D). In fact, the function C defined locally in the functor Residue-Domain is an isomorphism.

Of course, these two functors are pervasive in mathematics and the construction of algorithmic canonical simplifiers is one of the main methods for turning an inconstructive part of mathematics into a constructive one. Finding canonical simplifiers can be trivial, easy, non-trivial, difficult, or shown to be non-existent. For example, for the residue domain modulo 3, it is trivial to find a canonical simplifier. Just take

$$\sigma_Z\ z. := Mod(z, 3).$$

If we now call

$$Z3 = \text{Simplified} - \text{Domain}\ (Z)$$

then, for example, entering

$$2\ o_{Z3}\ 2$$

evaluates to

$$1.$$

For the residue domain of a multivariate polynomial ring modulo a polynomial ideal, as another example, the construction of a canonical simplifier is non-trivial (and it had been conjectured to be algorithmically unsolvable for quite some years) and needs the theory of Groebner bases. If P is some polynomial domain (over field coefficients or other suitable coefficient domains) and F is a (finite) set of polynomials in P,

$$\sigma_P \; p. := \text{Normal} - \text{Form} \; (p, \text{Groebner} - \text{Basis} \; (F))$$

is a canonical simplifier for the residue domain of P modulo the ideal generated by F and

$$Q = \text{Simplified} - \text{Domain} \; (P)$$

constructs an algorithmic isomorphic realization of this residue domain.

Note that the above sequences of statements are not only a mathematical description of the construction but are *executable* code in our system based on Mathematica. This is so mainly because of the fact that (the innermost kernel of) Mathematica is nothing else than an implementation of the equational part of higher order logic without extensionality and because of the following two features of Mathematica:

— Mathematica allows Currying. Thus, it is perfectly possible to handle terms like "$D(o)(s, t)$".
— The "Module" construct of Mathematica has exactly the semantics of the above "Functor" construct and, in fact, one can teach Mathematica the Functor construct by just defining "Functor = Module".

Of course, the functor construct is available also in a few other programming languages, notably in ML. However, in the above realization, Functor is much more flexible because it allows one to define (decision predicates for) arbitrary carrier sets. Also, the above realization of the functor concept is smoothly embedded into a full-fledged computer algebra system with its practical computational power. Unfortunately, neither the fact that Mathematica is essentially higher order equational logic nor the fact that Currying is allowed and Module can be used as Functor, is explicitly mentioned or observed anywhere in the manual (Wolfram 89) nor in the rich literature on Mathematica. One may also add that, surprisingly, the above implementation of the functor principle in Mathematica is quite efficient: Generating code for the operations in a domain by functors of the above kind is hardly slower than hand-coding the operations for a particular domain. The only practical disadvantage of the above realization of functors is that the tracing and debugging facilities of Mathematica, of course, are not tailored for this particular way of structuring code for towers of domains. Anyway, with some experience, it is satisfactorily possible to use the existing debugging tools of Mathematica for debugging mathematical software written in the functor style.

7. Overcoming the Proof Gap

At present, nearly no proving capabilities are available in general computer algebra systems. The "directed" equational proving capabilities of Mathematica are normally used only for computation. Sophisticated special provers, as for example Collins' prover for the theory of real closed fields, see (Collins 1975) and the collection of articles on his algorithm in Vol.5/1-2 of the Journal of Symbolic Computation, are stand-alone systems.

As a first step for combining algebra and logic on the system level, one should of course make these special provers available from within general computer algebra systems like Mathematica. This is not really a problem any more because we can use the MathLink facility for accessing independent systems. Thus, if we encounter a formula like

$$\forall x \in \mathbf{R} \; \exists y \in \mathbf{R} \; (y^2 = x)$$

we should be able to call Collins' algorithm from within Mathematica and obtain the answer

False.

Similarly, if we encounter

$$x \in \mathbf{R} \; \wedge \; \exists y \in \mathbf{R} \; (y^2 = x)$$

a call to Collins' algorithm would yield the equivalent quantifier-free formula

$$x \in \mathbf{R} \wedge x \geq 0.$$

Note that Collins' algorithm goes far beyond the present "simplification capabilities" in ordinary computer algebra systems and, hence, the available of this and other sophisticated special provers will greatly expand the potential of these systems.

Similarly, at present, universal theorem provers are not smoothly linked to computer algebra systems. Of course, it is very much desirable to have access to universal theorem provers, notably of the natural deduction type, from within computer algebra systems. For example if, in the development of some piece of mathematics, say in a "computerized" text book, we want to prove that

$\sim$ is an equivalence on $S \Longrightarrow S/\sim$ is a partition on S

then, given the present state of proving technology, it should be possible to produce the proof of this proposition automatically and to print the proof out in a way that closely resembles a proof generated by a human. In fact, the proof of such theorems hardly needs any special trick nor intuition and can basically be done by "expanding" the definition using the natural deduction rules for quantifiers and propositional connectives plus some "simplification", i.e. manipulation of quantifier-free terms and atomic formulae. Thus, given a proposition of the above type, we would wish to be able to call a natural deduction prover from within a computer algebra system and would expect that it produces just the answer

True

or, with the option "verbose", the answer

```
Let ~ be an equivalence on S.

By the definition of S\~ we have to show that
...
```

It seems that one major obstacle that still makes this desire unrealistic is the fact that

- mathematical knowledge must be well organized so that, in a given proof situation, only relevant knowledge and all relevant knowledge is available and
- in a given area of mathematics, in addition to the universal natural deduction proof techniques, relevant special proof techniques should be available.

One of the main points I would like to make in this paper is that I think that the functor principle is not only crucial for structuring mathematical knowledge and mathematical methods (including algorithms) but it is also the key for structuring proof techniques. Namely, the first part of any functor in the above realization is a definition of the elements in the carrier by describing its characteristic function (which is algorithmic in the case of algorithmic functors). The structure of the description of this characteristic function naturally suggests a special prover for properties of the objects in the domain generated by the functor. For example, if the characteristic function is described inductively, a corresponding inductive proof technique is naturally connected with the functor. If the characteristic function is defined using set braces then the usual set-theoretic proof techniques will naturally apply, etc.

We give an example which is trivial (sorry for that!) but, nevertheless, is surprisingly powerful: We take the parameterless functor that defines one particular realization of the natural numbers and define addition in the resulting domain:

Natural-Numbers() :=
Functor($\langle N, n, m \rangle$,

$0 \; \epsilon_N :\Longleftrightarrow$ True;
$s(n_.) \; \epsilon_N :\Longleftrightarrow n \; \epsilon_N$;
$n_. \epsilon_N :\Longleftrightarrow$ False;

$n_. +_N 0 := n$;
$n_. +_N s(m_.) := s(n +_N m)$;

N
).

Here, "0" and "s" are "arbitrary but fixed constants". The third clause in the inductive definition of the "ϵ_N" predicate is a clause that handles, the default case of any objects not having the format "$s(s(...s(0)))$". This is possible since Mathematica treats the clauses in equational inductive definitions successively.

Now we describe, in Mathematica, a simple inductive prover that can handle formulae of the form

$$\forall x_1 \forall x_2 \ldots \forall x_n \; l(x_1, \ldots, x_n) = r(x_1, \ldots, x_n)$$

where "l" and "r" are terms containing the free variables "x_1", ..., "x_n". The structure of the prover naturally reflects the induction scheme by which the above characteristic function for the natural numbers in the 0/s-representation is defined. In fact, at a higher level of the system, provers of this kind could be automatically generated from the inductive definition of the respective characteristic function. The prover proceeds by, first, trying to prove the formula by equational simplification. If this fails, "x_1" is taken as the induction variable. The base case for this induction is generated and, again, a proof by simplification is attempted. If it fails, recursively, an induction over "x_2" (with "x_1" replaced by 0) is started. If it succeeds, the induction hypothesis and the formula to be proved in the induction step for (arbitrary but fixed) "x_1" are generated and a proof by simplification is attempted. If it fails, recursively, an induction over "x_2" (with "x_1" arbitrary but fixed) is started. If it succeeds we are done.

In Mathematica (with Version 2.2 syntax), this recursive induction prover can be described in a few lines:

```
ProofByInduction[

   (* of the *) equality_,
   (* w.r.t. the induction variables *) { v1_, v2___},
   (* using the *) equalities_] :=


   ProofByInduction0[ equality, { v1, v2}, { }, equalities]


ProofByInduction0[ equality_, { v1_, v2___},

     (* arbitrary but fixed *) { w1___},

     equalities_] :=
Module[

   { ... (* local variables *)},


   equalitySimplified = equality //. equalities;


   If[ LeftHandSide[ equalitySimplified] ===

         RightHandSide[ equalitySimplified],
       Return[ True]
       ];


   equality0 = equality /. v1 -> 0;

   equality0Simplified = equality0 //. equalities;
   inductionHypothesis =

       Generalized[ LeftHandSide[ equality], { v1, w1}] ->

           RightHandSide[ equality];
   equalitiesForInductionStep =

       Append[ equalities, inductionHypothesis];
   equalitySv1 = equality /. v1 -> s[ v1];
```

```
equalitySv1Simplified =

    equalitySv1 //. equalitiesForInductionStep;


inductionBasisProved = False;
inductionStepProved = False;


If[ LeftHandSide[ equality0Simplified] ===

        RightHandSide[ equality0Simplified],
    inductionBasisProved = True
    ];
If[ LeftHandSide[ equalitySv1Simplified] ===

        RightHandSide[ equalitySv1Simplified],
    inductionStepProved = True
    ];


If[ ! inductionBasisProved,
    inductionBasisProved =
        ProofByInduction0[

            equality0Simplified,

            { v2}, { v1, w1}, equalities
            ]
        ];


If[ ! inductionStepProved,
    inductionStepProved =
        ProofByInduction0[

            equalitySv1Simplified,

            { v2}, { v1, w1},

            equalitiesForInductionStep,
```

```
                        ]
                  ];


            inductionBasisProved && inductionStepProved
            ]


ProofByInduction0[

    equality_, {}, { w1___},  equalities_, options___
    ] := False
```

In this description of the recursive induction prover we left out all the lines for printing the proof. The prover is so simple that it hardly needs an explanation. "Generalized" is a procedure that replaces a universally quantified variable by an "arbitrary but fixed constant". "/." is the Mathematica operator for applying, once, rewrite rules (i.e. directed equalities) to terms. "//." is repeated application of "/." until the term does not change any more. Correspondingly, as a technical detail, the equalities that constitute the inductive definitions of operations like, for example, "+" must be presented in the form of "rules". For example, if we want to apply the prover to the domain generated by the above functor, we must extract the defining equalities from the functor so that they are available, say, as the value of some constant "NN", which (in Version 2.2 syntax) could be done by an instruction of the kind

```
    NN =

      { sum[ x_, 0]] :> 0,

        sum[ x_, s[ y_]] :> s[ sum[ x, y]]}
```

Now, if we enter

```
    ProofByInduction[
        sum[ x, sum[ y, z]] == sum[ sum[ x, y], z],
        { x, y, z},
        NN
        ]
```

we obtain the following inductive proof of the associativity of addition over the natural numbers:

```
Simplification proof of:
for all( x, y, z, ( ( x + ( y + z)) = ( ( x + y) + z))).
```

```
. .Simplification of left-hand side:
. .( x0 + ( y0 + z0))


. .Simplification of right-hand side:
. .( ( x0 + y0) + z0)
Not proved  by simplification.


Induction proof of:

for all( x, y, z, ( ( x + ( y + z)) = ( ( x + y) + z))).
Induction variable: x.
****************************


.Prove induction basis (i. e. formula with x -> 0):
.for all( y, z, ( ( 0 + ( y + z)) = ( ( 0 + y) + z))).
.Simplification proof of:
.for all( y, z, ( ( 0 + ( y + z)) = ( ( 0 + y) + z))).
.-----------------------------------


. . .Simplification of left-hand side:
. . .( 0 + ( y0 + z0))


. . .Simplification of right-hand side:
. . .( ( 0 + y0) + z0)
.Not proved  by simplification.


.Induction proof of:

.for all( y, z, ( ( 0 + ( y + z)) = ( ( 0 + y) + z))).
.Induction variable: y.
.****************************
```

. .Prove induction basis (i. e. formula with y -> 0):
. .for all(z, ((0 + (0 + z)) = ((0 + 0) + z))).
. .Simplification proof of:
. .for all(z, ((0 + (0 + z)) = ((0 + 0) + z))).
. .------------------------------------

. . . .Simplification of left-hand side:
. . . .(0 + (0 + z0))

. . . .Simplification of right-hand side:
. . . .((0 + 0) + z0)
. . . . = (0 + z0)
. .Not proved by simplification.

. .Induction proof of:

. .for all(z, ((0 + (0 + z)) = ((0 + 0) + z))).
. .Induction variable: z.
. .****************************

. . .Prove induction basis (i. e. formula with z -> 0):
. . .for all(((0 + (0 + 0)) = ((0 + 0) + 0))).

. . . .Simplification of left-hand side:
. . . .(0 + (0 + 0))
. . . . = (0 + 0)
. . . . = 0

. . . .Simplification of right-hand side:
. . . .((0 + 0) + 0)
. . . . = (0 + 0)
. . . . = 0
. . .Proved induction basis with z -> 0.

. . .Let z0 be arbitrary but fixed.

```
. . .Induction hypothesis (i. e. formula with z -> z0):
. . .for all( ( ( 0 + ( 0 + z0)) = ( ( 0 + 0) + z0))).



. . .Prove induction step formula (i. e. formula with z ->



z0'):
. . .for all( ( ( 0 + ( 0 + z0')) = ( ( 0 + 0) + z0'))).



. . . . .Simplification of left-hand side:
. . . . .( 0 + ( 0 + z0'))
. . . . . = ( 0 + ( 0 + z0)')
. . . . . = ( 0 + ( 0 + z0))'
. . . . . = ( ( 0 + 0) + z0)'
. . . . . = ( 0 + z0)'



. . . . .Simplification of right-hand side:
. . . . .( ( 0 + 0) + z0')
. . . . . = ( ( 0 + 0) + z0)'
. . . . . = ( 0 + z0)'
. . .Proved induction step formula with z -> z0'.
. .Proved by induction over z.
. .Proved induction basis with y -> 0.



. .Let y0 be arbitrary but fixed.
. .Induction hypothesis (i. e. formula with y -> y0):
. .for all( z, ( ( 0 + ( y0 + z)) = ( ( 0 + y0) + z))).



. .Prove induction step formula (i. e. formula with y -> y0'



):
. .for all( z, ( ( 0 + ( y0' + z)) = ( ( 0 + y0') + z))).
. .Simplification proof of:
. .for all( z, ( ( 0 + ( y0' + z)) = ( ( 0 + y0') + z))).
. .-------------------------------------
```

```
. . . .Simplification of left-hand side:
. . . .( 0 + ( y0' + z0))


. . . .Simplification of right-hand side:
. . . .( ( 0 + y0') + z0)
. . . . = ( ( 0 + y0)' + z0)
. .Not proved  by simplification.


. .Induction proof of:

. .for all( z, ( ( 0 + ( y0' + z)) = ( ( 0 + y0') + z))).
. .Induction variable: z.
. .*****************************


. . .Prove induction basis (i. e. formula with z -> 0):
. . .for all( ( ( 0 + ( y0' + 0)) = ( ( 0 + y0') + 0))).


. . . .Simplification of left-hand side:
. . . . . . . . . .
```

In many more lines, the proof works its way completely automatically
through the recursion over the variables. Finally, it will return through all
levels so that the last few lines look like this:

```
. . . .Simplification of left-hand side:
. . . .( x0' + ( y0' + z0'))
. . . . = ( x0' + ( y0' + z0)')
. . . . = ( x0' + ( y0' + z0))'
. . . . = ( ( x0' + y0') + z0)'
. . . . = ( ( x0' + y0)' + z0)'
. . . .Simplification of right-hand side:
. . . .( ( x0' + y0') + z0')
. . . . = ( ( x0' + y0') + z0)'
. . . . = ( ( x0' + y0)' + z0)'
. . .Proved induction step formula with z -> z0'.
. .Proved by induction over z.
. .Proved induction step formula with y -> y0'.
.Proved by induction over y.
```

```
.Proved induction step formula with x -> x0'.
Proved by induction over x.
```

Of course, the crucial simplifications steps in the proof are exactly the ones that would be produced by the well-known test case method, see for example (Kapur *et al.* 1991). However, it is interesting that this prover evolves from a completely natural approach that does nothing more than reflecting the inductive definition of the particular representation of the natural numbers, it also allows to produce a natural language easy verbose presentation of the proof and it gets along without any human interaction. Most importantly, this prover is fully integrated and in fact programmed in the language of a full-fledged computer algebra system and can hence be used in intimate interaction with computation. It also should be mentioned that, of course, the sequence of the universally quantified variables may drastically influence the length of the proof. This is a phenomenon that has often been reported in the literature. In fact, in the above example, considering the variables in the order "z", "y", "x" produces a very short proof that succeeds in the simplest possible way. This is of course supported by the heuristics that the variable which is the induction variable in the inductive definition of the functions involved should also be treated first in an inductive proof. In more complicated examples, the proof may be stuck at certain points producing an equality which is "not yet known". Such equalities are often the appropriate guess for a lemma that should be proved (by the above prover) before the main proof can be attempted once more. In fact, this side step can be called automatically so that quite impressive proofs can be handled completely automatically. More details can be found in the preliminary report (Buchberger 1995).

8. Conclusion

In this paper we argued that the combination of the potential of computer algebra and logic systems is the next natural step for enhancing the problem solving power of "symbolic computation" systems. For closing the gap between the "ideal" future system and the systems available at present we suggested to start from a computer algebra system and to add logic and, in particular, proving power. For quite a few reasons which we discussed in some detail, Mathematica seems to be particularly appealing as a starting point for such an approach. We tried to illustrate that the functor view seems to be crucial for structuring mathematical knowledge, mathematical methods (including algorithms) and mathematical proofs in future symbolic computation systems. Thus, the implementation of a well designed system of basic and advanced functors (encompassing both the computation and the prover details for each functor) is the essence of building up a future system. Currently, at the RISC institute, we work on a project that elaborates the details of the approach described in this paper.

Acknowledgement: This paper was written in the frame of a research contract with Fujitsu Labs, Numazu, Japan.

BRUNO BUCHBERGER
Research Institute for Symbolic Computation
Linz, Austria

References

P.B. Andrews: *An Introduction to Mathematical Logic and Type Theory: To Truth Through Proof.* Academic Press, London 1986.

B. Buchberger: Groebner Bases: An Algorithmic Method in Polynomial Ideal Theory. Chapter 6 in: *Multidimensional Systems Theory*, (N.K. Bose ed.). D. Reidel Publishing Company, Dordrecht, 1985.

B. Buchberger: Induction Proofs in Equational Logic: A Case Study Using Mathematica. *Internal Technical Report*, The RISC Institute, A4232 Schloss Hagenberg, Austria, 1995.

G.E. Collins: Quantifier Elimination for Real Closed Fields by Cylindrical Algebraic Decomposition. Proceedings of the Second GI Conference on Automata Theory and Formal Languages. *Lecture Notes in Computer Science*, **33**, pp. 515-532, Springer, Heidelberg, 1975.

R. Constable: *The Nuprl System.* Lecture Notes of the Summer School on *Logic of Computation*, Marktoberndorf 1995. Edited by Institut für Informatik, Technische Universität München, 1995.

G. Huet: *The CIC System.* Lecture Notes of the Summer School on *Logic of Computation*, Marktoberndorf 1995. Edited by Institut für Informatik, Technische Universität München, 1995.

D. Kapur, P. Narendran, H. Zhang: Automating Inductionless Induction using Test Sets, *Journal of Symbolic Computation*, Vol. 11, No. 1&2, February 1991, pp. 83-111.

N. Soiffer: *Mathematical Typesetting in Mathematica.* Proceedings of the ISSAC 1995 Conference, pp. 140-149.

S. Wolfram: *Mathematica: A System for Doing Mathematics by Computers.* Addison-Wesley Publishing Company, Redwood, 1988.

JACQUES CALMET AND KARSTEN HOMANN

CLASSIFICATION OF COMMUNICATION AND COOPERATION MECHANISMS FOR LOGICAL AND SYMBOLIC COMPUTATION SYSTEMS

Abstract. The combination of logical and symbolic computation systems has recently emerged from prototype extensions of stand-alone systems to the study of environments allowing interaction among several systems. Communication and cooperation mechanisms of systems performing any kind of mathematical service enable one to study and solve new classes of problems and to perform efficient computation by distributed specialized packages.

The classification of communication and cooperation methods for logical and symbolic computation systems given in this paper provides and surveys different methodologies for combining mathematical services and their characteristics, capabilities, requirements, and differences. The methods are illustrated by recent well-known examples.

We separate the classification into communication and cooperation methods. The former includes all aspects of the physical connection, the flow of mathematical information, the communication language(s) and its encoding, encryption, and knowledge sharing. The latter concerns the semantic aspects of architectures for cooperative problem solving.

1. Introduction

The design of general techniques to combine and integrate several systems has been initiated in many areas. For instance, the integration of theorem proving and symbolic mathematical computing has recently emerged from prototype extensions of single systems to the study of environments with interaction among distributed systems. However, there are no common languages, protocols, or standards for such interfaces.

Communication and cooperation mechanisms for logical and symbolic

221

F. Baader and K.U. Schulz (eds.), Frontiers of Combining Systems, 221–234.
© *1996 Kluwer Academic Publishers.*

computation systems enable to study and solve new classes of problems and to perform efficient computation through cooperating specialized packages. On the one hand, computer algebra systems (*CAS*) offer an extensive collection of efficient mathematical algorithms which could improve the efficiency of theorem proving systems (*TPS*). On the other hand, they ignore AI methods (e.g. theorem proving, planning of proofs and computations, machine learning) and their capabilities, e.g. verification of properties of mathematical objects using a TPS.

Basic architectures for performing communication among TPS and CAS are introduced in (Homann & Calmet, 1994). The classification given here is a result of generalizations and extensions of communication and cooperation mechanisms for software systems performing any kind of mathematical computation. We call such systems *mathematical services* (*MS*) which cover CAS and other symbolic computation packages, TPS, proof checkers and verification tools, numerical computation systems, visualization and type-setting applications, and format converters. This classification is illustrated by well-known recent examples of communication and cooperation mechanisms for both logical and symbolic computation systems. It provides and surveys different methodologies for combining such systems and their characteristics, capabilities, requirements, differences, and may guide the developments and selection of methods in this ongoing research. However, it must be pointed out that some of the presented architectures and communication methods are not specific to mathematical information and could be applied to combine other systems as well.

We separate the mechanisms into *communication* and *cooperation* methods. The former include all aspects of the physical connection, the flow of mathematical information, the communication language(s) and its encoding, encryption, and knowledge sharing. Communicating *MS* send and retreive mathematical information and messages. The aspects of the "semantics" of these interactions are specified according to the level of cooperation among the distributed systems. Depending on their behaviour they can be classified into: master/slave, subpackage, black box, trust, extensible and exchangeable, consistency and closure.

As of today, there is no systematic investigation of the different possible methodologies to integrate heterogeneous mathematical systems. The goal of this paper is to fill this gap.

This paper is organized as follows. Section 2 gives an overview about architectures combining logical and symbolic computation systems. The advantages are illustrated by some recent well-known examples. The classification of such architectures based upon the features of the involved communication and cooperation methods is given in section 3 and section 4 respectively.

2. Combining Logical and Symbolic Computation Systems

The advantages of combining logical and symbolic computation systems are improved expressive power and more powerful inference capabilities. There are various applications for composing those systems, like multi-logic provers, hardware and software verification, proofs with arithmetics and constraints, program transformations.

There is a lack of languages and standards for interfaces between systems for mathematical computation. The reasons are manyfold: (i) CAS and TPS are designed, implemented and validated as stand-alone systems, (ii) many systems are copyrighted and allow neither communication nor external access to internal methods, (iii) they do not provide interfacing.

Several communication and cooperation methods have already been examined. The basic level of cooperation is just to exchange mathematical information. To enable mathematicians, TPS or CAS to pass proofs, theorems, functions, algorithms or any kind of mathematical objects offline by electronic mail, cut & paste or ftp requires communication in terms of a common language. Open Mechanized Reasoning Systems (Giunchiglia *et al.*, 1996) and OpenMath (Abbott *et al.*, 1995) introduce general languages suitable for specifying and communicating mathematical objects in theorem proving and symbolic mathematical computing respectively.

Higher levels of online cooperation can be achieved by adding links to interactive tools. The interfaces between HOL and Maple (Harrison & Théry, 1993) and Isabelle and Maple (Ballarin *et al.*,) introduce the powerful arithmetics of a computer algebra system into a tactical theorem prover to reason about numbers or polynomials much more efficient. Maple (Char *et al.*,) acts as a slave to the prover which controls external calls by evaluation tactics. (Jackson, 1994) presents an interaction to provide expressive algebra of constructive type theory in computer algebra. The theorem prover Nuprl is an algebraic oracle to the CAS Weyl. Analytica (Clarke & Zhao, 1992) is an example for cooperation within the language of another system. It is written in the Mathematica (Wolfram, 1991) environment and can solve sophisticated problems in elementary analysis.

CAS/π represents a sophisticated example of a powerful graphical system-independent common user interface (Kajler, 1992). It was designed so that expert users can set up connections to alternative CAS or visualization tools easily and at runtime. An architecture for proof planning in distributed theorem proving is given in (Denzinger & Fuchs, 1994). TPS compete and then cooperate using completion in pure equational logic using team work. The advantage of distribution is to profit from heuristics of several systems to reduce the typically immense search spaces.

3. Communication Methods

Prerequisite for distributed mathematical problem solving is communication. This section examines the communication language, its encoding and encryption, the flow of mathematical information, and the exchange of mathematical objects by common knowledge representation.

3.1. COMMUNICATION LANGUAGE AND ENCODING

A communication language defines how mathematical information can be exchanged among services. Such a language must be recognized by each system in order to to translate the information into their internal representation. Appropriate languages are the input language or internal encoding of one of the involved systems or standardized communication languages.

To select the syntax of the input language of one of the systems is natural and allows the straightforward interaction with this system. Currently, most interfaces between $\mathcal{MS}$ are built as prototypes designed to demonstrate the advantages of combining heterogeneous systems. Thus, the communication language has not been chosen according to general protocols. The prototypes described in (Harrison & Théry, 1993; Ballarin et al.,) communicate in terms of Maple expressions. The theorem provers HOL and Isabelle are extended both by adding syntax translations and evaluation tactics. In case of common knowledge representation (3.4) or subpackages (4.2) it is a good option to use the internal encoding of one specific system as communication method. The interaction with the Analytica (Clarke & Zhao, 1992) package is implemented with a common representation of the objects and in expressions of Mathematica's language.

Communicating in terms of the input language of one system is generally not a good choice because

systems are tough to interchange. (Ballarin et al.,) selects the input and output object representation of Maple as its communication language. To replace Maple by any other CAS requires to define a new syntax (4.3).

the input language differs from the representation language. The input object representations must be encoded into the internal application specific representations. Some types of cooperation gain efficiency at run time by communicating these internals (4.1,4.2).

services are based upon different semantics. There is no standardized semantics to expressions of mathematical objects. Some systems request case sensitive input, *d, diff* or *differentiate* may represent different functions, and the mathematical notions differ, like ^ or **.

Several communication languages for interfaces between software systems exchanging mathematical information have been developed, Camino-Real (Arnon *et al.*, 1988), ASAP, CC and central control (Dalmas *et al.*, 1994), Posso/XDR (Abbott, 1994), MP (Gray *et al.*, 1994), CAS/π (Kajler, 1992), and MathLink (Wolfram, 1993). OpenMath (Abbott *et al.*, 1995) classifies these projects according to the framework given in the basic Open-Math model as illustrated in figure 1.

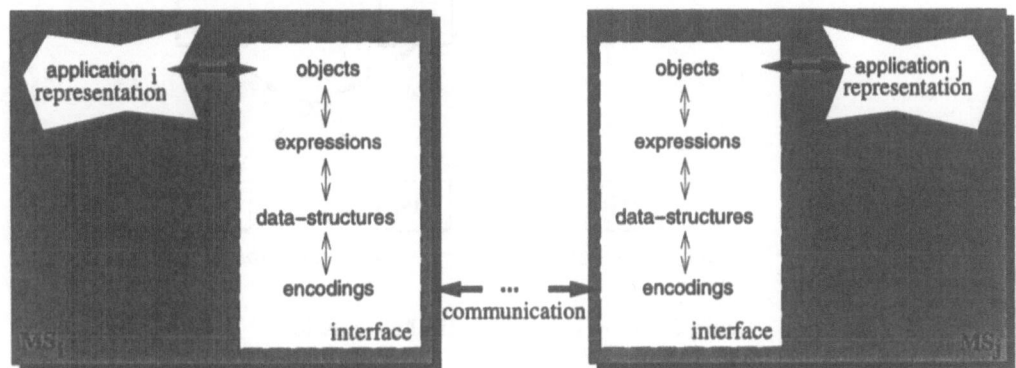

Figure 1.   The communication model

The communication is not implemented as the input language to one of the involved systems but as an interface compiling the service specific representation into a standardized encoding. This encoding is either a stream of bytes or an extended Lisp-like representation suitable for transmission via files, cut & paste, email, ftp, and broadcasting like Unix sockets. Thus, the communication language can be described by specifying the different levels in the model: objects, expressions, data structures, and encodings.

3.2. ENCRYPTION

Current interfaces between $\mathcal{MS}$ do not consider system security aspects since the interaction often only involves packages running in the same local network. Because of the wish to transmit mathematical information via files, cut & paste, email or ftp (see above) future encodings must be designed to provide connections with identification and encryption.

3.3. BIDIRECTIONAL COMMUNICATION

Cooperation among several software systems can be achieved with indirect, unidirectional and bidirectional communication. According to the flow of mathematical information several architectures are illustrated in figure 2.

Although there are no links between the services with indirect communication, interaction is possible if both systems can communicate with

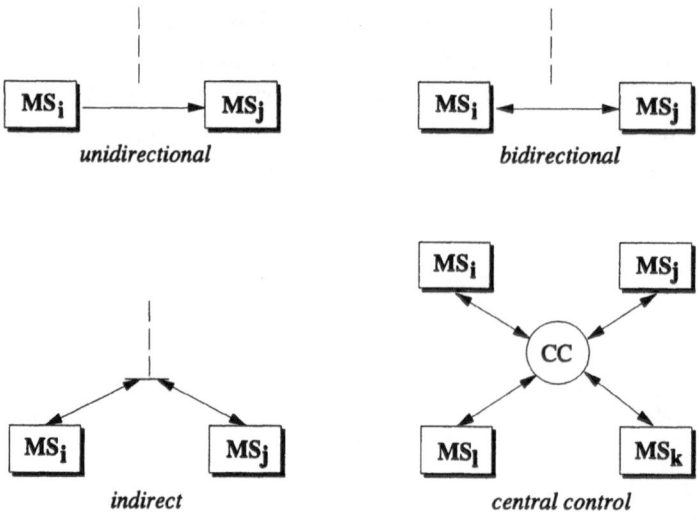

Figure 2.   Flow of mathematical information

a common user interface, central unit, mediator or evaluator. Such an interface provides links to some $\mathcal{MS}$. A user can access the systems and can apply (symbolic or numerical) algorithms or theorems to solve a given problem, depending on the class of the problem. Such a simple type of interaction allows already the use of arbitrary CAS and ATP. However, the systems do not interact directly and a user must be familiar with both systems. Such an architecture combines the advantages, but also the drawbacks. CAS/π represents a sophisticated example for such an architecture (Kajler, 1992).

To manage the communication and to hide the control from the user interface leads to an architecture with common evaluator or central control. The evaluator controls the selection of the modules by meta-knowledge on all functions and predicates. It also controls the application of algebraic algorithms and exchange of data and theorems in the $\mathcal{MS}$. The mathematical knowledge is represented separately in each module. The Central Control project (Dalmas *et al.*, 1994) is a typical representative for this architecture. The tools are mainly independent: they can perform their tasks without the help of other tools.

Unidirectional links can most often be found when communicating with input or output devices like math editors, visualization tools, graphical interfaces, SGML, in case of master/slave cooperation (4.1), or subpackages (4.2). As mentioned previoulsy, typically such interfaces do not support general encodings as communication language but the input or output language of one system (Ballarin *et al.*, ; Jackson, 1994;

Harrison & Théry, 1993; Clarke & Zhao, 1992).

The first environments providing bidirectional links have been studied recently. Such a communication requires to exchange common mathematical objects or relies on a common knowledge representation. At any step, arbitrary combinations of algorithms and theorems can be applied to solve a given problem. This combines the advantages of all involved mathematical services. The uni- and bidirectional communication is generalized to a software bus of mathematical services in (Calmet & Homann, 1995) as illustrated in figure 3. The highlighted connection between Maple and Isabelle is described in (Ballarin *et al.*,).

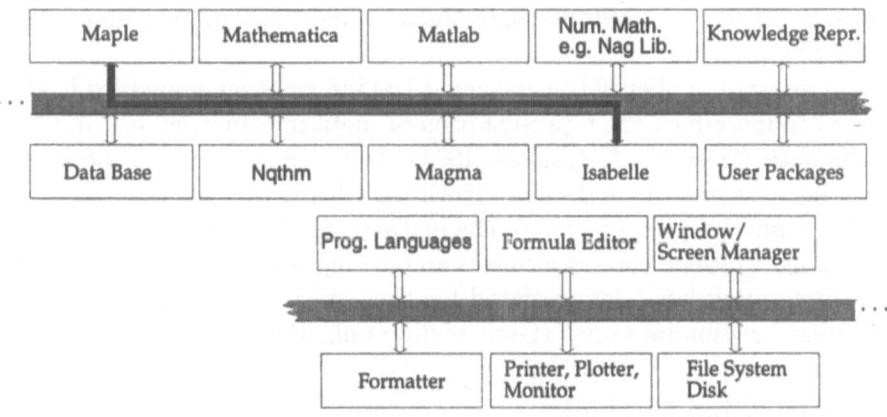

Figure 3.   Software bus of mathematical services

3.4. COMMON KNOWLEDGE REPRESENTATION

Many applications require several $\mathcal{MS}$ to share their knowledge about mathematical objects. In many cases, communicating this information is neither efficient nor practical, because it may not be explicitly known which knowledge is required.

Some cooperation mechanisms obviously benefit from sharing their knowledge, i.e. communication with subpackages or direct function calls in foreign packages (4.2, Analytica (Clarke & Zhao, 1992)). A software bus (figure 3) may include a knowledge representation system suitable for representing the common knowledge. Both architectures are illustrated in figure 4.

Recent communication methods are not restricted only to exchange of function calls, theorems, numerical data, polynomials or basic mathematical information. For example, OpenMath (Abbott *et al.*, 1995) provides the exchange of mathematical objects with a defined semantics derived from

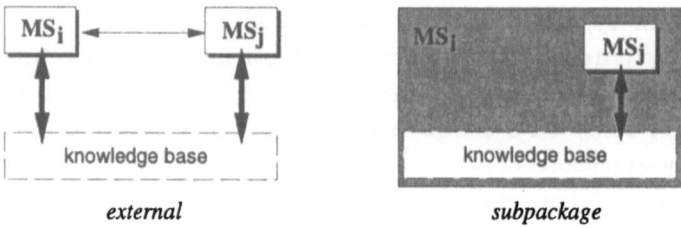

Figure 4.   Common knowledge representation

its associated lexicons. However, there are no protocols to provide meta-knowledge about the systems algorithms or type information about their arguments.

To represent explicitly the mathematical information embedded in CAS requires to introduce the representation of meta-information, e.g. in terms of schemata (Homann & Calmet, 1994). Different schemata contain this knowledge as type schemata and algorithm schemata.

The explicit representation of the mathematical information of systems performing symbolic computations is an ongoing research project. The corresponding work has been initiated for theorem provers, e.g. by the open mechanized reasoning group (Giunchiglia *et al.*, 1996).

4. Cooperation Methods

Communicating services exchange mathematical information and messages. This section introduces features and architectures according to the level of cooperation of these distributed systems. We illustrate interaction among systems playing different roles in the cooperation. Additionaly, some semantic issues and their resulting limitations are discussed.

4.1. MASTER/SLAVE

Mathematical problems are typically solved by dividing the problem into subproblems, solving subtasks by suitable CAS or TPS, and combining solutions to get the final result. Usually, one system is not sufficient for computing the solution. There are often more efficient special packages, some algorithms are not implemented, or the subproblem does not fit the scope of one $\mathcal{MS}$. However, it is often sufficient to solve the problem in the environment of one single system with the aid of other $\mathcal{MS}$. This is one reason why nowadays interfaces among CAS and TPS are typically restricted to master/slave cooperation (Clarke & Zhao, 1992; Harrison & Théry, 1993; Jackson, 1994; Giunchiglia *et al.*, 1996; Ballarin *et al.*,). The use of a $\mathcal{MS}$ is limited to some specialized tasks (algebraic simplifications, numerical computations) within the overall control of another $\mathcal{MS}$ (proofs, algebraic algorithms). The master acts as server to some client service.

Master/slave interfaces are easier to design. The master can act as a common control, the user interface of the master can act as GUI, the communication language can be chosen as the input and output language of the master, and the internal object representation of the master is the common knowledge representation. Master/slave communication typically occurs as unidirectional links (3.3) or with an intermediate bridge (Harrison & Théry, 1993). Some mathematical services act only as computational engine, decision procedure or oracle.

Additionaly, many groups improve the power of their own CAS or TPS by allowing external calls to other $\mathcal{MS}$. CAS are extended by links to TPS to verify certain conditions or type restrictions (Jackson, 1994), and TPS are extended by links to CAS or numerical software to deal with arithmetics (Harrison & Théry, 1993), mathematical objects, or to guide their proofs.

4.2. SUBPACKAGE

To avoid communication and common knowledge representation some $\mathcal{MS}$ are designed to work within the environment or language of another service (figure 4).

An example for a subpackage of CAS is Analytica (Clarke & Zhao, 1992) written in the framework of Mathematica (Wolfram, 1991). Another example is Otter (McCune, 1994) which allows external function calls out of proofs. User-defined algorithms are introduced with an identification by a special character (e.g. $GCD). The extension of the prover requires the recompilation of the whole system and each algorithm has to be implemented in C. CAS provide an extensive collection of very efficient mathematical algorithms, thus reimplementation is neither necessary nor meaningful.

4.3. EXTENSIBILITY AND INTERCHANGEABILITY

General interfaces are to a certain extent system-independent and may be connected to another or many other $\mathcal{MS}$ as illustrated in figure 5. A general communication language must be adopted by each of the involved systems (3.1).

In case of master/slave cooperation, it is typically easy to change the slave by replacing the syntax translations and if necessary the evaluation tactics. To replace the master is difficult because it hosts the complete interface. One example is the cooperation between Maple and Isabelle (Ballarin et al.,) where the CAS remains unchanged and is exchangeable. The interface is part of Isabelle's extended simplifier. Extending the interface to provide communication with other $\mathcal{MS}$ is also usually easy.

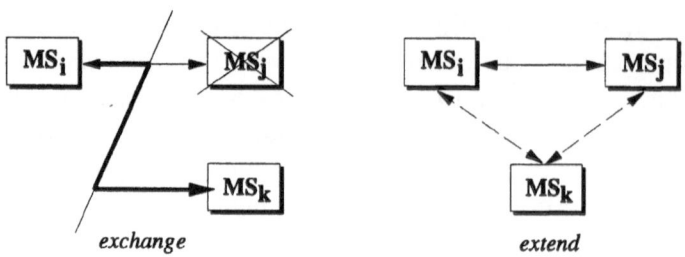

Figure 5.   Extending and interchanging mathematical services

To extend or change one $\mathcal{MS}$ in subpackage cooperation is more difficult. Since one system must provide the language and representation of both systems, extension or change is only meaningful between compatible packages. For example, it is impossible to replace Mathematica by Maple to run the Analytica package.

4.4. BLACK BOX

To cooperate among each other mathematical services must be able to interact in non-trivial ways. Deciders and black boxes (yes/no/result type packages) are not adequate in general because they do not provide internal mathematical information. Cooperating $\mathcal{MS}$ may need to accept information and produce results incrementally. Black boxes – often called oracles – are commonly implemented in interfaces among $\mathcal{MS}$ as they allow to combine systems as tools (clients) in master/slave cooperations (Ballarin *et al.*, ; Harrison & Théry, 1993).

Components like *handles* for open mechanized reasoning systems (Giunchiglia *et al.*, 1996) give access to proofs and derivation structures of TPS and open the black boxes. They can be extended to *contexts* (Calmet & Homann, 1995) which provide the necessary intermediate information to incremental and cooperative problem solving. One example of symbolic calculators providing internal information is by schemata ((Homann & Calmet, 1994), 3.4).

4.5. CONSISTENCY AND CLOSURE

Systems exchanging terms face two problems: does another service understand all transmitted terms and are the resulting terms defined in the signature of the service.

To guarantee that a mathematical service *understands* transmitted terms the term algebra must provide **consistency** w.r.t. the signature of that service.

Definition 1

The term algebra $T_\Sigma(X)$ is called consistent w.r.t. the signature A, *iff for all* $f \in A$ *any term* $f(a_1, \ldots, a_n) \in T_\Sigma(X)$ *already lies in* $T_A(X)$.

In a consistent term algebra all subterms of a term are in the subsignature $T_A(X)$, provided that the outermost connective belongs to the subsignature A. It is thus easy to recognize terms that can be passed to a service during cooperation. One has to verify that the outermost connective lies in A. Consistent term algebras are a strong limitation because they prevent to generate terms containing objects of foreign mathematical services. However, nowadays systems were designed as stand-alone systems and typically cannot handle such objects, e.g. they cannot reduce parts of expressions which also contain unknown objects like logical connectives between polynomial expressions of CAS.

The corresponding problem to understand the result computed by another service occurs in bidirectional communication and master/slave cooperation. To ensure that a term returned by a mathematical service (viewed as an operator E) does not contain unknown symbols is ensured by closure w.r.t. signatures. Even if $T_\Sigma(X)$ is consistent w.r.t. A the result returned by the service may lie in a signature $B \not\subseteq \Sigma$.

Definition 2

The operator $E : T_A(X) \to T_B(X)$ *is* closed w.r.t. A and B *if* $A \subseteq B \subseteq \Sigma$.

The OpenMath model introduces a common communication language for CAS which can be transformed into OpenMath objects. The mathematical service tries to compile these objects into the application specific representation. Objects containing unknown terms are rejected and can not be represented. In the case of communication in terms of input syntax of one of the systems problems with consistency and closure can be avoided by restriction to common subsets of both signatures. The examples illustrated in this paper require both consistency and closure.

4.6. TRUST

(Harrison & Théry, 1993) introduces levels of trust between CAS and TPS which can be generalized to classify any cooperation among $\mathcal{MS}$. Depending on the confidence on the accuracy of the answers given by another service a system trusts completely, partially, or not at all.

No trust at all may force one system to verify some conditions or results by another, or – if possible – use the results as an aid to compute the result by itself. This is especially useful in guiding a proof where verification is much less computationally complex than computation. Additionaly, $\mathcal{MS}$ may sometimes generate incorrect answers which is often unacceptable, i.e. in theorem proving.

Partial trust can occur when accepting results are accepted during interactive or temporary computations but these results have to be checked before being recorded. Another technique is to mark each result in the computation and communication with a rational representing its confidence.

Complete trust ((Jackson, 1994; Ballarin *et al.*,)) means that the result given to any request is accepted as truth. This is commonly implemented in current prototypes among $\mathcal{MS}$ and numerical packages because of its simplicity and efficiency. For example, the advantage of fast computations by symbolic or numerical software may be jeopardized by the slow arithmetics of theorem provers.

5. Conclusion

The development of general techniques for the integration of systems performing mathematical computations has not yet led to the definition of common languages, protocols, and standards. The classification of communication and cooperation methods given in this paper provides and surveys methodologies for combining $\mathcal{MS}$ and their characteristics, capabilities, requirements, differences. Its purpose is to guide the selection of methods and developments in this ongoing research.

Cooperation by distributing tasks between mathematical services is a subject of ongoing research. Among the arising problems is the black box behaviour of almost any current system. To plan and control such environments requires to represent meta-knowledge in local or global bridges or supervisors.

Among the work in progress is the design of an intelligent assistant – an environment whose semantics allows a consistent treatment of algorithms and theorems. A result of this work is the integration of the tactical theorem prover Isabelle and Maple (Ballarin *et al.*,). The extension of contexts (Calmet & Homann, 1995) is another step towards environments performing distributed mathematical problem solving.

There are obviously different approaches to what can be seen as interoperability of heterogeneous systems. For instance, we are presently investigating the feasibility of designing communication protocols based upon the types of the objects to be exchanged. This can be set in the framework of an agent approach to software engineering.

JACQUES CALMET AND KARSTEN HOMANN
Institut für Algorithmen und Kognitive Systeme
Universität Karlsruhe, Germany

References

J. ABBOTT
PossoXDR Specifications. In Posso project internal report, 1994.

J. ABBOTT, A. VAN LEEUWEN, A. STROTMANN
Objectives of OpenMath. Submitted to Journal of Symbolic Computation, 1995.

D. ARNON, R. BEACH, K. MCISAAC, C. WALDSPURGER
CaminoReal: an Interactive Mathematical Notebook. In Proceedings of International Conference on Electronic Publishing, Document Manipulation and Typography, Cambridge University Press, 1988.

C. BALLARIN, K. HOMANN, J. CALMET
Theorems and Algorithms: An Interface between Isabelle and Maple. In A.H.M. LEVELT (Ed.), Proceedings of International Symposium on Symbolic and Algebraic Computation (ISSAC'95), pp. 150–157, ACM Press, 1995.

J. CALMET, K. HOMANN
Distributed Mathematical Problem Solving. In E. SHAMIR, M. KOPPEL (Eds.), Proceedings of 4th Bar-Ilan Symposium on Foundations of Artificial Intelligence (BIS-FAI'95), pp. 222–230, 1995.

B.W. CHAR, K.O. GEDDES, G.H. GONNET, B.L. LEONG, M.B. MONAGAN, S.M. WATT
Maple V Language Reference Manual, Springer-Verlag, 1992.

E. CLARKE, X. ZHAO
Analytica – An Experiment in Combining Theorem Proving and Symbolic Computation. Carnegie Mellon University, Technical Report CMU-CS-92-147, 1992.

J. DENZINGER, M. FUCHS
Goal oriented equational theorem proving using team work. In B. NEBEL (Ed.), Proceedings of 18th German Annual Conference on Artificial Intelligence (KI'94), pp. 343-354, Lecture Notes in Artificial Intelligence 861, Springer-Verlag, 1994.

W.W. MCCUNE
OTTER 3.0 Reference Manual and Guide, Technical Report ANL-94/6, Argonne National Labaratory, 1994.

S. DALMAS, M. GAËTANO, A. SAUSSE
Distributed Computer Algebra: the Central Control Approach. In H. HONG (Ed.), Proceedings of 1st International Conference on Parallel Symbolic Computation (PASCO'94), pp. 104–113, Lecture Notes in Computing 5, World Science, 1994.

F. GIUNCHIGLIA, P. PECCHIARI, C. TALCOTT
Reasoning Theories – Towards an Architecture for Open Mechanized Reasoning Systems. In Proceedings of First International Workshop Frontiers of Combining Systems (FroCoS'96), Kluwer, 1996.

S. GRAY, N. KAJLER, P. WANG
MP: A Protocol for Efficient Exchange of Mathematical Expressions. In Proceedings of International Symposium on Symbolic and Algebraic Computation (ISSAC'94), ACM Press, 1994.

J. HARRISON, L. THÉRY
Extending the HOL Theorem Prover with a Computer Algebra System to Reason About the Reals. In J.J. JOYCE, C.-J.H. SEGER (Eds.), Higher Order Logic Theorem Proving and its Applications (HUG'93), pp. 174–184, Lecture Notes in Computer Science 780, Springer-Verlag, 1993.

K. HOMANN, J. CALMET
Combining Theorem Proving and Symbolic Mathematical Computing. In J. CALMET, J.A. CAMPBELL (Eds.), Integrating Symbolic Mathematical Computation and Artificial Intelligence (AISMC-2), pp. 18–29, Lecture Notes in Computer Science 958,

Springer-Verlag, 1995.

P. JACKSON
Exploring Abstract Algebra in Constructive Type Theory. In A. BUNDY (Ed.), Automated Deduction (CADE-12), pp. 590–604, Lecture Notes in Artificial Intelligence 814, Springer-Verlag, 1994.

N. KAJLER
CAS/PI: a Portable and Extensible Interface for Computer Algebra Systems. In Proceedings of International Symposium on Symbolic and Algebraic Computation (ISSAC'92), pp. 376–386, ACM Press, 1992.

S. WOLFRAM
Mathematica: a System for Doing Mathematics by Computer, Addison-Wesley, 1991.

WOLFRAM RESEARCH INC.
MathLink Reference Guide Version 2.2, Mathematica Technical Report, Wolfram Research, 1993.

ENRICO DENTI, ANTONIO NATALI, ANDREA OMICINI, MARCO VENUTI

LOGIC TUPLE SPACES FOR THE COORDINATION
OF HETEROGENEOUS AGENTS

Abstract. This work presents $\mathcal{ACLT}$, a coordination model aimed to combine and coordinate heterogeneous agents by means of a communication abstraction inspired to the Linda model, but rooted in a logic framework. The twofold interpretation of a logic tuple space both as a message and as a knowledge repository naturally induces a categorisation of agents as logic and non-logic. While non-logic agents adopt the basic Linda kernel, logic agents exploit the full power of the $\mathcal{ACLT}$ model, which supports deduction and reasoning over the content of the tuple space. By providing a conceptual framework where logic inference and temporal tuple space evolution coexist, $\mathcal{ACLT}$ provides a suitable environment to build heterogeneous multi-agent systems, where hybrid agent architectures can be designed, integrating reasoning capabilities together with reactive behaviours.

1. Introduction

When combining independent and possibly heterogeneous agents to build a multi-agent system, one of the most critical issues concerns the definition of a proper coordination and synchronisation model which adequately supports agent interaction. In fact, the notion of agent is usually intended to be a higher-level concept than just a system process, particularly emphasising problem solving and coordination aspects, as well as autonomous behaviour, which ask for high-level communication metaphors. A well-known approach to this issue consists of defining some form of shared memory abstraction to be used as a message repository, and a communication protocol allowing agent competition and cooperation in a multi-agent framework. Several such *coordination models* have been proposed in literature, based on abstractions like blackboards (Englemore *et al.*, 1988) and tuple spaces, like Linda (Gelernter *et al.*, 1985; Gelernter *et al.*, 1992; Gelernter *et al.*, 1989), Shared Prolog (Brogi *et al.*, 1991), and Polis (Ciancarini, 1994;

F. Baader and K.U. Schulz (eds.), Frontiers of Combining Systems, 235–248.
© 1996 *Kluwer Academic Publishers.*

Ciancarini, 1993).

However, the design and the implementation of multi-agent systems raise case-specific issues, including the development of suitable coordination strategies for effective problem solving, protocols to support reasoning activities (also concerning inter-agent communication), negotiation and conflict detection, resolution mechanisms, etc. These requirements call for agents with an enhanced capability of symbolic computing, which be able to manipulate knowledge and perform inference operations.

As a result, there is a growing need for enriched communication abstractions, which allow coordination and deduction aspects to coexist, supporting higher-level forms of interaction than raw information exchange (like message-passing or signals). This is the case, for instance, when agents are used to (incrementally) build and store in a knowledge repository a (possibly partial) representation of a given application domain, where the ability of deducing new facts which have not been explicitly generated, exchanged, or stored, becomes relevant.

The $\mathcal{ACLT}$ model (Agents Communicating through Logic Theories, first proposed in (Omicini et al., 1995)) is founded on a first-order logic-based approach, whose key idea is to assume the logic theory as a communication abstraction, supporting (and combining) Linda-like communication operations as well as logic operations based on don't know non-determinism. As a consequence, reactive behaviours may be naturally integrated with high-level symbolic activities based on reasoning and planning, leading to those hybrid architectures which currently represent one of the main goals in fields like robotics (Lyons et al., 1992; Zanichelli et al., 1994).

In fact, while low-level agents handling basic tasks and/or real-time jobs such as sensor data acquisition will just exploit the $\mathcal{ACLT}$ communication abstraction as a synchronisation device, high-level agents, encapsulating higher-level coordination policies and system strategies, will be able to interpret the tuple space also as a logic theory, performing logic inferences and deductions based on the available knowledge.

The main contribution of this work is to show how the interpretation of the communication abstraction both as a simple communication device, and as a logic theory containing knowledge over the application domain, suggests a natural criterion to conceptually categorise two kinds of agents, based on the way such abstraction is used. As a result, each kind of agent may interact with the others at its own abstraction level, achieving all advantages of coordination and synchronisation while maintaining its autonomy from the viewpoint both of its implementation technology and of its computation model.

This work is structured as follows. Section 2 introduces the basics of the $\mathcal{ACLT}$ coordination model, by discussing the assumptions, the language scheme, and the main problems related to the coexistence of time-related changes and logic-based computation model. In particular, this Section introduces the agent categorisation induced by the twofold interpretation of a logic tuple space. Section 3 discusses the role of logic agents in the $\mathcal{ACLT}$ framework, by presenting the basic hybrid primitives, which combine logic proofs based on don't-know non-determinism with the time-dependent evolution of logic tuple spaces. Finally, Section 4 is devoted to final remarks and conclusions, as well as to some comparisons with related works and similar models.

2. The Basic $\mathcal{ACLT}$ Coordination Model

2.1. MULTIPLE LOGIC TUPLE SPACES

In order to support the coordination of heterogeneous agents, an interaction model is required which is flexible enough to allow virtually any useful policy to be built upon it, yet provides a very limited number of basic primitives, which can be seamlessly inserted in any practical language framework. Among the models providing these features, we started from the Linda model (Gelernter *et al.*, 1985) as a general-purpose coordination language which intrinsically supports heterogeneity.

In fact, according to (Gelernter *et al.*, 1992), by providing a coordination model which is intentionally orthogonal to any computation model, Linda can be freely combined with any sequential language (such as C (SCA, 1990), Prolog (SICS, 1994), Modula-2 (Borrman *et al.*, 1988)).

In Linda, agents communicate through a *tuple space* which is managed by three basic operations: *in*, *out*, and *read*, which respectively remove, insert, and access tuples in the tuple space, by means of a pattern-matching mechanism.

The support for agent heterogeneity in Linda raises the question of type compatibility. Agents implemented in different languages would generate tuples of different types according to their type system. But in order to let them communicate through a shared data structure, a common type system is required. In principle, two choices are possible:

- a basic type system is defined as a sort of least common denominator, which is meant to accommodate any (sequential) host language;
- the type of the tuple space is chosen based on the required expressive power; then, a scheme for the distinct host languages is provided accordingly.

The former option clearly privileges the *uniformity* of the coordination model, by providing any kind of agent with exactly the same level of perception of the communication abstraction. The latter option, instead, emphasises the richness of *heterogeneity*, since agents based on different technologies may have different perception of the tuple space, so as to exploit the communication abstraction at their best.

Since we insist on the role of the high-level coordination abstraction, the $\mathcal{ACLT}$ coordination model adopts a logic-based approach, where the communication devices are *logic tuple spaces* à la Shared Prolog (Brogi *et al.*, 1991). By mapping the notion of tuple into the concept of (atomic) logic formula, an $\mathcal{ACLT}$ tuple space consists of an evolving (unitary) clause database representing the *current coordination state* (as defined in (Gelernter *et al.*, 1992)) as a first-order logic theory.

According to the idea that different kinds of agent may have different perceptions of the same coordination abstraction, the *pattern matching* mechanism may be replaced whenever possible by the more powerful *unification*mechanism, whenever an agent is able to deal with it (this is the case, for instance, of an $\mathcal{ACLT}$-Prolog agent). Moreover, since a logic tuple space can be interpreted as a knowledge repository (instead of a simple message repository), reasoning activities over the coordination process state may be combined in principle with the typical reactive behaviours induced by Linda synchronisation mechanisms, thus leading to hybrid agent architectures.

Like in Polis (Ciancarini, 1994; Ciancarini, 1993) and in Gelertner's Multiple Tuple Spaces (Gelernter *et al.*, 1989), $\mathcal{ACLT}$ tuple space is not unique. Instead, many communication abstraction instances, each denoted by a name, can be defined and used independently. However, unlike the Polis approach, an $\mathcal{ACLT}$ tuple space needs not be created dynamically by some tsc-like primitive, but exist simply by consequence of being referenced as such by some $\mathcal{ACLT}$ primitive. Consequently, $\mathcal{ACLT}$ operations have to be labelled with the name of the proper tuple space, with a syntax which obviously depends on the particular sequential language which an ACLT agent is built with.

For instance, an *out* operation performed by an $\mathcal{ACLT}$-Prolog agent has the following syntax, which adopts the same @ syntax used by Shared Prolog (Brogi *et al.*, 1991):

```
out(p(a))@world
```

where world denotes a logic tuple space.

2.2. HETEROGENEOUS AGENTS EXPLOITING A LOGIC TUPLE SPACE

Agents may be completely heterogeneous, both from the viewpoint of the technology and the language used to implement them, as well as for the abstraction level at which they operate: the only constraint is that their host language provides the basic Linda communication primitives adapted to the notion of logic tuple space. Obviously, low-level agents (such as $\mathcal{ACLT}$-C agents), performing basic tasks, will typically use the tuple space just as a synchronisation device, being unaware of its logic theory nature. Instead, higher-level agents (such as $\mathcal{ACLT}$-Prolog agents) will be able to exploit the information of the tuple space as knowledge which can be taken as the base for deductions. In this way, different categories of agents can exploit the tuple space according to the highest level of perception available to them.

Actually, one of the main features of our model comes from fully exploiting the twofold interpretation of the communication abstraction, which suggests a natural criterion to conceptually distinguish two kinds of agents, operating at different abstraction levels and featuring different capabilities:

- *non-logic agents*, exploiting the logic tuple space as a simple message repository, by means of the basic Linda primitives;
- *logic agents*, aware that the logic tuple space is a full-fledged logic theory, and thus able to exploit this feature to perform logic inferences over the current coordination state.

Typically, non-logic agents will be low-level agents, written in a language close to the machine, and aimed to handle basic tasks and/or real-time jobs such as sensor data acquisition. For this purpose, an ad-hoc $\mathcal{ACLT}$-C library has been implemented, which enables C-based agents to access the tuple space, intended as a pure message repository. Since the communication mechanisms provided by $\mathcal{ACLT}$ for non-logic agents are those of the basic Linda model, such category of agents is not a matter of discussion here.

On the other side, logic agents, encapsulating higher-level strategies and coordination policies, will likely be implemented by extending a sequential logic language such as Prolog with the $\mathcal{ACLT}$ primitives. Such agents will be able, when required, to interpret the shared information constituting the tuple space(s) as knowledge over the application domain provided by the whole multi-agent system. According to this view, any agent may be read as a knowledge source, providing its own (partial) description of an application domain.

The main issue is then how to extend the basic coordination model so as to allow logic agents to fully exploit the logic theory interpretation of $\mathcal{ACLT}$ tuple spaces.

3. Logic Agents in $\mathcal{ACLT}$

3.1. DEDUCTION OVER EVOLVING LOGIC TUPLE SPACES

Allowing agents to reason over the tuple space content yields a conceptual problem, due to the side-effect nature of the communication primitives.

In fact, logic tuple spaces evolve in response to *actions*, represented by *in* and *out* communication primitives. As a result, the interpretation of tuple spaces as communication devices is bound to a vision of clause databases representing dynamic, evolving information.

At the same time, logic agents are meant to perform *deductions* over the content of tuple spaces. But the logic theory reading of tuple spaces is founded over a notion of platonic, non-modifiable, universal truth, which is enforced, from a computation viewpoint, by the requirement of *correct* and *complete* deductions (as it is better explained below).

Since the two readings are in principle non-compatible, the main issue is to devise a solution to make them coexist in the same conceptual framework. However, our approach is not intended to reach the cleanness of other models, such as Polis[1], which try to integrate Linda primitives in a logic language while preserving its clean semantic characterisation. Instead, the $\mathcal{ACLT}$ approach is meant to emphasise the richness of *hybridisation*, by exploiting the conflict between reasoning activities and reactive behaviours as a feature rather than as a source of problems (at least, as far as this is possible).

Therefore, the main problem is deduction over an evolving logic theory. Fundamentally, this raises two kind of problems:

- how to ensure correctness, when assumptions on which a deduction is based may be retracted during the computation as an effect of performing a side-effect operation?
- even assuming that knowledge is never retracted, which form of deduction should we consider as complete, given that the information available is in principle always subject to grow?

At the very end, the basic problem to be solved is which information to consider for logic proofs in such an evolving knowledge space. This problem can be split in two (somehow independent) issues, which directly corresponds to the problems of correctness and completeness of the deduction process cited above:

- If knowledge may be retracted, how should we handle demonstrations in progress based on retracted knowledge? Should such deductions be permitted, or should they not?
- Conversely, even assuming that knowledge is never retracted, should new information produced during a demonstration be considered for the current logic proof itself?

With respect to the above issues, the first problem regards whether to charge the user for dealing with any inconsistency-related problem, or to

---

[1]In Polis (Ciancarini, 1994), sequential activities retain their own semantics by means of the special agent structure {test; consume; localEval; out} which encapsulates the side-effect Linda operations in separate sections, enveloping the sequential activity (localEval) so as to leave it autonomous and untouched.

provide some (controlled) form of (tuple space) knowledge access which guarantees some idea of consistency. The $\mathcal{ACLT}$ model disciplines then access and modification of the tuple space content, by providing tools for the support of deduction processes over evolving logic tuple spaces.

3.2. CLASSIFYING KNOWLEDGE

Generally speaking, with respect to the problem of conjugating retractable knowledge with "safe" logic inferences, the point is how to ensure that only logically valid (correct) demonstrations are performed, avoiding inconsistencies due to retraction of axioms underlying a deduction process. For instance, if some agent withdraws (with an *in* operation) from the tuple space one axiom on which the current line of reasoning of some logic agent is based, the correctness of the corresponding deduction would be undermined. Even though a trivial solution would be to prevent knowledge from being retracted at all, such an approach would defeat the effectiveness of the basic coordination model.

In order to face this problem without intentionally entering the field of non-monotonic reasoning (Reiter, 1987), $\mathcal{ACLT}$ introduces a classification scheme for the different sorts of knowledge, distinguishing between stable, although uncomplete, knowledge, such as the location of a fixed object like a door in a room, and transient knowledge, such as the current position of a human observer in a room.

The former category (called *extendible* knowledge) refers to that part of the application domain which, even though it may be practically considered as fixed and non-mutable, is only partially known, so that information about it is subject to grow during a computation. In other terms, while extendible knowledge refers to time-independent elements of the domain of discourse, its availability to the multi-agent system in form of logic tuples is time-dependent, in that it may increase over time, as a result of a domain exploration process.

The latter category (called *transient* knowledge), instead, refers to that part of the world which evolves during agent life, and cannot therefore be safely taken as a basis for logic inferences. As a result, transient knowledge can be retracted but, for this very same reason, not taken as a base for deductions, while stable knowledge can be used for logic inferences, but, consequently, never retracted.

To distinguish the two kinds of knowledge, a first possible approach might consist of exploiting the multiplicity of "knowledge containers" (the tuple spaces) by associating a unique kind of knowledge with each tuple space, thus constraining the access according to the container category. Instead, a finer-grained solution may be built upon the basic feature of first-order logic languages, which exploits predicate symbols as a modularity source. In particular, given predicate p of arity n, the collection of the facts of the form $p(\bar{t})$ (where $\bar{t}$ represents a generic n-uple of terms) constitutes a unique knowledge chunk, which may be naturally bound to represent either transient or partial knowledge. Because of the greater flexibility of the latter approach, information categories in the $\mathcal{ACLT}$ model are associated to each predicate symbol occurring in a tuple space, either explicitly, through suitable static declarations, or implicitly, by the primitives dynamically used to access the corresponding axioms (see (Omicini *et al.*, 1995) for more details). In other terms, each predicate in a tuple space is used to represent either transient or extendible knowledge, alternatively.

As a result, no deduction process can be performed over retractable knowledge, as well as no retraction is allowed over knowledge which may be involved in a logic proof. This in principle guarantees the correctness of the demonstration processes over the tuple spaces - at least, as far as only non-negative information is considered.

3.3. THE $\mathcal{ACLT}$ SNAPSHOT MODEL

As a consequence of the above classification, reasoning over the tuple space knowledge is meaningful (thus, allowed) only when stable (although partial, uncomplete) knowledge is considered. However, even in this case (that is, assuming that knowledge is never retracted), dealing with uncomplete information raises some further issues. For instance, if new knowledge is produced during a logic inference, should it be considered for the current logic proof? And how can synchronisation mechanisms be fruitfully combined with deduction processes?

These issues concern the nature of the demonstration primitives which the coordination model should provide logic agents with. These primitives, in their turn, determine the logic agent behaviours by settling how they can perform deductions over logic tuple spaces. Since we are trying to define a general model for hybrid logic agents, we have no elements for a-priori privileging one particular choice rather than another.

As a result, $\mathcal{ACLT}$ provide a family of *demo* meta-predicates (Bowen *et al.*, 1982), each one allowing different relevant behaviours to be achieved, and differing from one another for the choice they make with respect to the following two issues:

  i) in case the proof of a *demo* backtracks, should the newly-available knowledge (that is, those logic tuples inserted in the tuple space after the *demo* primitive has been first performed) be considered to provide further solutions?

ii) if there is no suitable knowledge when the *demo* starts, should we wait for such knowledge to become available, or just fail?

Since the two issues (*i*) and (*ii*) are basically orthogonal with respect to each other, there are (at least) four meaningful *demo* behaviours, that the $\mathcal{ACLT}$ model tries to capture.

In particular, issue (*i*) concerns completeness: how can we draw every logic consequence from a logic theory which grows during computations? Of course, we cannot. In order to give some (obviously, very weak) notion of completeness, we should fix the set of axioms, from where logic consequences may be drawn. To this aim, the best approximation we can provide is to associate an instantaneous *snapshot* of the logic tuple space content to any deduction process.

By means of such a snapshot, which represents precisely the coordination state (as defined in (Gelernter *et al.*, 1992)) referred to a given instant of the computation, we may think to perform somehow complete deductions, aware that the truth value of a logic formula in a tuple space is no longer absolute in time, but refers trivially to the specific instant when the snapshot is taken.

Obviously, the real issue becomes *when* such a snapshot should be taken. This choice represents a degree of freedom which is exploited in $\mathcal{ACLT}$ to model a variety of hybrid agent behaviours. Therefore, the $\mathcal{ACLT}$ *demo* primitives may differ for the instant assumed as the *snapshot time*, which triggers different deduction processes over the current coordination state. In this way, different choices for the snapshot time lead to different relevant logic agent behaviours.

More precisely, if the request for a *demo* operation to a given tuple space is performed at time τ, first served at τ', and (possibly) fails at τ'' after all admissible alternatives have been taken into account, *demo* primitives differ for the relationships between the three instants, as well as for which of the three instant is conceptually used as the snapshot time.

Actually, the $\mathcal{ACLT}$ model relies on a set of four *demo* meta-predicates, each one taking an atomic formula as its unique argument, and possibly associated to a given tuple space: `demo`, `demoWait`, `demoLast`, and `demoWaitLast`. The above instant denotation will be used in the following in order to give the intuitive semantics of the four primitives.

3.3.1. *The* `demoLast` *primitive*

The first problem to be solved (issue (*i*) on Page 242), then, concerns the (obviously, relative) completeness of the deduction process: should *demo*s take into account only knowledge available when they are first performed, or should they try to use as much knowledge as they can?

The simple demo primitive adopts the first option, by behaving as if the tuple space were frozen at the time the operation is invoked.

More precisely, by using the conventions introduced above for instants τ, τ', and τ'', its semantics may be described as follows. When an operation like demo(p(t̃))@world is invoked, the p(t̃)-snapshot of world (that is, those axioms of world unifying with p(t̃)) is taken at time τ, and the call is immediately served ($\tau = \tau'$). If this snapshot is empty, then the call immediately fails ($\tau' = \tau''$); otherwise ($\tau' < \tau''$), a logic derivation is performed according to the specific computation rule of the host language. If no branch of the derivation succeeds, the computation finally fails, and $\Delta\tau = \tau'' - \tau'$ is the non-null computation cost of the exploration of the p(t̃) axioms. Needless to say, any information produced after τ (τ') will be ignored by the demo primitive.

This may be useful, for instance, when synchronisation between an exploring agent (possibly, a non-logic agent) and a reasoning agent (typically, a logic agent) ensures that enough (possibly all) information about a previously unknown domain has been already accumulated in a tuple space by the explorer, so that the reasoner can fruitfully exploit available knowledge.

Like *demo*, *demoLast* is immediately served, too, so that $\tau = \tau'$ always holds. If there are no facts in world unifying with p(t̃) at time τ, the operation demoLast(p(t̃))@world behaves exactly as demo(p(t̃))@world: i.e., it immediately fails ($\tau' = \tau''$). In case, instead, that there are such facts, the deduction processes induced by the two primitives turn out different.

In particular, demoLast tries to exploit all the suitable world knowledge, including that one generated after the operation was first served (i.e., after τ'), by taking the p(t̃)-snapshot of world as late as possible. Thus, axioms taken into account are not only those existing in world at time τ, but also those possibly inserted in world before the computation finally fails (τ''), since the cost of computing all solutions available is not null. Thus, when a failure forces demoLast to backtrack, all new information can be fruitfully exploited to provide new branches for the deduction process.

Such an operation is repeated until all suitable facts in world have been tried (with failure) and no further facts unifying with p(t̃) have been added. Then, and only then, demoLast(p(t̃))@world fails. Since failure occurs only when none of the axioms which can be found in world at failure time has produced a successful branch, the p(t̃)-snapshot can be thought as taken at τ''.

For instance, if a reasoning agent is meant to work on the data produced by an exploring agent, while exploration is still going on, demoLast allows deduction to start while continuing to store the incoming knowledge in the tuple space, so as to have it available in case the *demo* operation backtracks.

3.3.2. *The* demoWait *primitive*

The second problem to be faced (issue (*ii*) on Page 243) concerns the synchronisation of the deduction process: in case no suitable information is available in the tuple space of choice, should *demos* wait for such information, or just fail immediately? Conceptually, the distinction arises when having a partial description of an application domain abstraction is enough to start a deduction process, while having no information at all is not acceptable. In other terms, this extends the notion of deduction with a synchronisation mechanism, by introducing the question "Is it too early to deduce?"

Again, the first option corresponds to the simple demo primitive: when the p($\bar{t}$)-snapshot of world is empty, demo(p($\bar{t}$))@world simply fails ($\tau = \tau'$).

Instead, when a demoWait(p($\bar{t}$))@world operation is performed, the calling logic agent is suspended until some suitable knowledge is found in world. This conceptually corresponds to taking the p($\bar{t}$)-snapshot of world at time τ', when the operation is first served. In fact, no suspension takes place if at time τ some axioms unifying with p($\bar{t}$) are already available, and demoWait(p($\bar{t}$))@world is immediately served ($\tau = \tau'$). In this case, demo and demoWait semantics perfectly match. Instead, in case no suitable facts are found in world at time τ, the snapshot is taken at $\tau' > \tau$, when some suitable knowledge is finally made available by some other agent.

This primitive may useful in those cases where any amount of suitable information available in the tuple space is enough for inferences, provided that it is non-null. Suppose, for instance, that an agent has to move from a room to another, passing through doors. In order to find the best path, every information available about doors should be used. However, no reasoning over a path can even be started if at least one door has not been discovered in the room: deduction should then be synchronised with the availability of some information about doors, or it would be "too early to deduce".

3.3.3. *The* demoWaitLast *primitive*

The orthogonality of the two issues (*i*) and (*ii*) allows the introduction of a fourth *demo* primitive, demoWaitLast, which combines the semantics of demoLast and demoWait. In fact, like demoWait, demoWaitLast never fails immediately ($\tau'' > \tau'$ always holds), even when no suitable knowledge is found in world. However, when world contains some facts unifying with p($\bar{t}$), demoWaitLast behaves exactly like demoLast, trying to exploit all suitable knowledge made available in world at any time before the primitive fails.

For instance, suppose that a logic agent controls a bunch of identical sensors. Obviously, sensors will likely output data in the tuple space nearly at the same time. Therefore, the logic agent could wait for the first sensor data available, and then start its deduction, knowing that all sensor data would likely arrive in time for being considered on backtracking.

3.3.4. *Other primitives*

Many other different primitives could be conceived, and added to the $\mathcal{ACLT}$ model. However, we choose to limit the discussion to the small family of *demo* primitives presented above, since in our opinion it allows a quite interesting number of hybrid agent architectures to be designed, by providing logic agents with concepts and tools for different kinds of safe deduction processes to be performed over evolving knowledge bases.

3.4. SKETCH OF THE CURRENT $\mathcal{ACLT}$ IMPLEMENTATION

The $\mathcal{ACLT}$ model has been implemented over the SICStus Prolog system (SICS, 1994) by providing a distributed support which allows agents to be spread over a network of machines, ensuring transparency of the communications: no information about physical allocations of either agents or tuple spaces is needed for an agent to access any tuple space. This goal has been achieved through the design of a proper network-layer interface built around two specific Unix daemons (written in Prolog themselves), which make $\mathcal{ACLT}$ a standard Unix system service. The general architecture of the system, as well the implementation details will be soon the subject of a further paper.

A set of SICStus Prolog libraries allows $\mathcal{ACLT}$ logic agents to be build in Prolog. Moreover, by means of an *ad-hoc* C library, non-logic $\mathcal{ACLT}$ agents can be written in C, too, obviously with no demo capability. As a result, the current $\mathcal{ACLT}$ implementation allows heterogeneous multi-agent systems to be actually built and experimented.

Although the SICStus programming environment is available on most hardware platform (UNIX, PC, Macintosh), the current $\mathcal{ACLT}$ implementation works only on a network of Sun and HP workstations, and on PCs running Linux. Further work will be devoted to extend the number of the supported platforms.

4. Related Works and Conclusions

$\mathcal{ACLT}$ is a model for the coordination of ensembles of heterogeneous agents. Based on the Linda (Gelernter *et al.*, 1985; Gelernter *et al.*, 1989) model, it adopts the idea of the separation between the coordination and the computation model [3], so as to ease cooperation of agents implemented with different sequential languages (e.g. C, Prolog). From this point of view, it is not comparable with other models and languages based on (generative) communication through shared memory abstractions too, such as Gamma (Ban *et al.*, 1990), and Tao (Porto *et al.*, 1995), which are built instead as unique languages or models integrating both computation and coordination aspects.

In order to provide a higher level of abstraction, $\mathcal{ACLT}$ is founded on the notion of multiple logic tuple spaces (like Shared Prolog, and ESP (Brogi *et al.*, 1991; Ciancarini, 1994; Ciancarini, 1993)), as the devices for agent communication/synchronisation. The main difference with respect to those logic-based approaches lays in the full exploitation of the twofold interpretation of the notion of logic tuple space both as a repository for message exchange, and as a logic theory representing evolving models for the objects of the application domain.

The latter interpretation, in fact, determines in $\mathcal{ACLT}$ a natural categorisation for modelling heterogeneous agents and their behaviours. Non-logic agents simply adopt the classical reading of the tuple space as a collection of messages, or signals (and performs only Linda-like operations), while logic agents are meant to perform reasoning activities over the logic tuple space representing the current coordination state.

However, when considering the dynamicity of information in non-trivial environments, such as multi-agent systems, the conventional idea of deduction from platonic, non-modifiable worlds has to be revised. Starting from this consideration, $\mathcal{ACLT}$ provides a knowledge classification criterion, along with a set of disciplined primitives for deduction over evolving logic theories, which allow the design of hybrid architectures for logic agents.

It may be observed, however, that the small set of different kinds of deduction processes discussed in this work may be even considered too rich, in that, given the lack of an adequate semantic characterisation, hybridisation may result in a source of confusion instead of clarification. In our opinion, these deduction processes have necessarily to be taken into account when dealing with real application environments, where high-level logic-based agents have to cooperate and compete with low-level reactive agents. Such environments, such as robotics, typically ask for hybrid systems (such as heterogeneous multi-agent systems), as well as for hybrid agent architectures, where reactive behaviours have to be combined with high-level symbolic activities, such as planning.

ENRICO DENTI, ANTONIO NATALI, ANDREA OMICINI AND MARCO VENUTI

LIA – DEIS
Università di Bologna, Italia

References

J.-P. Banâtre and D. le Métayer. The Gamma model and its discipline of programming. *Science of Computer Programming*, 15(1):55–77, November 1990.

L. Borrman and M. Herdieckerhoff. Linda integriert in Modula-2 – ein Sprachkonzept für portable parallele Software. In *Proceedings 10 GI/ITG-Fachtagung Architektur und Betrieb von Rechensystemen*, Paderborn, March 1988.

K.A. Bowen and R.A. Kowalski. *Amalgamating Language and Metalanguage in Logic Programming*, pages 153–172. Academic Press, 1982. Clark, Tarnlund (eds.).

A. Brogi and P. Ciancarini. The concurrent language, Shared Prolog. *ACM Transactions on Programming Languages and Systems*, 13(1), January 1991.

P. Ciancarini. Coordinating rule-based software processes with ESP. Technical Report UBLCS-93-8, Laboratory of Computer Science, University of Bologna, April 1993.

P. Ciancarini. Distributed programming with logic tuple spaces. *New Generation Computing*, 12, 1994.

R. Englemore and T. Morgan, editors. *Blackboard Systems*. Addison-Wesley, Reading, Mass., 1988.

D. Gelernter. Generative communication in Linda. *ACM Transactions on Programming Languages and Systems*, 7(1), January 1985.

D. Gelernter. Multiple tuple spaces in Linda. In *Proceedings of PARLE*, volume 365 of *LNCS*, 1989.

D. Gelernter and N. Carriero. Coordination languages and their significance. *Communications of the ACM*, 35(2):97–107, February 1992.

D.M. Lyons and A.J. Hendriks. Planning for reactive robot behavior. In *Proc. of the IEEE Int. Conf. on Robotics and Automation*, Nice, France, May 1992.

A. Omicini, E. Denti, and A. Natali. Agent coordination and control through logic theories. In *Topics in Artificial Intelligence - 4th Congress of the Italian Association for Artificial Intelligence, AI*IA'95*, volume 992 of *LNAI*, pages 439–450, Firenze, Italy, October 11–13 1995. Springer-Verlag.

A. Porto and V. Vasconcelos. Truth and action osmosis (the Tao computation model). http://theory.doc.ic.ac.uk/theory/COORDINATION, January 17 1995.

R. Reiter. Non-monotonic reasoning. *Ann. Rev. Computer Science*, 2:147–186, 1987.

Scientific Computing Associates, Inc., New Haven, CT, USA. *C-Linda Reference Manual*, 1990.

Swedish Institute of Computer Science, Kista, Sweden. *SICStus Prolog User's Manual*, 1994.

F. Zanichelli, S. Caselli, A. Natali, and A. Omicini. A multi-agent framework and programming environment for autonomous robotics. In *Proceedings of the International Conference on Robotics and Automation (ICRA'94)*, pages 3501–3506, S. Diego, CA, USA, May 1994.

CHRISTOPHER LANDAUER AND KIRSTIE L. BELLMAN

INTEGRATION SYSTEMS AND INTERACTION SPACES

Abstract. Integration systems are abstract methods of combining theories, computational paradigms, algorithms, and other rule-driven symbolic processing systems. This paper discusses our hopes and our progress in creating a principled approach to a theory of integration for large systems.

First, we discuss two types of integration systems, distinguished by the flexibility of the integration process and the types of infrastructure that they retain. In the next section of the paper, we present our approach to creating an integration infrastructure through the processing of explicit meta-knowledge.

In the following section, we introduce some work on combining wrappings with text-based Virtual Reality environments to form a new kind of integration infrastructure, and in the last section, we discuss the requirements for formalisms that have emerged from working in these testbeds.

This research was supported in part by the Aerospace Corporation's Sponsored Research Program, by the Federal Highway Administration's Office of Advanced Research, and by the Advanced Research Projects Agency's Software and Intelligent Systems Technology Office.

1. Introduction: Integration Systems

Our interest in building an "Integration Science" comes from working in designing, building, and analyzing large military space systems (Bellman, Gillam, 1988) (Bellman, Gillam, 1990) (Bellman *et al.*, 1993). These large complex systems are characterized by having a long design and development stage, extending over decades, with hundreds of researchers and developers coming and going. They also have an enormous number of different types of components, such as solar panels, propulsion systems, heat radiators, command and control software and hardware, communication devices, sensors, and so forth.

F. Baader and K.U. Schulz (eds.), Frontiers of Combining Systems, 249–266.
© *1996 Kluwer Academic Publishers.*

The computer-based support for these systems: models, databases, numerical analyses, graphics, rulebases and other computational resources reflect the enormous diversity of system components through the long life cycle of a space program. These resources vary from the earliest conceptual design and analysis programs through the much more detailed design and analysis models used for engineering and manufacturing, to the software controlling the on-board processing and ground stations and analysis program needed during the space programs operational life.

With many others, we became concerned about the need for developing a better, more principled approach to integrating this diversity of computational resources so that we can better develop, design and analyze large systems. Many of the large systems today are built in an overly costly, inflexible, and unanalyzable manner. By unanalyzable we mean that although the system can be and is often tested as a whole for certain overall performance characteristics, when faults, failures, or requirement changes occur, there is little ability to analyze the system or to apply more than limited testing strategies to it. We believe that part of this new integration science will include new formal foundations and languages for representing and reasoning about these large, complex systems. One of the goals of our research has been to help develop these formalisms, by first clearly defining what the requirements are for these formalisms and what processing capabilities must accompany them.

However, we also believe that there can be no one language or mathematical formalism or modeling technique that is adequate for the development, design and analysis of these large and complex systems. Hence, we believe that "integration science" must include not only techniques to put components together, as if once and for all, but also the additional techniques and infrastructure that allows us to select among and integrate different integration formalisms and techniques themselves. In other words, integration is not a one-shot job to be done with a given language, formalism, or technique. Rather, integration is a process that needs itself to be selected and fine tuned for a given application, and it occurs at many different levels of a complex system. These different level usually require different integration strategies. Lastly, we think that to bring together components to form a combined system is of limited use unless it promotes our ability to understand and manage how well each individual component is functioning within that combined system and how well each individual component relates to other components, and it enhances our ability to change or adapt both the functioning of the individual component and the combined system.

A term that was quite popular in the United States, especially in commercial and military software circles is "seamless integration". The goal of "seamless integration" was to bring together components into some new overall combinations, perhaps functionally a "new system", without any problems in interfacing these components or in the functioning of those components. The implications were that the user could remain naive about the components (hence the term "seamless") used to achieve the new functionality. In such a system, the user could for example, reach for a combination of graphic and editing tools without concern over compatibility of formats, hosts, or file system services. This ideal was picked up by many of the developers of the software environments, and for many of the more well-known commercial environments, users indeed are given a large array of tools that they may almost mix and match with impunity. However, even for the limited complexity of providing a single user a rich computing environment for a home computer, there are limitations for how these commercial systems are integrated: although the environments contain increasingly rich repertoires of services, they are not of the users' selection and "foreign" services (those services not invented and developed by or for that commercial vendor) are often difficult or impossible to use fully in these environments. Furthermore, the individual user of these tools has little help in integrating the services offered in the environment at a semantic rather than syntactic level. Both the rigid inflexibility revealed in the inability to import new computing resources into these environments (or perhaps more revealing, the inability to embed the environment or tools in some other environment) and the inability to integrate the services semantically undermine the ability of these systems to scale up to the multiple users and very demanding requirements of the type of complex systems we described above.

We need to invent better methods of integration: ones that are at once more permissive and flexible and yet also more powerful semantically and ones that allow more formality in the analyses of the processes and products of integration.

We therefore start by drawing a few distinctions between different types of integration systems, based on the flexibility of the integration process and the types of infrastructure that these systems retain. In the next section of the paper, we argue for the importance of retaining infrastructure for integration based on our experience with very large complex systems, and describe our work on *wrapping*.

In the following section, we introduce some work on combining wrappings with text-based Virtual Reality environments to form a new kind of integration infrastructure, and in the last section, we discuss the requirements for formalisms that has emerged from working in these testbeds.

252   C. LANDAUER AND K. L. BELLMAN

1.1. WHAT ARE INTEGRATION SYSTEMS?

An integration system is a system of rules for: identifying some basic objects or other entities and combining them and their behaviors into larger entities (the entities may be active, as in logical inference rules or programs).

We include in the integration system not only the particular components and integrative mechanisms, but also all of the processes required to determine and describe the semantic connections between and among components, to analyze proposed conceptual structures for consistency and appropriateness for the integration task, and to produce the integrated system.

Particular systems need their own definitions for the kinds of entities and processes considered. For example, different kinds of integrating systems integrate data structures, process behaviors, or even logics.

We distinguish between two classes of integration systems: combining systems and coordination systems. We take "combining" systems to be those for which the coordinative information structure is "compiled in", so that the integration scaffolding, i.e., the techniques and processes used to select and combine the components, is removed. We generally expect this kind of integration to run faster, because there are fewer choices to make at run time. We call a system a "coordination" system when the amount of coordinative information structure that remains in the system is itself selectable as part of the integration process. These latter systems can be used as the integration scaffolding to produce systems at many levels of flexibility. It is for this reason that we concentrate on coordination systems. Our wrapping approach, presented in the next section, is an example of a coordination system.

Without attempting even a modest review, we believe that the following research areas would benefit from a theory of coordination systems, and are providing some of the basic research that will enable coordination systems.

Integrating knowledge-based systems and their corresponding ontologies (knowledge structures) is a difficult problem area (Genesereth, Fikes, 1991) (MacGregor, 1991) (MacGregor, Burstein, 1991) (Gruber, 1994), and one of the outstanding research problems in this area is to devise appropriate integration mechanisms for different ontologies, even when they are viewed simplistically as object class hierarchies.

Integrating different formal models of complex systems (Walter, Bellman, 1990) brings up a difficult mathematical question: if we have two different mathematically formal systems (i.e., models) that purport to describe important aspects of some external entity or system, then how can we learn anything in one model from some knowledge in the other? The two models depend on different assumptions and simplifications, and there is no reason *a priori* to assume that anything can be transferred. This integration problem is one of the several research questions discussed in our New Mathematical Foundations for Computer Science Initiative (Landauer, Bellman, 1994), which is intended to stimulate basic research in several areas of mathematics relevant to complex heterogeneous systems.

The integration of large simulation programs is an important problem currently facing many large organizations (Widman *et al.*, 1989) (Zeigler, 1990) (Fishwick, Modjeski, 1991) (Landauer, Bellman, 1992). The main difficulties lie in the integration of interaction timings and object behaviors. We believe that the wrapping approach, described below in Section 2, helps to organize the processing appropriate for these problems, and to make explicit the meta-knowledge required about each object behavior.

2. Wrappings as Coordination Systems

In this section, we give a brief overview of the technical details of the wrapping approach; all of them are expanded elsewhere. In a series of papers (Landauer, 1990) (Bellman, 1991) (Landauer, Bellman, 1992) (Landauer, Bellman, 1993) (Landauer, Bellman, 1995) (and others), we have argued for a new approach to the development, integration, and management of heterogeneous software systems, based on two kinds of software entities: *Wrapping Knowledge Bases* (WKBs) and *Problem Managers* (PMs). The WKBs contain explicit, machine-processable, qualitative information (called *wrappings*) about the system components, architecture, and all the computational or information processing elements (called *resources*) in the system: not just how to use them, but also whether and when and why and in what kinds of combinations they should or can be used. The PMs are algorithms that use the wrapping descriptions to determine which resources to use and how to combine them to apply to problems.

These ideas have proven to be useful, even when implemented and applied in informal, *ad hoc* ways in some of the applications that have made or are making use of wrappings even at this preliminary stage of development (Miller, Quilici, 1992) (Bellman, Reinhardt, 1993).

The wrappings provide an example of our notion of a coordination system, since the scaffolding (which we usually call infrastructure) remains with the program at run-time.

In a similar vein, there has recently been a strong call for retaining the scaffolding used to build theorem proving and other deductive systems (Talcott, 1994) (Giunchiglia *et al.*, 1994), to combat the difficulties with their construction and (especially) modification. We believe that these systems are also amenable to our methods, even though they are not usually described as collections of resources that share context and other information to effect a common purpose.

2.1. WRAPPING OVERVIEW

The wrapping theory has four basic features:

1. ALL parts of a system architecture are *resources*, including programs, data, user interfaces, and everything else;
2. ALL activities in the system are *problem study*, (i.e., applying a resource to a problem), including user interactions, information requests and announcements within the system, and service or processing requests;
3. *Wrappings* are explicit machine-processable descriptions of the various ways to use the resources by performing the five *Intelligent User Support* functions (Bellman, 1991) with them: Selection, Assembly, Integration, Adaptation, and Explanation.
4. *Problem Managers*, including the Study Managers and the Coordination Manager, are algorithms that process those wrappings. Since they are also resources, they are also wrapped.

The wrapping information and processes form expert interfaces to all of the different ways to use resources in a heterogeneous system. The most important features that wrapping brings to this approach are the the uniformity of treating everything in the system as resources, the uniformity of treating everything that happens in the system as problem study, and the reflection provided by treating the PMs as resources themselves. These uniformities are the coordinative processes for our desired flexibilities of resource selection and problem posing (it is a fundamental philosophical principle of ours that every flexibility must have a corresponding coordinative mechanism to manage it).

First, every part of the system is a *resource* that provides some kind of *information service*. This includes tools, functions, ordinary files, databases, programs, data, user interfaces, other communication interfaces, interconnection architectures, symbolic formula manipulation systems, scripts that refer to other resources (e.g., plans), and analysis tools that refer to other resources (e.g., parametric study), and everything else (Everything!). We think about *applying resources* instead of "invoking tools" because the resource being applied might not be the active part of that process.

Second, everything that happens in the system is the response to a *posed problem*. Since not all problems can be solved, we think of *studying* problems rather than solving them. Moreover, that allows the system to do more or less undirected explorations as it studies certain kinds of problems, so it can treat some problems as suggestions for study when appropriate, not as strict goals. Our notion of problems deals with context as an explicit part of the problem study process: there must be a problem context before posing a problem even makes sense. Therefore, problem study always occurs after a context is chosen and a problem is posed (we allow these choices to be made either by human users or by other programs as users). Therefore, instead of thinking about "issuing commands" to the system, we think about *posing problems* for the system. Then the wrapping processes find resources that can deal with the problems by studying them directly or decomposing them into collections of simpler problems. This *problem posing* interpretation of programs and systems allows the wrapping processes to mediate all problem study using the Wrapping Knowledge Bases. Instead of having direct calls between resources, we have the resources pose problems that correspond to service requests. Other resources announce information services that they provide and the interactions are all mediated through the Wrapping Knowledge Base.

Third, every resource has one or more *wrappings*, which are explicit machine-processable descriptions of the different ways to use the resource, including the different roles played by the Intelligent User Support functions in their use:

— *Selection* (which resources to apply to a problem),
— *Assembly* (how to let them work together),
— *Integration* (when and why they should work together),
— *Adaptation* (how to adjust them to work on the problem), and
— *Explanation* (why certain resources were or will be used).

A wrapping is not simply an interface "to" a resource; it is an interface to the "use" of a resource. There will be separate wrappings for different common uses of certain complex tools, especially analysis tools that have grown by accretion. Similarly, combinations of resources that often work together may have a single wrapping for the combination, in addition to separate wrappings for separate ways to use the resources by themselves.

Fourth, the *Problem Managers* (PMs) are algorithms that use the wrapping descriptions to collect and select resources to apply to problems. There is a distinguished class of PMs called the *Study Managers* (SMs) that coordinates the basic problem study process, and a specialized PM called the *Coordination Manager* (CM), which is a kind of basic "heartbeat" that drives all of the processing. The SMs mediate between the problem at hand and the wrappings to select and apply resources to the problem, and the CM cycles between posing problems and using the SM to study them.

2.2. WRAPPING PROCESSES

We concentrate on the processes that read and interpret wrappings in this paper, since they organize the information services provided by the resources. Problem study always occurs after a context is chosen and a problem is posed; these choices may be made either by human users or by other programs as users. The alternation between problem definition and problem study is organized by the *Coordination Manager* (CM), which is a special resource that coordinates the wrapping processes.

The basic problem study sequence is monitored by a resource called the *Study Manager* (SM). The Study Manager organizes problem solving into a sequence of basic steps, which we believe represent a fundamental part of problem solving. It is the default resource for the problem *Study Problem*, which is our version of what we do with problems that are already posed (Landauer, Bellman, 1992) (Landauer, Bellman, 1993). The SM performs the sequence of steps in Figure 1.

Interpret problem :
  Match resources : get list of candidate resources
  Resolve resources : reduce the list, make some bindings
  Select resource : choose one resource to apply
  Adapt resource : finish the bindings
  Advise poser : describe resource and bindings chosen
Apply resource : go do it
Assess results : evaluate

Figure 1.   Study Manager (SM) Step Sequence

To "Match resources" is to find a set of resources that might apply to the current problem in the current context. It is intended to allow a superficial first pass through a possibly large WKB. To "Resolve resources" is to eliminate those that do not apply. It is intended to allow negotiations between the problem poser and each wrapping of the resource to determine whether or not it can be applied, and make some initial bindings of formal parameters of resources that still apply. To "Select resource" is simply to make a choice of which of the remaining candidate resources (if any) to use. To "Adapt resource" is to set it up for the current problem and problem context, including finishing all required bindings. To "Advise poser" is to notify the problem poser (who could be a user or another part of the system) of what resource was chosen and how it was set up to be applied. To "Apply resource" is to use it for its information service, which either does something, presents something, or makes some information or service available. To "Assess results" is to determine whether the application succeeded or failed, and to help decide what to do next.

Find context : get a containing context from the user

loop:

> **Pose problem** : determine the current problem
>
> **Study problem** : use SM to do something about the problem
>
> **Present result** : to user

Figure 2.   Coordination Manager (CM) Step Sequence

The original context and problem come from the CM mentioned above, which has the steps listed in Figure 2, of which the middle one of the cycle is studied by the SM. This way of providing context and tasking for the SM is familiar from many interactive programming environments: the "Find context" part is usually left implicit, and the rest is exactly analogous to LISP's "read-eval-print" loop, with rather different processing at each step mediated by the SM. In this sense, the CM is a kind of "heartbeat" that keeps the system moving.

Up to this point in the description, the SM is just a (very) simple type of planning algorithm. Several additional design features make it a framework for something more. First, all of the wrapping processes are themselves wrapped. Second, the processing is completely recursive: "Match resources" is itself a problem, and is studied using the same SM steps, as are "Resolve resources", "Apply resource", and ALL of the other steps listed above. The simple form we described above is just the bottom of the recursion. These two things mean that every new planning idea that applies to a particular problem domain (which information would be part of the context) can be written as a resource that is selectable according to context; it also means that every new mechanism we find for adaptation or every specialization we have for application can be implemented as a separate resource and selected at an appropriate time.

This SM recursion is unusual, since it is a recursion in the "meta-" direction, not within the problem domain or within the planning process. The result is that planning the planning process itself is an integral part of the SM behavior, not a separate kind of function. The SM recursion also means that there are many layers of study context, each with its own problem context, problem poser, and problem specification.

This recursion has profound implications in other applications of wrapping (Landauer, Bellman, 1995), but for its application to software development, the fact that the SM can be selected is a key. Since the SM steps above that interact with the Wrapping Knowledge Base (WKB) are themselves posed problems, we can use completely different syntax and semantics for different parts of the WKB in different contexts, and select the appropriate processing algorithms according to context. In particular, whether one writes about one WKB or several is a matter of taste and viewpoint. Finally, the SM is only one of a family of processes called *Problem Managers* (PMs), each of which is wrapped and selectable according to context, and each of which can pose problems and organize their study in different ways. For example, the original context and problem are problems posed by the CM, since the CM steps: "Find context", "Pose problem", "Study problem", and "Present result", are all problems posed to the system and studied by the SM. The kinds of resources that might be chosen by the SM to apply to the first two problems would be menus or other requests to the human user to make a choice of context and problem, and the SM is the main resource that applies to the "Study problem" problem.

The recurrence of posing and studying problems is managed by the CM. The poser (i.e., any of the resources applied by the SM to the "Pose problem" problem) reads expressions from somewhere, as determined by context, and the SM interprets them. The poser has a parser (different problem posers may have different parsers), which reads text and makes symbol structures, within a particular context defined by the "Find context" step of the CM.

3. Multi-User Domains as Interaction Spaces

In this section, we describe our new vision for Multi-User Domains (MUDs), combining the infrastructure flexibility that a wrapping architecture provides with the processing styles of current MUDs to form a new kind of integration system called an *interaction space*.

Imagine reading a story set in some time or place. If the story is well-written, it will take the reader to a different situation, maybe a different character or even a different kind of character, regardless of whether the story is fiction or non-fiction. Stories can present information about a situation that is not apparent, or that can only be learned through experience of the situation. Even primarily visual media like movies present this kind of experiential information through words. If a MUD is well-designed, it is like a well-written story in its power to transport the player to a different situation, but it has three other important features. It is interactive, which means that the reader can affect the behavior and outcome of the story, so in particular, the reader can explore the story in many different ways or directions. It is also multiple-person, so that the reader can interact with other readers, and moreover, it has the added feature that parts of the story are still being written. Both humans and computer programs enter these worlds and act within them as a distinct characters with names,

description and behaviors. A simple command language is provided by all
the client programs for MUDs that allows one to move around and act and
talk within the virtual world. There is also a simple construction language
in many MUDs that makes it easy for a player to immediately become a
builder (a creator) in that world.

MUDs are one approach to providing some richness of interaction among
programs: import people into the program, so that they can generate interesting interaction behaviors. These MUDs have become enormously popular
as games and as educational support tools over the last few years because they get the human interactions right in some fundamental sense, and
because they engage our sense of "place" (Bartle, 1990) (O'Brien, 1992)
(Riner, Clodius, 1994). MUD clients and servers are easy to obtain and
run (most servers and clients are free), but they usually only provide text
worlds; there is little interaction with existing tools that are outside the
MUD; though some have construction languages that allow complex programming, it is the usual kind of programming; and it is not very easy to
access large volumes of information.

3.1. MULTI-USER VIRTUAL ENVIRONMENTS

A Multi-User Virtual Environment (MUVE) is an interaction-based virtual
reality program, in which multiple users share a common "Virtual World",
for conversation, cooperation, and other activities. The users are both humans and computer programs.

A sense of location is provided directly by the *Virtual World*, which
is divided into a set of locations called *locales*. The primitive locales are
called *locations*. They correspond to rooms in MUDs, and provide explicitly
bounded scope for interactions.

Characters are essentially the same as in MUDs. An *agent* is an active
object that can respond to and issue commands. All characters are agents,
but the converse is not true.

The search path for trying to match commands issued by an agent is
defined by the set of locales containing the agent's current location. The set
of available commands comes from the objects in those locales. Neither the
matching algorithm nor the particular phrases to match are fixed. Every
object in a locale that contains the agent's location provides its parser
to the matching and selection process. Thus the matching algorithm is
heterogeneous, not just variable, since different objects can use different
styles of parser.

Everything is an object: things, agents, characters, rooms, connections,
exits, groups, etc., and it is assumed (according to the processing algorithms
described later) that each parser has both a preliminary matching part and

a final resolving part. If there are not very many resources, then the matcher can be trivial.

3.2. OBJECT BEHAVIOR IN MUDS

Many styles of MUD now allow interesting object behavior, but all of them are fundamentally limited by their choice of behavior definition language. The one used by TinyMUDs (Applegate *et al.*, 1991) (DMAuthors, 1989) (O'Brien, 1992) (Riner, Clodius, 1994) is much less a programming language in the usual sense than it is an application domain specific notation for the domain of building TinyMUD objects. This fact may seem like a disadvantage compared to other MUD languages (especially from the language technology point of view), but this simplicity is its great strength. People who want to build can, regardless of their conventional programming experience. Most servers have a more powerful programming language than TinyMUDs do, and some servers, particularly MOOs (for MUD Object-Oriented (Curtis, 1992) (Curtis, Nichols, 1993), of which the most prominent example is LambdaMOO (Curtis, White, 1995)), and also CoolMUD (White, 1994) and ColdMUD (Hudson, 1994) (Gillespie, 1995), have much more powerful object-oriented programming languages, with all attendant advantages in expressiveness and behavioral richness and variety, and all attendant disadvantages in debugging, unexpected behavior, and inertia.

3.3. OBJECT BEHAVIOR IN MUVES

Our proposed change is to make the entire recognition process much more explicit and to separate out the basic steps of the algorithms, so we can deal with them explicitly also. We treat every object as a resource, and every interaction item as a posed problem, with the source location as part of the context that limits the applicability of resources.

The computational steps are shown in Figure 3. They should look familiar, since they are almost the same as the steps of the simplest wrapping SM. We implement the step sequence as an alternative SM, called the *MUVE* SM to distinguish it from the other SMs. As with all of the SMs, each step is in turn a posed problem, which is studied by a wrapping SM according to context. Therefore, we can invent many different selectors, resolvers, and other resources for each of these steps. As long as this MUVE SM keeps track of enough context information to know which one to use when it matters, we have an enormous gain in flexibility right at the heart of the MUVE server processing.

This simple change of replacing the matching algorithm with wrappings provides a far-reaching change in the nature of the servers: independence from a particular construction language. There is still a difficult problem

Interpret problem :

> **Match resources** : use a filter for preliminary object matching, including source location, to reduce the time spent matching (parsing)
>
> **Resolve resources** : use the object recognizers for parsing, to see which ones can recognize the posed problem
>
> **Select resources** : select appropriate resources
>
> **Adapt resources to problem** : set parameter bindings for resource application
>
> **Advise problem poser** : note selected resource applications and reasons for selection

Apply resources : feed the structure that results from parsing to the resource and let it do what it does (if there is interference among resources here, then select needs to be more interesting)

Assess results : evaluate

Figure 3.   MUVE SM Step Sequence

of defining how the different semantics of different construction languages should interact, but the wrapping mechanism and its use of context allows expressions to be disambiguated, so there need be no syntactic confusion. This independence means that different locations may imply different kinds of interaction and behavior, so that programs with very different assumptions about their environment can be integrated.

3.4. INTERACTION SPACES

With the MUVE servers, we can define a new way for computer programs to be integrated. *Interaction spaces* are computer-mediated virtual places in which computer programs and humans meet. The basic idea is that when computer programs are *embodied*, there is an entirely different level of processing required (Bellman, Goldberg, 1984) (Bellman, Walter, 1984) (Maes, 1990), and that explicit attention to using co-experience as an integration mechanism can lead to very different styles of interaction among computational agents and humans, and therefore possibly much more interesting program structures and behaviors. There has been some progress on these kinds of programs, which are generally called 'bots (short for robots) in the MUD communities, and softbots in the computer science community, but we know much less about these embodied programs than we do about making programs interact as objects.

The main feature of MUDs that we use here is that users can interact with each other directly through the primitive "say" command, which is uninterpreted by the server. That fact allows new user and agent collections with new interaction protocols to be integrated with no changes to the server at all. These external users that are programs may be written in languages that do not have parsers in the MUVE, and they need only a rudimentary ability to interact with the MUVE (we believe that the basic metaphor of movement and action will be common across most construction languages).

We use wrappings to combine humans and computer programs as users, all of whom share a common virtual environment. In other words, we are combining the *interactions* among the different kinds of users. To construct these wrappings, we need to define the information services provided by the various programs, and the kinds of information and protocols used. This information is generally available for interesting programs, though it is not usually in an explicit machine-processible form. We believe that MUVEs will help us create the testbeds in which protocols, man-machine interfaces, coordination among multiple users and resources, and the processing of contextual information may be studied in a more explicit and enriched fashion.

4. Conclusions

Integration systems are general mechanisms for allowing many individual systems to cooperate in a shared context. Our research has developed a particular approach to coordination systems called "wrapping", and combined it with MUDs to make a new way of integrating software: interaction spaces. Wrappings allow many special case processes to be combined with general case methods, and MUDs allow multiple programs to be connected together with humans as users.

We are only beginning to study interaction spaces as a style of integration, but we believe that they are widely applicable. The MUVE advances our work by creating the testbeds where we can explicitly study the relationship between different contexts (rooms) and different resources (programs personified as character objects). It also allows us to explore fascinating new strategies for integrating resources and human users.

Integration is importantly an issue of defining suitable relationships among components under new contexts. Above all else, an "integration science" must have the formal basis and the techniques for dealing with the representation and processing of context information. Context takes the initial component to be integrated into a system and reinterprets, changes, and biases it, whatever its processing, use, or goal.

What is context? For now let us call the operational setting, the manner of use, and the services required by a component all parts of its context. The word "context" like the word "relationship" is a slippery term because it is so dependent upon the type of component and what its use will be within a system. The way that a system is partitioned into components by its developers or users often reflects implicit assumptions about the "context" for those components, i.e., who will use them for what. One of the lessons we have learned from our wrapping approach is that this context information must be made explicit, so that we can suitably "integrate" system components for their purposes. Our wrapping experiences have also taught us a great deal about what type of context information needs to be made available and how one can process that information to help the users and developers of the systems to select, integrate, and adapt components.

The wrappings approach may seem more like modeling than integration, because "integration" in the information sciences has long been associated with "construction" processes, rather than the development of conceptual structures as in modeling. Indeed, the goal for integration in the United States is often likened to providing "tinkertoys" or "LEGO blocks" for system construction. Our point here is that "integration" involves more than the process of choosing the appropriate components, parameterizing them as possible, and checking for violations of construction and relational constraints among components. These construction activities are very important to integration, but in addition, integration does involve the formation of new concepts, and this is akin to modeling activities in general. Integration involves two key modeling activities: one is forming the overall concept for some new combination of components, and the second, related to the first by the requirements for semantic integration, has to do with reconciling and creatively integrating the multiple views – assumptions, goals, meanings for terms, reasons for using certain methods – that accompany each of the different resources (both components in the usual sense and also the different languages and mathematical formalisms that may have been used in developing those components). The implications of these admittedly philosophical points for the integration formalisms currently under development are far-reaching: the techniques must be able to represent and reason about much more than searching limited solution spaces and glorified constraint checking. Rather, they must be able to represent and handle a diversity of logics and methods; they must be able to handle truly different levels of abstraction; they must be able to help in the formations of new "theorems" or concepts about the overall system, creatively fused from the rules for the different components; and they must be able to help analyze the correctness and consistency of the underlying multiple viewpoints. We are hoping that our wrapping approach, with its emphasis on the collection and processing

of explicit metaknowledge for different integration goals, will be useful in a
growing dialog about how we as a community can build towards a better
and more principled approach to integrating our large systems. The need
is certainly there and the challenge is before us.

CHRISTOPHER LANDAUER
The Aerospace Corporation
Herndon, Virginia, USA

AND

KIRSTIE L. BELLMAN
Advanced Research Projects Agency
Arlington, Virginia, USA

References

David Applegate, James Aspnes, Timothy Freeman, and Bennet Yee, *TinyMUD*,
source code available via anonymous ftp from ftp.math.okstate.edu, in directory
/pub/muds/servers/, in file tinymud-1.5.4.tar.Z (availability last checked 16 October 1995)

Richard Bartle, "Interactive Multi-User Computer Games", MUSE, Ltd. (December 1990), available via anonymous ftp from parcftp.parc.xerox.com, in directory
/pub/MOO/papers/, in files mudreport.* (availability last checked 16 October 1995)

Kirstie L. Bellman, "The Modelling Issues Inherent in Testing and Evaluating Knowledge-based Systems", *Expert Systems With Applications Journal*, **Vol. 1**, pp. 199–215 (1990)

Kirstie L. Bellman, "An Approach to Integrating and Creating Flexible Software Environments Supporting the Design of Complex Systems", pp. 1101–1105 in *Proceedings of WSC '91: The 1991 Winter Simulation Conference*, 8-11 December 1991, Phoenix, Arizona (1991); revised version in Kirstie L. Bellman, Christopher Landauer, "Flexible Software Environments Supporting the Design of Complex Systems", *Proceedings of the Artificial Intelligence in Logistics Meeting*, 8-10 March 1993, Williamsburg, Va., American Defense Preparedness Association (1993)

Kirstie L. Bellman and April Gillam, "A knowledge-based approach to the conceptual design of space systems", pp. 23–27 in *Proceedings of the 1988 SCS Eastern Multi-Conference*, March 1988, The Society for Computer Simulation (1988)

Kirstie L. Bellman and April Gillam, "Achieving Openness and Flexibility in *VEHICLES*", pp. 255–260 in *Proceedings of the SCS Eastern MultiConference*, 23-26 April 1990, Nashville, Tennessee, Simulation Series, **Vol. 22, No. 3**, SCS (1990)

Kirstie L. Bellman, April Gillam, and Christopher Landauer, "Challenges for Conceptual Design Environments: The VEHICLES Experience", *Revue Internationale de CFAO et d'Infographie*, Hermes, Paris (September 1993)

Kirstie L. Bellman and Lou Goldberg, "Common Origin of Linguistic and Movement Abilities", *American Journal of Physiology*, **Vol. 246**, pp. R915–R921 (1984)

Dr. Kirstie L. Bellman, Captain Al Reinhardt, USAF, "Debris Analysis Workstation: A Modelling Environment for Studies on Space Debris", *Proceedings of the First European Conference on Space Debris*, 5-7 April 1993, Darmstadt, Germany (1993)

Kirstie L. Bellman and Donald O. Walter, "Biological Processing", *American Journal of Physiology*, **Vol. 246**, pp. R860–R867 (1984)

Richard Bellman, P. Brock, "On the concepts of a problem and problem-solving", *American Mathematical Monthly*, **Vol. 67**, pp. 119–134 (1960)

David D. Clark, "The Design Philosophy of the DARPA Internet Protocols", pp. 106–114 in *Proceedings ACM SIGCOMM 1988 Symposium on Communications Architectures and Protocols*, 16-19 August 1988, Stanford, California (1988)

John P. Crane, *DragonMUD*, accessible via telnet (or other MUD clients) at tinylondon.ucsd.edu 4201 (availability last checked 20 December 1995)

Pavel Curtis, "Mudding: Social Phenomena in Text-Based Virtual Realities", in *Proceedings 1992 Conference on Directions and Implications of Advanced Computing*, Berkeley, May 1992, also available as Xerox PARC technical report CSL-92-4, also available via anonymous ftp from parcftp.parc.xerox.com, in directory /pub/MOO/papers/, in files DIAC92.* (availability last checked 16 October 1995)

Pavel Curtis and David A. Nichols, "MUDs Grow Up: Social Virtual Reality in the Real World", available via anonymous ftp from parcftp.parc.xerox.com, in directory /pub/MOO/papers/, in files MUDsGrowUP.* (availability last checked 16 October 1995)

Pavel Curtis, Stephen White, *LambdaMOO*, accessible via telnet or other clients at lambda.parc.xerox.com 8888, source code available via anonymous ftp from ftp.parc.xerox.com, in directory /pub/MOO/, in files LambdaMOO-1.7.8p4.tar.Z and others (availability last checked 16 October 1995)

Paul A. Fishwick, Richard B. Modjeski (eds.), *Knowledge-Based Simulation*, Springer (1991)

Michael R. Genesereth, Richard E. Fikes, "Knowledge Interchange Format Version 2 Reference Manual", Logic-91-3, Stanford U. Logic Group (1991)

Brandon Gillespie *ColdX*, source code available via WWW at URL ftp://sticky.usu.edu/ pub/brandon/coldmud/drivers/ (availability last checked 16 October 1995)

F. Giunchiglia, P. Pecchiari, C. L. Talcott, "Reasoning Theories: Towards An Architecture for Open Mechanized Reasoning Systems", IRST Technical Report 9409-15 and Stanford University Technical Note STAN-CS-TN-94-15 (December 1994); also available by anonymous ftp from sail.stanford.edu, file /pub/MT/94omrs.ps.Z

Tom Gruber, "Knowledge Sharing Effort public library", abstracts on World-Wide Web page, http://www-ksl.stanford.edu/knowledge-sharing, papers accessible via anonymous ftp, from ksl.stanford.edu, in directory /pub/knowledge-sharing/papers (availability last checked 16 October 1995)

Greg Hudson, *ColdMUD*, source code available via anonymous ftp from ftp.math.okstate.edu, in directory /pub/muds/servers/, in file coldmud-0.10.2.tar.gz (availability last checked 16 October 1995); newer versions available via WWW at URL http://web.mit.edu/ afs/sipb/project/coldmud/ release/coldmud-0.10/coldmud-0.10.tar.gz, documentation at URL http://web.mit.edu/ afs/sipb/project/coldmud/ doc/coldmud.ps (availability last checked 16 October 1995)

Christopher Landauer, "Wrapping Mathematical Tools", pp. 261–266 in *Proceedings of the 1990 SCS Eastern MultiConference*, 23-26 April 1990, Nashville, Tennessee, Simulation Series, **Vol. 22, No. 3**, SCS (1990); also pp. 415–419 in *Proceedings of the 22nd Symposium on the Interface (between Computer Science and Statistics)*, 17-19 May 1990, East Lansing, Michigan (1990)

Christopher Landauer, Kirstie L. Bellman, "Integrated Simulation Environments" (invited paper), *Proceedings of DARPA Variable-Resolution Modeling Conference*, 5-6 May 1992, Herndon, Virginia, Conference Proceedings CF-103-DARPA, published by RAND (March 1993); shortened version in Christopher Landauer, Kirstie Bellman, "Integrated Simulation Environments", *Proceedings of the Artificial Intelligence in Logistics Meeting*, 8-10 March 1993, Williamsburg, Va., American Defense Preparedness Association (1993)

Christopher Landauer, Kirstie L. Bellman, "The Role of Self-Referential Logics in a Software Architecture Using Wrappings", *Proceedings of ISS '93: the 3rd Irvine Software Symposium*, 30 April 1993, U. C. Irvine, California (1993)

Christopher Landauer, Kirstie L. Bellman, "New Mathematical Foundations for Computer Science", Initiative announcement available via anonymous ftp from aero-space.aero.org, in directory /pub/newmath, in file workbook.html, and from WWW at URL http://www.cs.umd.edu/~cal/newmath.html (original July 1994), revision 1.4 (February 1995) (availability last checked 16 October 1995)

Christopher Landauer, Kirstie L. Bellman, "The Organization and Active Processing of Meta-Knowledge for Large-Scale Dynamic Integration", pp. 149–160 in *Proceedings 10th IEEE International Symposium on Intelligent Control, Workshop on Architectures for Semiotic Modeling and Situation Analysis in Large Complex Systems*, 27-30 August 1995, Monterey (August 1995)

Pattie Maes (ed.), Special Issues of *Robotics and Autonomous Systems*, **Vol. 6, Nos. 1 and 2** (June 1990); reprinted as Pattie Maes (ed.), *Designing Autonomous Agents: Theory and Practice from Biology to Engineering and Back*, MIT / Elsevier (1993)

Pattie Maes, D. Nardi (eds.), *Meta-Level Architectures and Reflection, Proceedings of the Workshop on Meta-Level Architectures and Reflection*, 27-30 October 1986, Alghero, Italy, North-Holland (1988)

Robert MacGregor, "The Evolving Technology of Classification-Based Knowledge Representation Systems", pp. 385–400 in John F. Sowa (ed.), *Principles of Semantic Networks: Explorations in the Representation of Knowledge*, Morgan Kaufman (1991)

Robert MacGregor, Mark H. Burstein, "Using a Description Classifier to Enhance Knowledge Representation", *IEEE Expert*, **Vol. 6, No. 3**, pp. 41–46 (June 1991)

Lawrence H. Miller, and Alex Quilici, "A Knowledge-Based Approach to Encouraging Reuse of Simulation and Modeling Programs", in *Proceedings of SEKE'92: The Fourth International Conference on Software Engineering and Knowledge Engineering*, IEEE Press (June 1992)

Mike O'Brien, "Playing in the MUD", Ask Mr. Protocol Column, *SUN Expert*, **Vol. 3 No. 5**, pp. 19–20, 23, 25–27 (May 1992)

R. Riner and J. Clodius, "Simulating Future Histories", Anthropology and Education Quarterly (Fall 1994)

Carolyn L. Talcott, "Reasoning Specialists Should Be Logical Services, Not Black Boxes", pp. 1–6 in *Proceedings of CADE-12 workshop on Theory Reasoning in Automated Deduction*, 26 June - 1 July, Nancy (1994); also available through anonymous ftp from sail.stanford.edu, in file /pub/MT/94cade-tr-work.ps.Z

Donald O. Walter, Kirstie L. Bellman, "Some Issues in Model Integration", pp. 249–254 in *Proceedings of the SCS Eastern MultiConference*, 23-26 April 1990, Nashville, Tennessee, Simulation Series, **Vol. 22, No. 3**, SCS (1990)

Stephen White, *CoolMUD*, source code available via anonymous ftp from ftp.math.okstate.edu, in directory /pub/muds/servers/, in file coolmud2.1.4.tar.gz (availability last checked 16 October 1995)

Lawrence E. Widman, Kenneth A. Loparo, Norman R. Neilson (eds.), *Artificial Intelligence, Simulation and Modeling*, Wiley (1989)

Bernard P. Zeigler, *Object-Oriented Simulation with Hierarchical, Modular Models*, Academic Press (1990)

EVELINA LAMMA, MICHELA MILANO, AND PAOLA MELLO

COMBINING SOLVERS IN A META CONSTRAINT LOGIC PROGRAMMING ARCHITECTURE

Abstract. We present a general technique for the combination and the integration of different Constraint Logic Programming (CLP) solvers. The main idea behind the work concerns the possibility of building meta CLP architectures by adding CLP solvers in a natural and effective manner. In the meta architecture, levels are constraint solvers each reasoning on constraints of the underlying system. The architecture presented starts from a meta Constraint Logic Programming general scheme. A distinguishing feature of the architectural scheme concerns its operational semantics which can be seen as a general combination method for data and control of two constraint solvers. A set of linking rules define how systems exchange data, while a set of transition rules define how systems combine their control flow. We propose a specialization of a meta CLP architecture on finite domains. The specialization concerns the possibility of combining qualitative and quantitative reasoning in a CLP framework. This combination can be useful, for example, in the field of temporal reasoning.

1. Introduction

Many Artificial Intelligence fields such as temporal reasoning, scheduling and planning require the combination of different kinds of constraint solving techniques. On one hand, the Constraint Logic Programming (CLP) paradigm[1] provides efficient constraint handling mechanisms, while maintaining the natural declarative semantics of Logic Programming (Kowalski, 1979; Lloyd, 1987). On the other hand, CLP solvers manage only one class of constraints. Therefore, it can be useful to integrate various constraint solving techniques.

---

[1] For a survey on CLP see (Jaffar & Maher, 1994).

F. Baader and K.U. Schulz (eds.), Frontiers of Combining Systems, 267–283.
© 1996 *Kluwer Academic Publishers.*

In the field of Logic Programming (LP) (Lloyd, 1987) several efforts have been performed in order to develop meta-systems dealing with the LP provability relation (Aiello *et al.*, 1986; Bowen & Kowalski, 1982). Thank to this technique, expressive power of LP is augmented in a very flexible and modular way. While in the LP setting meta-programming techniques are now well-known and widely accepted, this is not true for CLP. The aim of this paper is to apply meta-programming to the CLP framework, and show that meta-programming is a suitable technique for combining various constraint solvers. In LP, the meta-level system treats the clauses of the object-level program as data structures. In our approach to meta CLP, we focus on the constraint solving mechanism and each meta-system treats the constraints of the underlying system as data structures in order, for instance, to change the propagation method or to combine different constraint solving techniques.

This idea greatly extends the expressive power of CLP, and allows us to face those applications where a single CLP solver expressiveness does not suffice. For example, CLP uses constraints in order to restrict the domain of variables (performing what is called quantitative reasoning), but it is not able to reason and to infer information on constraints such as determine the tightest constraint which holds between two variables, or compose and intersect constraints (what we call qualitative reasoning). We show that a meta CLP approach can solve these problems, thanks to the capability of reasoning on object-level constraints.

The way we combine different constraint solvers (levels) is described by the operational semantics. When combining two systems we focus on two problems: exchanging the data, i.e., find a way of combining knowledge and information, and synchronize the control, i.e., when and how the two systems exchange information. The first problem is solved by defining a set of linking rules for exchanging data (constraint stores), while the second is solved by defining a set of transition rules that specify the control flow between two levels. In this way, we maintain the identity of the constraint solvers and overcome the problems that derive from the combination of constraint domains in a flat, single level architecture.

An attractive feature of the architecture concerns the very easy implementation of specializations. In this paper, we propose a specialization of a meta CLP on finite domains (CLP(FD), see (Jaffar & Maher, 1994)) architecture: it concerns the possibility of combining in a CLP framework the quantitative and qualitative reasoning. In this specialization, the object level is a constraint solver on finite domains of integers performing quantitative reasoning, while the meta level is a constraint solver on finite domain of sets of constraints of the underlying system. By suitably handling these constraints, the meta-level is able to perform qualitative reasoning. For example, if we have at the object level the constraints $X < Y$, $Y < Z$ and $X \leq Z$, the meta-level treats these constraints as data structures, infers the tightest constraint linking X and Z and then enforces the constraint $X < Z$ at the object level.

The paper is organized as follows: in Section 2 the Meta Constraint Logic Programming general scheme is introduced. Section 3 describes the operational semantics of the object-level, the meta-level and their combination. In Section 4, we present a specialization of the general scheme concerning qualitative reasoning. In Section 5 a simple example is discussed. Related works are presented in Section 6. Conclusions and future works follow.

2. Meta CLP General Scheme

Each CLP language can be seen as an instance of a general CLP(X) scheme where each instance can be specified by instantiating the parameter X (for more details see (Jaffar & Maher, 1994)). This parameter X represents a 4-tuple $(\Sigma, \mathcal{D}, \mathcal{L}, \mathcal{T})$ where Σ is a signature, $\mathcal{D}$ is a Σ-structure, $\mathcal{L}$ is a class of Σ-formulas and $\mathcal{T}$ is a first-order Σ-theory. The signature Σ represents the predefined predicates and function symbols, $\mathcal{D}$ is the domain structure, $\mathcal{L}$ is a class of constraints which can be expressed in the domain and $\mathcal{T}$ is the axiomatization of some properties of $\mathcal{D}$. Each CLP language can be defined by specifying all the 4-tuple components. Each CLP system is determined by the constraint domain $(\mathcal{D}, \mathcal{L})$ and by the operational semantics.

We are interested in a general technique for combining homogeneous or heterogeneous constraint solvers. A particular way of combining systems is to organize them in a hierarchy where higher levels reason and act upon lower levels. This organization leads to the development of meta-systems. A meta-system (Maes, 1987) is a computational system that has as its domain another computational system, called its object-system. Therefore, a meta-system is a system reasoning and acting upon another computational system which represents the object-system in its data. Meta-level architectures can be classified in two categories: *amalgamated* and *separated* (Aiello *et al.*, 1995). In *amalgamated* systems there is no distinction between object and meta systems: they are based on a single theory. In *separated* systems, sentences of the object and meta-levels are kept separated. The object theory is a first order theory while the meta theory contains sentences on the object theory. In this paper we present a separated approach to meta systems. One of the main advantages of separating meta and object levels concerns the avoidance of paradoxes originating from the self-reference.

In this paper, we are interested in *separated* CLP meta-systems where each system treats the constraints of the underlying level as data-structures.

In our meta-architecture, each meta-level reasons and applies propagation mechanisms on constraints of the underlying level. The constraint store of each meta level maintains a representation of the object level constraint stores but the two levels are kept separated.

We propose a meta CLP general scheme composed by a number $n \geq 1$ of modules. Each module presents the structure $(\Sigma, \mathcal{D}, \mathcal{L}, \mathcal{T})$. In our scheme, each level reasons on the constraints of the underlying level. In particular, the constraint store of the $n + 1\text{-}th$ level contains an *implicit* or *explicit* representation of constraints of the n-th level. For each constraint at level n, we have a meta-variable at level $n + 1$ which represents (explicitly or implicitly) this relation.

The representation of the meta-signature should be defined according to the application we have to solve. In Section 4 we specialize this general framework in order to integrate two constraint solvers on finite domains. The meta-signature symbol representation is given in terms of constraint symbols of the underlying system, thus providing an implicit representation.

3. Operational Semantics

The operational semantics of a CLP system can be defined in terms of transition rules[2]. These transitions can be specialized for a particular domain, by defining the operations that some transitions involve.

In our meta architecture, for each level we maintain the operational semantics of CLP solvers as presented in (Jaffar & Maher, 1994) (see section 3.1). For the interaction among levels, the combination of data (constraint stores) is defined in terms of *upward* and *downward* linking rules (see section 3.2), while the combination of control flows is defined in terms of transition rules (see section 3.3).

In this paper, we focus on two levels[3] but the semantics can be generalized for a number $n > 0$. In fact, some applications can benefit by a n level architecture, see (Lamma et al., 1995c).

3.1. OPERATIONAL SEMANTICS OF A SINGLE COMPONENT

The operational semantics of a constraint system can be defined in terms of transition rules which allow the system to switch from different states, i.e., the system performs the so called "top-down execution" (Jaffar & Maher, 1994).

---

[2]There are other definitions for the operational semantics of CLP, see (Hentenryck & Deville, 1990).

[3]In the following sections we will refer to two related components as the object and the meta system.

States are represented by tuples $\langle A, C, S \rangle$ where A is a multiset of atoms and constraints, C is the constraint store of active constraints, i.e., constraints that are *awake*, and S is the constraint store of passive constraints, i.e., constraints that are *asleep*. The failure of the system is represented by a state called *fail*. The transition system is very general but can be parameterized by defining the set of active and passive constraints, a predicate *consistent* and a function *infer* tailored on the particular domain.

The first transition is denoted by $\rightarrow_r$, where r stands for resolution:

$$\langle A \cup a, C, S \rangle \rightarrow_r \langle A \cup B, C, S \cup (a = h) \rangle$$

if a is an atom selected by the computation rule and the program P contains the rule $h \leftarrow B$, renamed to new variables, and h and a have the same predicate symbol. Otherwise, $\rightarrow_r$ leads to a failure[4].

The second transition is denoted by $\rightarrow_c$, where c stands for constraint, and introduces constraints into the passive constraint store:

$$\langle A \cup c, C, S \rangle \rightarrow_c \langle A, C, S \cup c \rangle$$

where c is a constraint selected by the computation rule.

The third transition rule, denoted by $\rightarrow_i$, where i stands for infer, infers new active constraints and modifies (generally simplifies) the passive constraints starting from the current collections of constraints C and S with the function $(C', S') = infer(C, S)$:

$$\langle A, C, S \rangle \rightarrow_i \langle A, C', S' \rangle.$$

The last transition rule, denoted by $\rightarrow_s$, where s stands for satisfiability, tests whether the active constraints are consistent with the predicate $consistent(C)$. If $consistent(C)$ holds the transition is:

$$\langle A, C, S \rangle \rightarrow_s \langle A, C, S \rangle$$

otherwise, the system fails.

3.2. INTERACTIONS BETWEEN TWO COMPONENTS: KNOWLEDGE COMBINATION

An interesting feature of the system concerns the interaction between two levels in terms of knowledge combination. A change in the meta-level constraint store should be reflected in the object-level and vice versa. Therefore, some *linking rules* between the two systems should be defined.

---

[4]The system switches to the state *fail*.

In each Constraint Logic Programming system we identify active and passive constraint stores. The operation $consistent(C)$ analyses only the active constraint store. Passive constraints are used for inferring new information. Some constraint systems (e.g., CLP(FD)) do not reason on passive constraints but many application, e.g., qualitative reasoning, need this kind of reasoning. Our system is organized in such a way that the meta system reasons on passive (and possibly active) constraints of the underlying system. Therefore, the active meta constraint store contains a representation of the passive object constraint store. For example, in a finite domain solver, if the passive object constraint store contains the constraint $X \leq Y$, the meta active constraint store should contain one possible representation in terms of a meta variable XY whose domain contains the constraint symbols linking the object level variables X and Y. We combine the constraint stores of the two systems, i.e., the object- and meta-active and passive constraint stores (referred to as C_{obj}, C_m, S_{obj}, S_m respectively) relying on *upward* and *downward* linking rules.

The *downward* linking rule assures that knowledge in the meta system is reflected in the object system. Each change in the meta constraint stores concerns the constraints of the object system. Therefore, the object level constraint store has to be modified (extended or simply changed) each time a meta-level active constraint changes. This propagation is formalized by the following linking rule which reflects down the constraints of the meta level and produces new object level passive constraints:

$$\frac{C_m \cup S_m}{C_{obj} \cup S_{obj} \cup S_{obj-m}}.$$

The constraint store $S_{obj-m} = reflect(C_m, S_m)$ is derived from the translation of the meta constraints in object level constraints. As happens for *infer*, the function *reflect* should be specialized on a particular domain. Generally speaking, *reflect* has the following structure. First, it builds a representation of constraints $C_m \cup S_m$, and selects a constraint c from this store. The constraint c is transformed in a set containing object constraints $\{c_1 \ldots c_n\}$ which are finally added to S_{obj-m}. In order to specialize the operation *reflect*, we have to define precisely the selection rule and how each meta constraint is transformed in a (possibly empty) set of object constraints.

The *upward* linking rule concerns the propagation in the meta system when object level constraints are modified. This transition generates meta-constraints starting from object level constraints and adds the new meta constraints to the meta constraint store

$$\frac{C_{obj} \cup S_{obj}}{C_m \cup S_m \cup S_{m-obj}}.$$

The constraint store $S_{m-obj} = reify(S_{obj}, C_{obj})$ is obtained by the reification of the object level constraints into meta constraints. The function $reify$ should be specialized on a particular domain. The operation $reify$ has the very similar structure of $reflect$. In fact, it builds a representation of constraints $C_{obj} \cup S_{obj}$, and selects a constraint c from this store. The constraint c is transformed in a set of meta-constraints $\{c_1 \ldots c_m\}$ which are finally added to S_{m-obj}.

It is worth mentioning that in both transitions the new active and passive constraints S_{m-obj} and S_{obj-m} are first added to the passive constraint stores. The transition $\rightarrow_i$ of each system provides the movement of active constraints in the active constraint store. This choice is motivated by a uniform treatment of constraints. In a constraint solver during the transition $\rightarrow_c$ (section 3.1), when a new constraint is encountered, it should be first added to the passive constraint store. Then the transition $\rightarrow_i$ moves the active constraints from the passive to the active set and performs the propagation.

3.3. TOP-DOWN EXECUTION: CONTROL FLOW COMBINATION

In this Section we define our system behavior by providing one or more sequences of operators (transitions) to be applied to the states of the system which lead it to either a solution or a failure. These transitions describe how and when the two systems combine their control flows.

A derivation is a sequence of transitions:

$$\langle A_1, C_1, S_1 \rangle \rightarrow \langle A_2, C_2, S_2 \rangle \ldots \rightarrow \langle A_i, C_i, S_i \rangle \rightarrow \ldots$$

where $\rightarrow$s are one of the transitions described in Section 3.1. The initial state depends on the program to be executed. This meta architecture is transparent to the user which writes a program as if the architecture contains only one level. Usually, the syntax of the programs is the same as the first level while queries are much more powerful because they involve also meta-level computation. Therefore, the initial state of the first level is $\langle A, \emptyset, \emptyset \rangle$, where A is the initial multiset of atoms and constraints, and the initial state of each meta constraint solver is $\langle \emptyset, \emptyset, \emptyset \rangle$[5]. A state which cannot be rewritten further is called *final state*. The *final state* succeeds if it has the form $\langle \emptyset, C, S \rangle$. A derivation fails if it is finite and the final state is *fail*. A derivation flounders if it is finite and the final state has the form $\langle A, C, S \rangle$ where $A \neq \emptyset$.

---

[5]There is the possibility of writing special purpose programs for other levels and their initial state in this case is $\langle A_m, \emptyset, \emptyset \rangle$, where A_m is the initial multiset of meta-atoms and constraints. We have defined a possible syntax for meta systems. It is described in (Lamma *et al.*, 1995b).

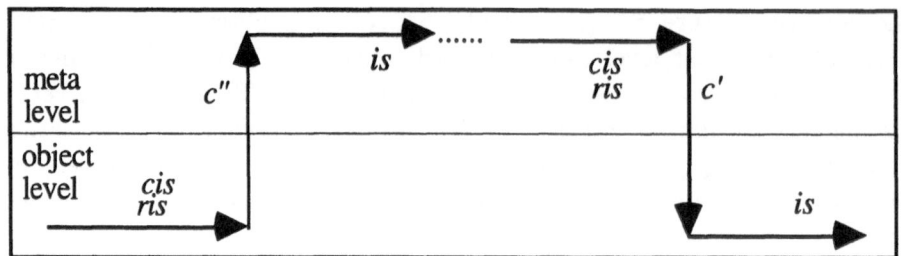

Figure 1.   Transitions between the object and the meta systems

In the meta system a final state succeeds if for each level it has the form $\langle \emptyset, C, S \rangle$. If a derivation fails at some level, the overall system fails. A derivation flounders if each level is in a floundering state.

An interesting point concerns the transitions from the meta to the object level and vice versa corresponding to the *upward* and *downdard* linking rules. Roughly speaking, transition rules allows to make a deduction step in the object-context by temporarily switching the attention to the meta-context (reification), then doing a proof (or a constraint propagation) at the meta-level and finally getting the result in the object-context (reflection).

When the meta level system performs an *infer* and a *consistent* transition, i.e., $\rightarrow_{is}$, it produces new consistent constraints. These constraints $(C_m \cup S_m)$ should be reflected at the object level via a transition, called $\rightarrow_{c'}$, and added to the object passive constraint store (in the set S_{obj-m}). After a $\rightarrow_{c'}$ transition the object level performs a $\rightarrow_{is}$ transition for processing the new constraints. The $\rightarrow_{c'}$ transition can be written as:

$$\frac{\langle A_m, C_m, S_m \rangle \rightarrow_{is} \langle A_m, C'_m, S'_m \rangle}{\langle A_{obj}, C_{obj}, S_{obj} \rangle \rightarrow_{c'} \langle A_{obj}, C_{obj}, S_{obj} \cup S_{obj-m} \rangle} \qquad (1)$$

where S_{obj-m} has been defined in Section 3.2 as $S_{obj-m} = reflect(S_m, C_m)$.

The transition $\rightarrow_{c''}$ refers to the *upward* linking rule. After an *infer* and *consistent* transition, i.e., $\rightarrow_{is}$ in the object level, the $\rightarrow_{c''}$ transition translates active and passive constraints stores $C_{obj} \cup S_{obj}$ in the meta passive constraint store $S_{obj-m} = reify(S_{obj}, C_{obj})$.

$$\frac{\langle A_{obj}, C_{obj}, S_{obj} \rangle \rightarrow_{is} \langle A_{obj}, C'_{obj}, S'_{obj} \rangle}{\langle A_m, C_m, S_m \rangle \rightarrow_{c''} \langle A_m, C_m, S_m \cup S_{m-obj} \rangle} \qquad (2)$$

Transitions $\rightarrow_{c'}$ and $\rightarrow_{c''}$ can be considered a sort of $\rightarrow_c$ transition with the only difference that the constraints added to the passive constraint store are not taken from the set A but come from another level. The partial order between the two level transitions is depicted in Figure 1.

The soundness and completeness of the scheme depends on the characteristics of the single levels and on the transitions between them. The soundness and completeness of single levels represent a necessary (but not sufficient) condition for the soundness and completeness of the whole architecture. The linking rules should also be sound and complete. Informally, the object and the meta constraint solver could be unified in a single level special purpose constraint solver, hereinafter referred to as C_{flat}. If the constraint solver C_{flat} is sound and complete with respect to its domain, i.e., the combination of the object and the meta domains, then the meta system is sound and complete. The operations $consistent_{C_{flat}}$ and $infer_{C_{flat}}$ in the flat constraint solver correspond to the composition of operation of $consistent$, $infer$, $reify$ and $reflect$ in the meta system. A more detailed study of these properties is a basis for future works.

4. Combining Quantitative and Qualitative Reasoning

In this Section, we show the real applicability of the general meta CLP scheme by presenting a possible specialization concerning the combination of quantitative and qualitative reasoning (Lamma et al., 1995a). Even if CLP has very efficient constraint propagation and handling mechanisms, it does not reason on passive constraints and, therefore, it is able to perform quantitative but not qualitative reasoning.

We specialize the general scheme in order to build a two-level architecture. The first level is a CLP(FD). The second level reasons on finite domains of sets of constraints of the underlying system, performs operations on them (composition and intersection) and propagates meta constraints (tighter relations[6]).

This specialization has been used in the field of temporal reasoning see (Lamma et al., 1995a), following Meiri's framework (Meiri, 1991). For defining the specialization, we have to specify the constraint domain, and the operational semantics.

4.1. THE DOMAINS

As already said, the object level constraint solver is a CLP(FD) component. Therefore, its domain, according with (Jaffar & Maher, 1994), is the following: let D=Z and

$$\Sigma = \{\{\in [m, n]\}_{m \leq n}, +, =, \neq, \leq\}.$$

_____
[6]A constraint C' is tighter than C" if every pair of values allowed by C' is also allowed by C".

The symbols in Σ have their usual meaning apart from $\{\in [m,n]\}_{m\leq n}$ which is defined as follows: for every pair of integers m and n, $\mathcal{D}$ interprets the unary constraint $x \in [m,n]$ as $m \leq x \leq n$. $\mathcal{L}$ is the class of constraints which is generated by the primitive constraints where each variable is subject to an interval constraint. $\mathcal{FD} = (\mathcal{D}, \mathcal{L})$ is the constraint domain.

The meta level domain contains the constraints of the underlying system and reasons on finite domains of relations thus performing qualitative reasoning. A meta-variable $R_{XY}{}^7$ is automatically generated each time a new qualitative constraint between object level variables X and Y (passive object-level constraint) is encountered. Qualitative constraints are disjunctions of primitive constraints between variables. Therefore, if between the variables X and Y we have a relation of the kind: $(X\ r_1\ Y)\vee\ldots\vee (X\ r_n\ Y)$ which can be written as $X\{r_1,\ldots,r_n\}Y$ where r_is are primitive qualitative relations[8], then the system creates a meta-variable R_{XY} whose domain D_{XY} is $\mathcal{P}(D)$, the powerset of $\{r_1,\ldots,r_n\}$. Now let V be a subset of $\mathcal{P}(D)$, i.e., $V\subseteq \mathcal{P}(D)$, the signature of the meta-level system is

$$\Sigma_m = \{\{\in V\}, +_m, =_m, <_m, \neq_m\}^9.$$

The symbols in Σ_m are interpreted by the Σ-structure $\mathcal{D}_m$ in the following way. The symbol $\{\in V\}$ is defined as follows: for each $V\subseteq \mathcal{P}(D)$, $\mathcal{D}_m$ interprets the constraint $r \in V$ as the usual instance of the set V, therefore, r can assume one value among all the possible elements of V (subsets of constraints). The advantages of using the powerset instead of the set of relations are the following: first, the expressive power of the meta-level is increased. In fact, we can say that $XY \in [\{>\}, \{<,=\}, \{\}]$, i.e., XY ranges on $\{<,=\}$ or on $\{>\}$, which cannot be translated in $XY \in [<,=,>]$, i.e., XY ranges on $<$ or $>$ or $=$. Second, we can define a partial order relation between the elements of the powerset. Finally, the element $\{\}$ in the powerset can be very useful in order to detect inconsistencies at the object level without failing at the meta-level. This object level failure can be either propagated in the meta system (as a failure) or handled by the meta-reasoner in order, for example, to retract some constraints.

The symbol $+_m$ is interpreted by $\mathcal{D}_m$ as the composition operator. The composition of two meta-variables R_{XZ} and R_{ZY} admits only those values r_{xy} for which there exists $r_{xz} \in D_{XZ}$ and $r_{zy} \in D_{ZY}$ such that $r_{xy} = r_{xz} \otimes r_{zy}$ where the symbol $\otimes$ represents the composition of primitive constraints and is defined by a transitivity table on the constraints of D.

[7]R_{XY} is derived from a naming relation linking this meta-variable with variables X and Y.

[8]Primitive relations are constraints of $\mathcal{L}$ in the finite domain framework.

[9]The subscript m in symbols of Σ has been introduced in order to distinguish these operation in the object system and in the meta system.

The symbol $=_m$ is interpreted by $\mathcal{D}_m$ as the intersection of constraints. It unifies two variable's domain of $\mathcal{P}(D)$. The intersection between two constraints $X\{r_1,\ldots,r_n\}Y$ and $X\{r_1',\ldots,r_n'\}Y$ is the powerset of the set theoretic intersection between the two vectors $\{r_1,\ldots,r_n\}$ and $\{r_1',\ldots,r_n'\}$. A meta-variable representing the possible relations between X and Y is associated with each constraint.

The symbol $<_m$ is interpreted as the relation "tighter" among constraints. A set of constraints C' is tighter than another set of constraints C" if each pair of values allowed by C' is also allowed by C". In our case, the relation tighter can be translated in the set inclusion. This relation is a partial order relation on the elements of the domain. Therefore, a constraint imposing a relation on non-comparable elements should fail.

The symbol $\neq_m$ is interpreted as usual. If two meta-variables are linked by a $\neq_m$ constraint, and one of them assumes a meta-value (a set of constraints), this set is deleted by the domain of the other meta-variable.

$\mathcal{L}_m$ is the class of constraints which are generated by the primitive constraints (i.e., $\neq_m, <_m, =_m$) where each meta-variable is subject to a set-inclusion constraint[10]. Then, $\mathcal{FD}_m = (\mathcal{D}_m, \mathcal{L}_m)$ is the constraint domain. A typical constraint which can be expressed in this domain is

$$r_{xy} \in [\{<,=\}, \{<\}, \{=\}, \{\}] \wedge r_{yz} \in [\{<\}, \{\}] \wedge r_{xz} =_m r_{xy} +_m r_{yz}$$

4.2. THE OPERATIONAL SEMANTICS

The operational semantics of the general scheme is specialized by defining the operations *infer* and *consistent* for the two solvers, and the operations *reify* and *reflect* for the interactions between the two.

In the finite domain solver (the object level), the *infer* operation performs an arc consistency algorithm (see, for instance, (Mackworth, 1977)). The arc consistency is a constraint propagation algorithm which ensures that every arc of the network is consistent. An arc, linking variables X and Y, is consistent iff for each value in the domain of X there exists at least one value in the domain of Y which is consistent with the constraint linking the two variables. The predicate *consistent* should be applied to the active constraint store and holds if no variable domain is empty.

In the meta constraint solver, the two predicates have the same structure: the predicate *infer* performs the arc consistency on the meta-variables by eliminating constraint symbols which cannot satisfy meta-constraints and operations. The predicate *consistent* holds if no meta-variable domain is empty, i.e., $r_{xz} \in [\ ]$[11].

---

[10]With an abuse of notation, we refer to $\{\in V\}$ as to a set-inclusion constraint.

[11]If $r_{xz} \in [\{\}]$, it means that a failure in the object system occurred.

The most interesting part of the operational semantics concerns the interactions between the two systems. The *downward* linking rule should be specialized by defining the operation $reflect$. In this case, $reflect$ works on a representation of C_m (S_m is not reflected in the object system[12]). For each unary constraint $(r_{xz} \in D_{xz}) \in C_m$, $D_{xz} \subseteq \mathcal{P}(D)$, $reflect$ creates a passive object constraint and adds it to S_{obj-m}.

$$S_{obj-m} = \bigcup_{r_{xz} \in D_{xz}} x \ lub(r_{xz}) \ z.$$

For instance, suppose that after the computation of $infer(C_m, S_m)$ the set of active meta constraints C'_m contains the constraint:

$$r_{xz} \in [\{<\}, \{=\}, \{<, =\}, \{\}].$$

The *lub* of the variable's domain is $\{<, =\}$, represents the constraints that hold between the object variables x and z in the object system, i.e., $x \leq z$.

As concern the *upward* linking rule, we should specialize the operation $reify$. This operation acts on a representation of the constraint stores S_{obj} and C_{obj} and creates active and passive meta constraints which are added to the set S_{m-obj}.

$$S_{m-obj} = (\bigcup_{\substack{\forall x, y \in D, \\ x\{r_1, ..., r_n\}y, \\ r_1, ..., r_n \in \mathcal{L}}} r_{xy} \in \mathcal{P}(r_1, ..., r_n)) \bigcup (\bigcup_{\forall x \in D_x, y \in D_y} r_{xy} \in C_{xy})$$

The first part of the set S_{m-obj} concerns the reification of passive object constraints. Suppose that the set of passive object constraints S_{obj} contains the constraint $x \leq y$ which can be written as $x\{<, =\}y$, the operation $reify$ creates a meta-variable r_{xy} ranging on the powerset of $\{<, =\}$.

The second part of the set S_{m-obj} concerns the reification of active object constraints. In particular it adds the constraints $r_{xy} \in C_{xy}$ where C_{xy} is the set of constraints supported by the domain values of the object system. For example, if the domains of X and Y contain respectively the values [1..10] and [14..24] the only constraint supported by the domain values is $<$. If the constraint between them is $X \leq Y$ the constraint $=$ is no longer entailed by the values in their domain.

The *upward* linking rule also connects meta variables that share an object variable with the composition operation and meta-variables that share two object variables with the operation intersection.

---

[12]The operation $reflect$ works on a representation of C_m and S_m but the selection rule selects only constraints of C_m.

5. An Example

We now present a simple example which shows the way in which we combine knowledge and control in the framework proposed. The example shows that, by using our meta architecture specialized for combining qualitative and quantitative knowledge, we can infer more information from a simple CLP(FD) program. The program considered is the following.

```
example([X,Y,Z]):-
    X::1..20, Y::5..14, Z::14..21, % object active constraints
    X ≤ Y, X < 5, Y < Z, X ≤ Z. % object passive constraints
```

The first three constraints state that variables X, Y and Z range on a domain (set of integer). The consistency check is performed only on active constraints. The constraint solver, in fact, fails if one of the variable domain becomes empty. The last four constraints are used by the predicate *infer* in order to reduce the domain of active constraints.

Now we describe which are the steps performed by our meta system when the query example([X,Y,Z]) is raised. The addition of the first three active constraints does not lead to any propagation at the object level because *infer* cannot use passive constraints which are not yet examined. The *upward* rule creates three meta-variables XY, XZ and YZ. The first two meta-variables range on the powerset of $\{<, =, >\}$ because there are couples of values in the object variables domain that support these constraints. The third meta-variable range on the powerset of $\{<, =\}$. No couple of values in the domain of Y and Z support the constraint $>$. Moreover the *upward* linking rule creates the constraint $XY +_m YZ =_m XZ$. The meta level performs an $infer_m$ which does not alter meta variables domain. The $consistent_m$ holds. The *downward* rule passes the control to the object level.

The addition of the constraint $X \leq Y$ at the object level restrict the domain of X to 1..14. At this point, the predicate *consistent* is checked and it holds. The *upward* linking rule creates a meta variable XY, whose domain is $[\{<, =\}, \{<\}, \{=\}, \{\}]$, i.e., the sets of possible passive constraints that hold between X and Y. This variable is intersected, by $infer_m$, with the one XY previously built whose domain is reduced by deleting the constraint sets containing the symbol $>$. The $infer_m$ reduces also the domain of variable XZ to the powerset of $\{<, =\}$ because of the composition of XY and YZ and the intersection with the old XZ.

The *downward* transition passes the control to the object level. The object level adds the constraint $X < 5$ to the passive constraint store. This constraint produces a restriction (by the application of *infer*) in the domain of X which becomes [1..4]. It is worth noting that the constraint $X \leq Y$ is no longer supported by the values of the variables. In fact, there is no couple of values for X and Y that satisfy the constraint =. Therefore, the constraint XY ∈ [{<}, {}] is added to the meta passive constraint store.

The $infer_m$ now intersects the new constraint and the one previously built for XY. This constraint is consistent and the information is passed to the object level through *downward* transition. Therefore, in the object level system passive constraint we find X < Y. The object level system then considers the constraint Y < Z which does not alter the domain values. This information, as before, is reified and the constraint YZ ∈ [{<}, {}] is added to the meta system passive constraint store. The $infer_m$ propagates the constraints and produces the new constraint XZ ∈ [{<}, {}].

The new constraint X < Z is added to the object passive constraint store and the subsequent *infer* does not alter the domain values. The last constraint of the program, X ≤ Z, does not produce any variation because a tighter constraint, X < Z, has already been produced by a meta level propagation. Thus the query example([X,Y,Z]) succeeds with X::1..4, Y::5..14, Z::14..21, X < Y, Y < Z, X < Z.

We can query the system with queries like, for example, which is the minimal constraint network, and the system answers, in this very simple case, with the above mentioned constraints. An answer to this query could not be possible with a single constraint solver. This example has shown that the expressive power of CLP(FD) is enhanced by this approach thanks to the combination of constraint solving mechanisms.

6. Related Works

One of the starting points of this work is the work by Tsang, (Tsang, 1987). The author proposes the use of a graph and a meta-graph where graph-nodes are variables of the problem, graph-arcs refer to (qualitative) constraints, meta-graph nodes refer to constraints of the underlying graph and meta-graph arcs refer to meta-relations between constraints (e.g., the tighter relation). We embed the concept of meta-graph in the meta-constraint solver which reasons on constraints of the underlying system (corresponding to the graph).

In the field of temporal reasoning several efforts have been performed in order to integrate qualitative and quantitative reasoning, see (Meiri, 1991; Vilain *et al.*, 1990). In particular, the Meiri's framework has been used as a starting point of this work. Meiri uses the composition and intersection operations on (temporal) qualitative constraints, and uses a path-consistency propagation method on quantitative constraints. In our meta architecture we apply the same propagation methods except for the path-consistency at the object level which is replaced by an arc-consistency algorithm.

Another related work is that by T. Frühwirth (Frühwirth, 1994) on

Constraint Handling Rules (CHRs). There are two important relations between our work and that of Frühwirth. First, the integration of CLP(FD) with qualitative and quantitative reasoning has been approached with CHRs in a general way. The main difference with our work is that the CHRs rely on a flat architecture while our approach benefits of modularity and flexibility of meta-systems. Moreover, in our approach CLP program syntax is unchanged while with CHRs we have to write special purpose programs.

The second relation concerns the fact that CHRs represent a very general architecture for prototyping and implementing constraint solvers. Our approach has been implemented by using the CLP(FD) library of ECL^{i}PS^e, but it can be easily implemented by using CHRs. In fact, we can use CHRs in order to write our meta-constraint solver and the linking rules between the object level and the meta level systems. The use of CHRs for writing our meta-constraint solver is described in (Lamma et al., 1995a).

A more general approach is that adopted by C.Kirchner, H.Kirchner and M.Vittek in (Kirchner et al., 1995). The authors present a framework for describing computational logics. Within the framework, equational programming, Horn clause programming and constraint solvers can be expressed as instances of the general scheme. The framework is based on rewriting techniques which can be used for reason about termination, confluence, combination of computational systems. Our approach is less general. We focus only on the combination of constraint solvers and we rely on the CLP general scheme.

Finally, the work of Codognet and Nardiello (Codognet & Nardiello, 1994) on m-constraints is related to our work in the sense that they use an explicit representation of constraints in terms of a matrix. Again they embed in the constraint solver this mechanism while we use two different constraint solvers and our representation is implicit.

7. Conclusions and Future Works

In this paper, we have presented a general method for combining constraint solvers. Our work leads to the definition of a general meta CLP architecture where each level is a constraint solver reasoning on constraints of the underlying level. The approach benefits of the efficiency of CLP and maintains all the advantages of meta architectures such as modularity, expandibility and flexibility. The architecture is flexible, and transparent to the user. We think that it could be a basis for prototyping and implementing new constraint solvers.

We define a general meta CLP scheme which can be specialized by defining the domain and the operational semantics of each level composing the architecture. In addition, two additional rules (*upward* and *downward*) should be defined, which link two levels.

We have presented a specialization of the scheme concerning the integration of quantitative and qualitative reasoning. Each meta level reasons on the constraints of the underlying object level. This specialization has been applied to the field of temporal reasoning, following Meiri's framework.

We have developed a prototype version of this specialization by using an extension of the ECL^{i}PS^e (ECLiPSe, 1992) finite domain library (see (Lamma *et al.*, 1995a; Lamma *et al.*, 1995b) for details). Our research is now concentrated on specializations on real numbers, boolean, sets and Herbrand Universe. We are also investigating another application of the system to qualitative reasoning by using a CLP(Intervals) (Benhamon *et al.*, 1994) as object-level and the Allen's Interval Algebra (Allen, 1983) as meta-level.

We are also investigating a way of embedding the capability of making some recovery actions on the lower level system in the meta constraint solver. In fact, the meta level is able to detect some inconsistencies of the low level system, and can therefore restore some previous consistent situations in order to recovery from a failure. We will investigate the possibility of using some Truth Maintenance System techniques (de Kleer, 1986) for handling this kind of problems.

8. Acknowledgments

This work has been supported by M.U.R.S.T. (Ministero dell'Università della Ricerca Scientifica e Tecnologica) Project 60%. We would like to thank the anonymous referees for their very useful comments and remarks on the first version of this paper.

EVELINA LAMMA AND MICHELA MILANO
DEIS
Università di Bologna, Italia

AND

PAOLA MELLO
Istituto di Ingegneria
Università di Ferrara, Italia

References

L.Aiello, M.Cialdea, D.Nardi, M.Schaerf, "Modal and Meta-Languages: Consistency and Expressiveness", in *Meta-Logics and Logic Programming*, K.Apt and F.Turrini eds., MIT Press.

L.Aiello, C.Cecchi, D.Sartini, "Representation and Use of Metaknowledge" in *Proceedings of the IEEE*, vol. 74, No. 10, 1986, pp. 1304–1321.

J.F.Allen, "Maintaining Knowledge About Temporal Intervals", in *Communications of the ACM*, vol. 26, 1983, pp. 832–843.

F.Benhamon, D.McAllester, P.Van Hentenryck, "CLP(Intervals) Revisited", *Tech. Report* CS-94-18 Computer Science Department, Brown University, 1994.

K.A.Bowen, R.A. Kowalski, "Amalgamating Language and Metalanguage in Logic Programming" in *Logic Programming*, K.Clark and S.Tarnlund Eds., Accademic Press NY, 1982, pp. 153–173.

P.Codognet, G.Nardiello, "Path Consistency in clp(FD)", in *Proceedings of the First International Conference Constraint in Computational Logics CCL94*, 1994, pp. 201–216.

J. de Kleer, "An Assumption-based TMS", in *Artificial Intelligence*, n.28, vol 3, 1986, pp. 127–162.

ECL'PS^e User Manual Release 3.3, ECRC 1992.

T. Frühwirth, "Temporal reasoning with constraint handling rules", *Tech. Report* ECRC-94-05, ECRC, 1994, Germany.

J.Jaffar, J.L.Lassez, "Constraint Logic Programming", in *Proceedings of the Conference on Principle of Programming Languages*, Munich 1987.

J.Jaffar, M.J.Maher, "Constraint Logic Programming: a Survey", in *Journal of Logic Programming on 10 years of Logic Programming*, 1994.

C.Kirchner, H.Kirchner, M.Vittek, "Designing Constraint Logic Programming Languages using Computational Systems", in *Principles and Practice of Constraint Programming*, The NewPort Papers, P.Van Hentenrick and V.Saraswat eds. MIT Press, 1995.

R.Kowalski, "Logic for Problem Solving", North-Holland,1979.

E.Lamma, P.Mello, M.Milano, "A Meta Constraint Logic Programming Architecture for Qualitative and Quantitative Temporal Reasoning", Technical Report DEIS-LIA-95-001, 1995.

E.Lamma, P.Mello, M.Milano, "A Meta Constraint Logic Programming Scheme", Technical Report DEIS-LIA-95-005, 1995.

E.Lamma, P.Mello, M.Milano, "A Multi-Level CLP Architecture for Consistency Techniques", submitted for publication.

J.Lever, B.Richards, R.Hirsh, "Temporal Reasoning and Constraint Solving", *Deliverable CHIC*, ESPRIT Project EP5291, IC-Park, London, 1992.

J.W.Lloyd, "Foundation of Logic Programming", Second Extended Edition, Springer-Verlag 1987.

P.Maes, "Computational Reflection", Tech. Report AI-lab VUB, Brussels, 1987.

A.K. Mackworth, "Consistency in Networks of Relations", in *Artificial Intelligence*, vol.8, 1977, pp. 99–118.

I.Meiri, "Combining Qualitative and Quantitative Constraints in Temporal Reasoning", in *Proceedings of AAAI91*, pp.260–267.

E.P.K.Tsang, "The Consistent Labeling Problem in Temporal Reasoning", in *Proceedings of AAAI87*, 1987, pp.251–255.

P.Van Hentenryck, Y.Deville "Operational Semantics of Constraint Logic Programming over Finite Domains", *Tech. Report* CS-90-23 Computer Science Department, Brown University, 1990.

M.B.Vilain, H.Kautz, P. Van Beek, "Constraint Propagation Algorithms for Temporal Reasoning: A Revised Report", in *Readings in Qualitative Reasoning about Physical Systems*, D.S.Weld, J.De Kleer, Morgan Kaufmann, 1990, pp.373–381.

MEMBERSHIP-CONSTRAINTS AND COMPLEXITY IN LOGIC PROGRAMMING WITH SETS

Abstract. General agreement exists about the usefulness of sets as very high-level representations of complex data structures. Therefore it is worthwhile to introduce sets into constraint logic programming or set constraints into programming languages in general.

We start with a brief overview on different notions of sets. This seems to be necessary since there are almost as many different notions in the field as there are applications such as e.g. rapid software prototyping and unification-based grammar formalisms.

An efficient algorithm for treating membership-constraints is introduced. It is used in the implementation of an algorithm for unifying finite sets with tails also presented here. Such a unification algorithm is useful in any logic programming language embedding sets.

Finally it is shown how a full set language including the operators $\in$, $\notin$, $\cap$, $\cup$ can be built on membership-constraints. The text closes with a reflection on the complexities of different algorithms – which are single exponential – showing the efficiency of our new algorithm.

Keywords. Constraint logic programming, logic programming with sets, set unification, data structure sets.

1. Introduction

Implementations of logic programming languages have terms as their main data type. But often it may be useful and more natural to represent objects by sets instead of simple terms. In particular sets can be conveniently used in rapid software prototyping. Therefore it is worthwhile to introduce sets into logic programming. Of course there are many more applications for sets; we will list them together with different notions on sets later on in the text.

F. Baader and K.U. Schulz (eds.), Frontiers of Combining Systems, 285–302.
© *1996 Kluwer Academic Publishers.*

1.1. THE DATA STRUCTURE SETS

We will first recall some definitions of the term *set* which are common in the literature and used in applications. Various systems of set constraints have been set up for the purpose of axiomatising (Zermelo-Fraenkel) set theory and proving theorems in it. However in practice it seems to be necessary to restrict discussion to a fragment of the theory such that problems involving set constraints remain computable and reasonably efficient procedures exist. We will now discuss several notions of sets informally and briefly.

In (Gervet, 1994) *ground sets* are considered. These are finite subsets of the Herbrand universe, i.e. sets of ground terms or integers. This restricted notion of set leads to efficient algorithms. Set domain variables are introduced, i.e. variables which are attached with a set of ground sets. Such a set is approximated by its greatest lower and least upper bound with respect to set inclusion. This allows us to exploit efficient constraint satisfaction techniques (arc consistency) (Van Hentenryck, 1989) for the basic set relation constraints inclusion and disjointness of ground sets. Application domains for ground sets are combinatorial problems based on sets, relations or graphs, e.g. set partition and bin packing.

A *hereditarily finite set* is a set of finite depth that is finite, its members are finite, the members of its members are finite, etc. Such sets are being considered e.g. in a French project on constraint logic programming with sets (Legeard *et al.*, 1993) which aims at rapid prototyping combinatorial problems with sets, multisets and sequences, writing executable specifications and software modelling which also makes use of partial constraint consistency techniques. – There are (at least) two syntactic variants for these kind of sets:

Firstly there are *finite set expressions* over arbitrary terms, i.e. including variables, that may be (finitely) nested, e.g. $\{x, \emptyset, f(\{a, b\})\}$. These sets are used in unification theory, set and associative-commutative-idempotent unification problems, theorem proving and logic programming with sets (Kapur & Narendran, 1992; Stolzenburg, 1993). Here e.g. the expression $\{x, y\}$ may denote a set consisting of one or two entities. This is in the nature of sets of course, but we cannot know in advance whether x and y become identical or not because they are variable terms.

Secondly (and in the sequel mainly) we will consider *sets with tails* $S = \{t_1, \ldots, t_k | T\}$, understood as the union $\{t_1, \ldots, t_k\} \cup T$; here T is a variable for a set, called the tail of S. These sets may be (finitely) nested too. They are well-suited for the extension of logic programming with sets (Dovier *et al.*, 1993; Dovier & Rossi, 1993; Stolzenburg, 1994a) by exploiting constraint logic programming techniques.

Hypersets (Aliffi *et al.*, 1993) are rational, i.e. possibly cyclic hereditarily finite sets. If this concept is combined with feature structures as done in (Pollard & Moshier, 1990; Manandhar, 1994) then it is possible to implement unification-based grammar formalisms. There sets are used in the treatment of so-called unbounded dependency constructions, relative clauses, questions, linguistic quantifiers and anaphora resolution.

In (Bruscoli *et al.*, 1994) it is shown that the combination between constraint logic programming with sets and constructive negation opens up the possibility of representing a general class of *intensionally defined sets*, i.e. sets that are defined by the properties of its elements, not *extensionally* by enumerating the elements of the sets. The presence of intensional sets leads to an increase in the expressive power and abstraction level offered by the host logic language.

1.2. NOTIONS AND NOTATIONS

In our context we are mainly concerned with finite sets with tails. We will make use of unification theory and standard constraint logic programming notions, see e.g. (Siekmann, 1989; Jaffar & Maher, 1994). Let Σ be a set of functional symbols together with their arity and V a denumerable set of variables $\{x, y, z, \ldots\}$. Then $\mathcal{T}(\Sigma \cup V)$ and $\mathcal{T}(\Sigma)$ denote the sets of terms $\{s, t, \ldots\}$ and ground terms, respectively.

Let the notions substitution (here written within square brackets), more general than (symbol $\leq$), equality, unifier, etc. used in this context be defined as in standard texts, e.g. (Siekmann, 1989). For two substitutions σ and τ we define their *composition* $\sigma\tau$ as the usual composition of mappings $(\sigma\tau)(t) = \tau(\sigma(t))$.

We assume that the signature Σ of our language is endowed with two functional symbols: the constant $\emptyset$ for the empty set, and $\{\cdot|\cdot\}$ – the (binary) set constructor where $\{s|T\}$ stands for $\{s\} \cup T$ in usual notation; here T is a set (variable). Thus $\Sigma \supseteq \{\emptyset, \{\cdot|\cdot\}\}$. Furthermore $\{s_1, s_2|T\}$ is an abbreviation of $\{s_1|\{s_2|T\}\}$, and $\{s\}$ of $\{s|\emptyset\}$. The following properties, i.e equational axioms hold:

(1) $\{x, y|T\} = \{y, x|T\}$ (permutativity)

(2) $\{z, z|T\} = \{z|T\}$ (absorption)

This set theory corresponds to standard finite set theory. For a detailed discussion of this, the interested reader is referred to (Dovier & Rossi, 1993). In that paper it also is shown how logic programming with sets fits into the framework of *constraint logic programming* (Jaffar & Maher, 1994). For this we fix the set of constraint predicate symbols of our scheme to be $\in$ (membership), $\notin$, $=$ and $\neq$. They form a subset of all predicate symbols Π in our language.

We take our set unification algorithm based on membership-constraints as a *constraint simplification algorithm* in this context. As interpretation domain we choose a hereditarily finite universe over the Herbrand domain $\mathcal{T}(\Sigma\backslash\{\emptyset, \{\cdot|\cdot\}\})$. Non-variable terms t except $\emptyset$, i.e. $t \in \mathcal{T}(\Sigma \cup V)\backslash(V \cup \{\emptyset\})$ (called *urelements*) are disallowed as set tails.

But another view is possible from the standpoint of *unification theory*. Then the interpretation domain can also be understood as the equivalence classes of the finest equivalence relation over the Herbrand domain $\mathcal{T}(\Sigma)$ that fulfils the above properties. Then we have to perform set *theory unification*. Our set theory is *finitary*, i.e. for two sets there may be (at most) finitely many different and most general unifiers in general. We will consider the *general unification problem* here where additional free function symbols of arbitrary arity may occur.

Finite sets (possibly) with tails shall be used on argument positions in (logic) programs. If we query the system e.g. $\{x, y\} = \{a, b\}$ the system will answer with two variable bindings (substitutions), namely $[x/a, y/b]$ and $[x/b, y/a]$ which represent the solutions of the corresponding set unification problem. It is also useful to introduce more operators in the programming language, namely $\in$ and $\notin$. Furthermore set union $\cup$ and intersection $\cap$ may be desirable.

There are constraint canonization algorithms like the ones stated in (Dovier & Rossi, 1993; Manandhar, 1994). Both need an algorithm for *set unification* but they do not develop algorithms that are reasonable efficient in the average case. In this paper we will concentrate on the implementation of an algorithm for set unification that is needed in any logic programming language embedding (finite) sets. Before we address this problem we have to give a precise definition of set unification.

Definition 1 A substitution σ is called a *unifier of two sets* $S = \{s_1, \ldots, s_m\}$ and $T = \{t_1, \ldots, t_n\}$ where $s, t \in \mathcal{T}(\Sigma \cup V)$ iff for every $s \in S$ there exists some $t \in T$ (and vice versa too) such that $s\sigma = t\sigma$ (Kapur & Narendran, 1986), or (stated differently) $S\sigma = T\sigma$, i.e. $S\sigma \subseteq T\sigma$ and $T\sigma \subseteq S\sigma$.

1.3. OVERVIEW OF THE PAPER

In the following we will present an algorithm for unifying sets with tails that behaves efficiently in many cases. Firstly, the case where the set tails are empty can be treated by so-called *membership-constraints* which can be seen as a tricky variant of the predicate *member/2* of Prolog using non-unifiability constraints (Stolzenburg, 1994b). Constraint techniques such as delayed execution and the first-fail principle lead to even more improvements. Secondly, it will be shown how the algorithm can be lifted to the general case with non-empty set tails.

The algorithm compares very well with other approaches since our algorithm avoids the computation of many redundant solutions and has a good run-time performance. Although it uses ideas from constraint logic programming with *finite domains* (Van Hentenryck, 1989) and *generalized propagation* (Le Provost & Wallace, 1993), it goes beyond that. The algorithm can be easily implemented in logic programming languages that provide delay mechanisms and explicit control of delayed goals. A prototype has been implemented in ECLiPSe-Prolog (ECLiPSe, 1995).

2. Membership-Constraints

In this section we want to define two primitive predicates, namely $memb/2$ and $wake/1$ that help us deal with *membership-constraints*. The semantics of $memb/2$ shall be the same as that of $member/2$ in Prolog. The difference is that $memb/2$ is treated as a constraint. Membership-constraints are a means for unifying sets without tails.

2.1. DEFINITIONS

We will call $C = memb(s, L)$ a *membership-constraint* where $s \in \mathcal{T}(\Sigma \cup V)$ is a term that will be unified with (at least) one element of the list L which represents a set. Only a complete list L is admissible, i.e. its (final) tail must be the empty list $[]$. By $|L|$ we denote the length of the list L. If $L = [t_1, \ldots, t_n]$ we can also view C as the disjunction $(s = t_1 \vee \cdots \vee s = t_n)$. We look for a substitution σ that solves one of the disjuncts, i.e. $s\sigma = t_k\sigma$ for some $1 \leq k \leq n$. We need the following definition in the algorithm stated next.

Definition 2 The *most specific generalization* of two terms t_1 and t_2 is a term t that is more general than t_1 and t_2, and every other term with the same property is more general than t.

A membership-constraint $C = memb(s, L)$ where $L = [t_1, \ldots, t_n]$ can be simplified by the following **Algorithm A**. Its implementation is sketched in Figure 1. The natural language description of the algorithm follows immediately.

(1) If there is a $t \in L$ with $s = t$ (syntactical identity), then C is already solved by this t.

(2) Otherwise remove all $t \in L$ which (a) are not unifiable with s or (b) are identical duplicates of other terms in L. Let L' be the list of remaining elements.

(3) If $L' = [\,]$ then the computation is stopped here. Backtracking has to be initiated.

(4) If $L' = [t]$ then s and t have to be unified. Their most general unifier σ is applied to all relevant terms occurring in the problem.

(5) If $|L'| > 1$ then s is unified with the most specific generalization (also called *anti-unification*) of all $t \in L'$ yielding s'. It replaces each occurrence of s in the problem.

(6) After that the membership-constraint $C' = memb(s', L')$ is set up and will be solved later on. The old constraint C is discarded.

```
INPUT        constraint C = memb(s, L) where L = [t₁,...,tₙ]
OUTPUT       simplified constraint
VARS         t : term

BEGIN
(1)              IF ∃t ∈ L : s == t THEN
                         RETURN                    {C is already solved }
                 ELSE BEGIN
(2)                      FORALL t ∈ L DO BEGIN
    (a)                      IF s\= t      {s and t are definitely not unifiable }
    (b)                      OR t is a duplicate in L
                             THEN L := L\t          {remove element }
                         END
                         CASE |L| OF
(3)                          0 : STOP;                      {no solution }
(4)                          1 : [t] := L;      {t is the only element in L}
                                 unify(s,t);
(5)                          >: s := unify(s, generalize(L));
(6)                              delay(C);
                         END
                 END
END
```

Figure 1. Simplification procedure – Algorithm A.

Proposition 3 *Algorithm A is correct and complete with respect to the satisfiability of membership-constraints, i.e. the algorithm leads to an equivalent set of constraints where equivalent means to have the same set of most general unifiers.*

Proof. The correctness of the procedure is obvious. So we will only provide some remarks on the completeness of the procedure. Firstly, completeness is preserved in step (1) because we are only interested in the most general solutions of the constraint C, and the empty substitution ϵ which solves the disjunct $s = t$ is more general than any other substitution. Secondly, step (2b) is justified by a similar argument. Finally, since s must eventually be unified with one $t \in L'$ in order to solve the constraint, s may be unified by the above-mentioned generalization in step (5).   □

2.2. CONSTRAINT-BASED TREATMENT

Delayed membership-constraints have to be solved eventually. This is done when a constraint C is woken implicitly or explicitly. If a variable in C is instantiated or unified with another term then it is woken *implicitly*. Algorithm A is applied to C so that we get a better approximation of this constraint. The predicate $wake(C)$ shall allow the user to wake up a delayed membership-constraint *explicitly*. By this x is unified with one $y \in L$.

A call $wake(C)$ chooses one of the delayed constraints in a non-deterministic fashion and unifies it with C. The well-known first-fail principle can serve as a heuristic for that choice, i.e. the constraint $memb(x, L)$ with smallest $|L|$ should be chosen. Also it is a good idea to treat those constraints first which share the most variables with other constraints. This predicate is needed in order to compute concrete solutions and to assure correctness of the whole procedure. It corresponds to the labelling procedures used for programming with finite domain and related constraints. The reader should consult (Van Hentenryck, 1989) for details.

Now we want to present another effective optimization. A membership-constraint can be viewed as the disjunction shown in (1) below. We observe that (1) is equivalent with (2) where $\neq$ denotes the non-unifiability *constraint* that s and t shall never become identical. This can be implemented via the delaying predicate $\sim=/2$ of some Prolog dialects. It is clear that version (2) can avoid a lot of redundant solutions. In (3) it is shown how to code this optimization into a real logic program for membership-constraints. We will refer to it as **Algorithm B**.

(1) $s = t_1 \lor s = t_2 \lor \cdots \lor s = t_n$
(2) $s = t_1 \lor (s = t_2 \land s \neq t_1) \lor \cdots \lor (s = t_n \land s \neq t_{n-1} \land \cdots \land s \neq t_1)$
(3) $memb(x, [x|\_])$.
     $memb(x, [y|L]) \leftarrow x \neq y \land memb(x, L)$.

2.3. HOW TO ENCODE SET UNIFICATION

We can express set unification by means of membership-constraints as follows. If we want to unify the sets A and B (represented as the Prolog lists $[x_1, \ldots, x_m]$ and $[y_1, \ldots, y_n]$ respectively) then we have to solve the membership-constraints in (1). That follows directly from the definition of set unification. The code for automatically reducing unification of finite

sets to membership-constraints is shown in (2) and (3) which we will call
Algorithm C. It is sound and complete for set unification without tails.

(1) $memb(x_1, B) \wedge \cdots \wedge memb(x_m, B) \bigwedge memb(y_1, A) \wedge \cdots \wedge memb(y_n, A)$
(2) $unify\_sets(A, B) \leftarrow subset\_of(A, B) \wedge subset\_of(B, A)$.
(3) $subset\_of([], \_)$.
    $subset\_of([x|R], L) \leftarrow memb(x, L) \wedge subset\_of(R, L)$.

3. Sets with Tails

Now we will consider the unification of sets with tails and formulate a
generalized unification algorithm based on Algorithm C.

3.1. THE UNIFICATION ALGORITHM

The algorithm is based on a case distinction. The set tails may both be
the empty set (where it reduces to Algorithm C), or exactly one of them is
the empty set; if both tails are variables, then we treat the case separately
where both are identical in order to improve efficiency. The algorithm works
as follows:

Algorithm D. A unifier of two sets $\{s_1, \ldots, s_m|S\}$ and $\{t_1, \ldots, t_n|T\}$
where the set tails S and T may be variables or the empty set can be
computed by the following non-deterministic procedure:

(1) Let A, A', B and B' be finite sets without tails such that $A \uplus A' = \{s_1, \ldots, s_m\}$ and $B \uplus B' = \{t_1, \ldots, t_n\}$ where $\uplus$ denotes the union of
    disjoint sets. If $S = \emptyset$ then it must be $B' = \emptyset$; if $T = \emptyset$ then it must be
    $A' = \emptyset$.
(2) Let σ be a unifier of A and B, computed by the Algorithm C. If the
    sets are not unifiable then the computation stops here with failure.
    Backtracking is initiated then.
(3) Let A'', B'' and C be finite sets (without tails) such that $A'' \uplus B'' \uplus C = A\sigma$ (or equivalently $B\sigma$). If $S = \emptyset$ then $B'' = \emptyset$; if $T = \emptyset$ then $A'' = \emptyset$.
    If S and T are identical variables then this step is omitted. This step
    prunes a lot of branches because of the *disjoint* union.
(4) If S and T are identical variables then let σ' be defined by the unifi-
    cation equation system $[S\sigma = T\sigma = A'\sigma \cup B'\sigma \cup N]$. Otherwise σ' is
    defined by $[S\sigma = B'\sigma \cup B'' \cup N, \ T\sigma = A'\sigma \cup A'' \cup N]$. N is a new
    variable in this context. If $S = \emptyset$ or $T = \emptyset$ then it is $N = \emptyset$.
(5) If all steps can be executed successfully then $\tau = \sigma\sigma'$ is a unifier of the
    given sets.

In Figure 2 the algorithm is stated more formally, namely as pseudo-
code. We will soon present some examples that shall clarify how it works.
But before that we show the correctness and completeness of the algorithm.

```
INPUT      Two sets {s₁, ..., sₘ|S} and {t₁, ..., tₙ|T}
OUTPUT     unifying substitution τ
VARS       A, B, C and derivatives : finite sets without tails
           N : set (tail) variable

BEGIN
(1)        IF S = ∅ THEN B' := ∅;
           IF T = ∅ THEN A' := ∅;
           A ⊎ A' := {s₁, ..., sₘ};
           B ⊎ B' := {t₁, ..., tₙ};
(2)        σ := [A = B];
(3)        IF S ≠ T THEN
               IF S = ∅ THEN B'' := ∅;
               IF T = ∅ THEN A'' := ∅;
               A'' ⊎ B'' ⊎ C := Aσ(= Bσ);
(4)            σ' := [Sσ = B'σ ∪ B'' ∪ N, Tσ = A'σ ∪ A'' ∪ N]
           ELSE
               σ' := [Sσ = Tσ = A'σ ∪ B'σ ∪ N];
(5)        τ := σσ'
END
```

$S = \emptyset$; $B' := \emptyset$; $T = \emptyset$; $A' := \emptyset$; $A \uplus A' := \{s_1, \ldots, s_m\}$; $B \uplus B' := \{t_1, \ldots, t_n\}$; $\sigma := [A = B]$; $S \neq T$; $S = \emptyset$; $B'' := \emptyset$; $T = \emptyset$; $A'' := \emptyset$; $A'' \uplus B'' \uplus C := A\sigma(= B\sigma)$; $\sigma' := [S\sigma = B'\sigma \cup B'' \cup N, \; T\sigma = A'\sigma \cup A'' \cup N]$; $\sigma' := [S\sigma = T\sigma = A'\sigma \cup B'\sigma \cup N]$; $\tau := \sigma\sigma'$

Remarks: (a) $A \uplus B := C$ partitions the set C non-deterministically into the sets without tail A and B. If one of the latter sets has been defined earlier as the empty set $\emptyset$, then the other one becomes identical with C. $\uplus$ denotes disjoint union. – (b) $[s = t]$ denotes one of the solutions of the unification equation system containing $s = t$. – (c) steps (1) and (2) may be totally interleaved for efficiency reasons in order to avoid too much generating and testing. – (d) In step (2) any unification algorithm for finite sets without tails can be used.

Figure 2.   Main function – Algorithm D.

Theorem 4 *Algorithm D computes a correct and complete set of unifiers.*

Proof. We will only prove the most complex case where the tails S and T are *distinct* variables. The other cases can be shown in a similar manner. The correctness proof is straightforward and therefore omitted here.

The algorithm computes a complete set of unifiers, i.e. for every unifier τ of the given sets, there is a unifier $\tau' \leq \tau$ (i.e. τ' is more general than τ) that can be computed by the algorithm. Let τ be a unifier of the two sets $\{s_1, \ldots, s_m|S\}$ and $\{t_1, \ldots, t_n|T\}$. During step (1) put all $s \in \{s_1, \ldots, s_m\}$ for which there exists a $t \in \{t_1, \ldots, t_n\}$ with (∗) $s\tau = t\tau$ into A; similarly, put all $t \in \{t_1, \ldots, t_n\}$ for which there exists an $s \in \{s_1, \ldots, s_m\}$ with (∗) into B. For all other s and t, let $s \in A'$ and $t \in B'$, respectively. – Because of the completeness of Algorithm C, a substitution $\sigma \leq \tau$ can be computed in step (2) of Algorithm D with $s\sigma = t\sigma$.

If (∗) and at most *one* of the conditions (†) $s\tau \in T\tau$ and $t\tau \in S\tau$ for some s and t hold then we can choose $s\sigma \in B''$ or $t\sigma \in A''$, respectively in step (3). Otherwise, if *both* conditions (†) hold then $\sigma\sigma' \leq \tau$ is a unifier where σ' contains a solution of $[S\sigma = \{t\sigma|R_1\},\ T\sigma = \{s\sigma|R_2\}]$ where R_1 and R_2 are the remainders of the sets in question. However the more general solution $\tau' = \sigma\sigma'' \leq \sigma\sigma' \leq \tau$ can be computed by the algorithm where $\sigma'' \leq \sigma'$ contains a solution of $[S\sigma = R_1,\ T\sigma = R_2]$ in step (4). □

The algorithm is always terminating. We will just give some remarks on that. – If at all, the only source for infinite loops is in step (4) where the equation system $[S\sigma = B'\sigma \cup B'' \cup N,\ T\sigma = A'\sigma \cup A'' \cup N]$ has to be solved. The only critical case is where $S\sigma$ or $T\sigma$ are not variables. This can only happen if S or T are tails of (nested) sets occurring in the original problem. Since there can only be finitely many such tails and every computed σ in each recursive step eliminates at least one such tail variable, the computation will terminate.

3.2. EXAMPLE UNIFICATIONS

Let us now consider the two sets $\{x_1, x_2|T\}$ and $\{c_1, c_2|T\}$ with identical variable tails T where x_1 and x_2 are variables and c_1 and c_2 are constants. In step (1) of Algorithm D we choose $\{x_1, x_2\} = \{x_1, x_2\} \uplus \emptyset$ and $\{c_1, c_2\} = \{c_1\} \uplus \{c_2\}$. A unifier of $\{x_1, x_2\}$ and $\{c_1\}$ computed by Algorithm C is $\sigma = [x_1/c_1, x_2/c_1]$ in step (2). We omit step (3) because the tails of the sets are identical variables. In step (4) we have to solve the equation $T\sigma = \emptyset\sigma \cup \{c_2\}\sigma \cup N$ where N is a new variable. We get $\sigma' = [T/\{c_2|N\}]$ as its solution. Thus $\tau = \sigma\sigma' = [x_1/c_1, x_2/c_1, T/\{c_2|N\}]$ is one of the most general unifiers of the two given sets. If we take different choices we will find further solutions. They coincide with the minimal set of most general unifiers for this example.

We do not hesitate to give two more examples: The first is $\{x, y|S\} = \{z\}$ where all identifier denote variables. Here step (1) implies $A' = \emptyset$ and hence $A = \{x, y\}$. This enforces $B = \{z\}$ and $B' = \emptyset$ since otherwise in step (2) $A = B$ would not be solvable. Step (2) yields $\sigma = [x/z, y/z]$ then. In step (3) $A'' = \emptyset$ is constrained. Let us choose $B'' = \{z\}$ and thus $C = \emptyset$. Finally, step (4) leads to $\sigma' = [S/\{z\}]$. So the overall solution is $\tau = [x/z, y/z, S/\{z\}]$. There is another solution $\tau' = [x/z, y/z, S/\emptyset]$. Both τ and τ' are most general. – As last example we want to consider the problem $\{a|S\} = \{b|T\}$ where a, b are constants and S, T are set tail variables. Here in steps (1) and (2) there is nothing left for A and B but to become $\emptyset$ and hence $A' = \{a\}$ and $B' = \{b\}$. In step (3) it happens that $A'' = B'' = C = \emptyset$. The last steps (4) and (5) yield the only most general unifier $\tau = [S/\{b|N\}, T/\{a|N\}]$.

3.3. IMPLEMENTATION

The presented set unification algorithm is implemented in ECLiPSe-Prolog (ECLiPSe, 1995) as an extension of Prolog. All solutions can be enumerated via backtracking. In order to avoid a combinatorial explosion, constraint techniques are exploited. Sets on argument positions in Prolog predicates can be written as expected with curly brackets and transformed into a metaterm.

A *metaterm* is a variable with an associated attribute. It behaves like a normal variable, however when it is unified with another term, an event is raised and a user-defined handler specifies what the result of the unification will be. Thus rapidly implementing set unification is possible. Building a full set constraint language is easy then. The Algorithms A and B may be used for the treatment of membership-constraints as an efficient kernel of the language.

The overall architecture of the system works as follows: The Algorithm D calls in step (2) Algorithm C as a subprocedure for unifying sets without tails. This task is performed by means of (optimized) membership-constraints (Algorithm B). They are preprocessed by the constraint simplification procedure – Algorithm A.

Below it is shown how basic set-theoretic operations can be expressed in a clear and concise way within our language all using the efficiently implemented membership-constraints. – First of all the $\in$-relation can directly be expressed by means of membership-constraints (1). In (2) and (3) the definitions for $\notin$ and $\subseteq$ are stated. If we are able to treat *restricted universal quantifiers* of the form $\forall x \in s$ where s is a finite set (possibly) with tail then we can express the relation $\subseteq$ as shown in (4). In (Dovier *et al.*, 1993) an algorithm for transforming extended Horn clauses with restricted universal quantifiers into ones without them is shown. This allows us to express e.g. the operations intersection $\cap$ and union $\cup$ quite naturally; see (5) and (6).

(1) $x \in \{y_1, \ldots, y_n | S\} \leftarrow memb(x, [y_1, \ldots, y_n]) \vee S = \{x|\_\}$.

(2) delay $x \notin S$ if $var(S)$.

    $x \notin \emptyset$.

    $x \notin \{y|R\} \leftarrow x \neq y \wedge x \notin R$.

(3) $\emptyset \subseteq S$.

    $\{x|R\} \subseteq S \leftarrow x \in S \wedge R \subseteq S$.

(4) $R \subseteq S \leftarrow \forall x \in R : x \in S$.

(5) $intersection(S_1, S_2, R) \leftarrow \forall x \in R : (x \in S_1 \wedge x \in S_2)$

(6) $union(S_1, S_2, R) \leftarrow \forall x \in R : (x \in S_1 \vee x \in S_2)$

That means, we can really incorporate set constraints into logic programming by the above-stated definitions of set-theoretic operations which

are based on membership-constraints. Since unification is one of the main
ingredients in logic programming, an algorithm for set unification is absolutely necessary. In addition an extended constraint simplification algorithm
is useful such as the one in (Dovier & Rossi, 1993). For example, the following clash $x \neq x$ can be detected. Here T is a new set tail variable and $\leadsto$
means "simplifies to".

$$x \in S \wedge x \notin S \leadsto S = \{x|T\} \wedge x \notin S \leadsto S = \{x|T\} \wedge (x \neq x \wedge x \notin T) \leadsto \text{FALSE}$$

4. Comparison with Other Approaches

Let us first make an experimental comparison of different methods that are
applicable to set unification problems based on their implementations in
Prolog. Later on we will also give a more theoretical analysis. In any case
we will not consider nested sets, although most of the algorithms could
handle them.

4.1. EXPERIMENTS AND EXPERIENCES

Let us now draw our attention to the examples of set unification problems
shown below. Due to the lack of space we restrict discussion to only these
few examples. Of course most of them can be generalized. Problem (2) e.g.
could easily be turned into the generic one $\{x_1, \ldots, x_n\} = \{x_1, \ldots, x_n\}$ for
$n \in \mathbb{N}$. So each problem represents a whole class.

(1) $\{w, x, y, z\} = \{a, b\}$
(2) $\{x, y, z\} = \{x, y, z\}$
(3) $\{x, f(y_1), g(y_1), g(z_1)\} = \{x, f(y_2), g(y_2), g(z_2)\}$
(4) $\{x, y|T\} = \{z\}$
(5) $\{x, y|T\} = \{a, b|T\}$
(6) $\{u, v, w|S\} = \{x, y, z|T\}$

The results we get with Prolog implementations (done by the author)
are listed in the table shown in Figure 3. The first row indicates the *problem*
number from above. In an expression of the form n/t, n means the number
of solutions that are computed by the respective method and t the overall
time in *ms* to compute the complete solution set. This notation is used in
the middle rows. The last row shows the number of *minimal* solutions.

Firstly, the results of the *naïve* approach are listed where the usual
definition of *member*/2 is used instead of membership-constraints. In this
case quite a lot of redundant solutions are computed which can be avoided
by the other methods. This and the next algorithm are applicable only to
finite sets without tails, i.e. for examples (1), (2) and (3).

problem	(1)	(2)	(3)	(4)	(5)	(6)
naïve	48/2.2	729/20.7	969/38.2	–	–	–
propia[a]	12/68.8	1/17.8	9/141.7	–	–	–
unify[b]	14/13.9	73/61.2	93/222.0	2/0.6	?	1372/204.2
set[c]	14/3.0	15/4.4	17/9.8	2/0.5	9/1.1	829/46.9
sua[d]	14/28.4	1/17.4	6/65.6	2/7.5	9/37.9	652/784.3
memb	14/6.1	1/0.6	7/5.1	4/0.6	9/3.0	1900/165.1
minimal	14	1	3	2	9	652

[a](Le Provost & Wallace, 1993)
[b](Dovier & Rossi, 1993)
[c](Stolzenburg, 1993)
[d](Arenas-Sánchez & Dovier, 1995)

Figure 3.   Table with number of solutions and timings.

Secondly, generalized propagation (Le Provost & Wallace, 1993) applied to the predicate *member*/2 is used as it is implemented in the library *propia* of ECLiPSe-Prolog (ECLiPSe, 1995). The main idea of generalized propagation is to extract information from the definition of an arbitrary predicate which is common to *all* answers to a given goal, say $p(x_1, \ldots, x_n)$. This means anti-unification (generalization) is performed on all rules defining p/n. In case there remains information not yet extracted, the constraint goal must be delayed so that completeness is preserved.

When no more information can be extracted by constraint propagation further progress requires that the system makes some choices which can be made automatically by the goal *propia_labeling*. After that the solution may still contain some constraints saying that a variable may take one of several values; so solutions are bundled sometimes. This fact explains the number of 12 solutions for example (1).

Generalized propagation (as implemented in *propia*) behaves not as good as the other methods, e.g. the membership-constraints presented here. This indicates that the optimization via non-unifiability constraints is a good idea which is not incorporated in generalized propagation. The constraint propagation steps appear to be very time-consuming.

Thirdly, the results gained by a rapid Prolog implementation of the constraint simplification procedure named *unify* in (Dovier & Rossi, 1993) are listed. It is quite fast in computing a single solution, i.e. the ratio number of solutions / time is not too bad, but it produces a lot of redundant unifiers. The case where the set tails are identical variables is a bit complicated because the plain algorithm might go into an infinite loop. This is why we do not have a measuring for example (5).

Fourthly, the row labelled with *set* shows the number of solutions computed by Algorithm D where in step (2) the set unification algorithm presented in (Stolzenburg, 1993) is inserted instead of Algorithm C. This illustrates the fact that we can take an arbitrary algorithm for unifying finite sets without tails in Algorithm D. The results are optimal for the examples where there are no variable interdependencies among the elements of the sets and both set tails are empty. But for other cases it may perform badly. See e.g. example (2). The algorithm is quite fast because there are no expensive tests.

Fifthly, the set unification algorithm *sua* (Arenas-Sánchez & Dovier, 1995) is considered. In this paper the minimal number of unifiers are stated for some sample problems. The algorithm presented therein avoids some of the redundant solutions produced by the algorithm named set here. But these optimizations can also be built into Algorithm D. For example the values of the set tails S and T can be constrained not to take in variables occurring in some left-to-right or right-to-left fork, respectively. We speak of a *fork* iff an element of one set has been unified (i.e. chosen in a membership-constraint) with *two* or more elements of the other set. – The algorithm *sua* computes in most but not all cases a minimal solution set; look at example (3). However its run-time performance is not so good because of the many tests that are built into the algorithm.

Last but not least, the behaviour of our membership-constraints (with the proposed optimization) on the examples is shown (row *memb*), i.e. Algorithm D plus C. It turns out that its performance is quite reasonable in the average, but there seems to be a trade-off. On the one hand the non-unifiability constraints avoid redundant solutions if there are variable interdependencies among the elements of the sets. But on the other hand, if not, then the Algorithm D does not always find a minimal complete solution set.

However, we made an interesting observation on membership-constraints, changing the order in which the elements are chosen by the predicate *memb*/2 for each membership-constraint. The results suggest that we can almost always achieve that only the minimal solutions are computed, provided we take the right ordering. This point needs further investigation. For optimal orderings both the number of solutions and the run time decrease simultaneously of course.

Yet another approach can be found in (Shmueli *et al.*, 1992). It is dedicated to database applications. In that paper a compilation technique is proposed which in some cases unfortunately increases the code size exponentially. In addition only *matching* of sets without tails is considered. So it is more restricted than the other algorithms. – More analyses are stated in (Stolzenburg, 1993; Stolzenburg, 1994b; Arenas-Sánchez & Dovier, 1995).

4.2. COMPLEXITY ISSUES

The algorithm presented here of course is much better than the naïve algorithm that makes use of the ordinary predicate *member/2*. The latter approach leads to an explosion of (at most)

$$T_{\text{naïve}}(m,n) = m^n \cdot n^m$$

solutions, provided we want to unify the sets $A = \{x_1, \ldots, x_m\}$ and $B = \{y_1, \ldots, y_n\}$. (In this context T shall always denote the maximal number of unifiers of finite sets without tails computed by the respective algorithm.) In addition, if there are many interdependencies among the terms in the sets, then our algorithm outperforms the others in many cases because of the constraint techniques.

Nevertheless, the set unification (decision) problem is NP-complete, even if only sets without tails are considered as shown in (Kapur & Narendran, 1986). The algorithms for unifying finite sets with tails have "only" single exponential complexity, whereas in deduction with associative-commutative functors (possibly with idempotency and unit element) the number of unifiers may be double exponential (Kapur & Narendran, 1992) (although this could be reduced by constraint techniques in many cases). However, if we restrict ourselves to finite sets with tails, single exponential complexity is obtained. Look at the following theorem.

Theorem 5 *The complexity of Algorithm D is single exponential. That means, given two sets $A \cup S$ and $B \cup T$ where A and B are as above, and S and T are the set tails, the algorithm does not compute more than $c^{p(m,n)}$ solutions where c is a constant and $p(m,n)$ a polynomial in m and n.*

The most critical step is (2) where two sets without tails have to be unified. If we assume the worst case, i.e. A and B consists of variables that are pairwise distinct, then we have T_{min} minimal solutions which is the number of minimal left and right total relations between A and B. From the theory of exponential generating functions it follows that

$$T_{\text{min}}(m,n) = m!\,n!\cdot \frac{\partial f^{m+n}}{\partial x^m\,\partial y^n}(0,0) \quad \text{where} \quad f(x,y) = e^{x\,(e^y-1)+y\,(e^x-1)-xy}$$

which has been proven in (Stolzenburg, 1993). – Anyway, all algorithms for unifying finite sets without tails presented here have single exponential complexity, even the naïve one because of the following proposition:

Theorem 6 *For all $m,n \geq 1$ and $c \geq e^{2/e} \approx 2.087$ it holds:*

$$\min(m,n)! \overset{(1)}{\leq} T_{\text{min}}(m,n) \overset{(2)}{\leq} T_{\text{set}}(m,n) \overset{(3)}{\leq} T_{\text{memb}}(m,n) \overset{(4)}{\leq} T_{\text{naïve}}(m,n) \overset{(5)}{\leq} c^{m \cdot n}$$

Due to the lack of space we will not carry out the proof in detail but give only some remarks on the parts of the chain of inequations:

(1) $\min(m, n)!$ is a lower bound of the complexity because each permutation of the smaller set leads to a solution. – Since factorials increase faster than any power c^{m+n}, the degree of the polynomial in the exponent must be greater than 1.

(2) Since all considered algorithms compute a complete set of unifiers this relation is quite clear. – It holds $T_{\min} = T_{\text{set}}$ if there are no variable interdependencies among the terms in the sets.

(3) This holds because the solution set computed by the algorithm in (Stolzenburg, 1993) always is a subset of the solutions computed with membership-constraints.

(4) As in the previous item, also a subset-relationship holds here.

(5) The proof for this case requires standard mathematical analysis techniques. – It follows that the degree d of the polynomial in the exponent must be less than 2, i.e. we estimate the complexity by a function of the form $c^{(m+n)^d}$.

The lower and upper bounds in the chain of inequations above imply that for the degree d it holds $1 < d < 2$. Furthermore d tends to 1 for $m, n \to \infty$. But this only gives a rough estimation of the complexity. Of course Algorithm C whose analysis is given next is much better than the naïve algorithm.

Proposition 7 *Algorithm C produces (at most) the following number of solutions:*

$$T_{\text{memb}}(m, n) = \sum_{k=0}^{\min(m,n)} k! \left\{ {m \atop k} \right\} \left\{ {n \atop k} \right\}$$

Here the notation $\left\{ \begin{smallmatrix} \cdot \\ \cdot \end{smallmatrix} \right\}$ denotes Stirling numbers of the second kind.

5. Conclusion

The Algorithm D presented here is reasonably efficient and easily implementable. It uses constraint techniques and delaying mechanisms which avoid many redundant solutions and hence combinatorial explosion. It can be embedded in a full set constraint language (Dovier & Rossi, 1993; Bruscoli *et al.*, 1994) such that it is possible to use sets as first-class citizens in logic programming. – Furthermore we gave a complexity analysis which gives an asymptotic estimation of our algorithm as well as the problem itself. It shows the good performance of the algorithm presented here.

Acknowledgements. I would like to thank Peter Baumgartner, Jürgen Dix, Andreas Podelski, Jörn Richts, Gianfranco Rossi, Martin Volk, Graham Wrightson and some anonymous referees for comments on this paper or helpful discussions.

FRIEDER STOLZENBURG
Institut für Informatik
Universität Koblenz, Germany

References

D. Aliffi, G. Rossi, A. Dovier, and E. G. Omodeo. Unification of hyperset terms. In E. G. Omodeo and G. Rossi, editors, *Proceedings of the Workshop on Logic Programming with Sets, in Conjunction with the 10th International Conference on Logic Programming*, pages 27–30, Budapest, Hungary, June 1993.

P. Arenas-Sánchez and A. Dovier. Minimal set unification. In M. Hermenegildo and S. D. Swierstra, editors, *Proceedings of the 7th International Symposium on Programming Language Implementation and Logic Programming*, pages 397–414. Springer, Berlin, Heidelberg, New York, 1995. LNCS 982.

P. Bruscoli, A. Dovier, E. Pontelli, and G. Rossi. Compiling intensional sets in CLP. In P. Van Hentenryck, editor, *Proceedings of the 11th International Conference on Logic Programming, Santa Margherita, Ligure, Italy, June 1994*, pages 647–661. MIT Press, Cambridge, MA, London, England, 1994.

A. Dovier, E. G. Omodeo, E. Pontelli, and G. Rossi. Embedding finite sets in a logic programming language. In E. Lamma and P. Mello, editors, *Proceedings of the 3rd International Workshop on Extensions of Logic Programming, Bologna, Italy, February 1992*, pages 150–167. Springer, Berlin, Heidelberg, New York, 1993. LNAI 660.

A. Dovier and G. Rossi. Embedding finite sets in CLP. In D. Miller, editor, *Proceedings of the International Logic Programming Symposium*. MIT Press, Cambridge, MA, London, England, 1993.

ECRC GmbH, München. *ECLiPSe 3.5: User Manual – Extensions User Manual*, February 1995.

C. Gervet. Conjunto: Constraint logic programming with finite set domains. In M. Bruynooghe, editor, *Proceedings of the International Logic Programming Symposium, Ithaca, NY, November 1994*, pages 339–358. MIT Press, Cambridge, MA, London, England, 1994.

J. Jaffar and M. J. Maher. Constraint logic programming: a survey. *Journal of Logic Programming*, 19,20:503–581, 1994.

D. Kapur and P. Narendran. NP-completeness of the set unification and matching problems. In J. H. Siekmann, editor, *Proceedings of the 8th International Conference on Automated Deduction, Oxford, July 1986*, pages 489–495. Springer, Berlin, Heidelberg, 1986. LNCS 230.

D. Kapur and P. Narendran. Double-exponential complexity of computing a complete set of AC-unifiers. In *Proceedings of the 7th Annual Symposium on Logic in Computer Science, Santa Cruz, CA*, pages 11–21, 1992.

T. Le Provost and M. Wallace. Generalized constraint propagation over the CLP scheme. *Journal of Logic Programming*, 16(3&4):319–359, 1993.

B. Legeard, H. Lombardi, E. Legros, and M. Hibti. A constraint satisfaction approach to set unification. In *Proceedings of the 13th International Conference on Artificial Intelligence, Expert Systems and Natural Language*, pages 265–276, Avignon, May 1993.

S. Manandhar. An attributive logic of set descriptions and set operations. In *Proceedings of the 32nd Annual Meeting of the Association for Computational Linguistics*, 1994.

C. J. Pollard and M. D. Moshier. Unifying partial description of sets. In P. Hanson, editor, *Information, Language, and Cognition*, pages 285–322. University of British Columbia Press, Vancouver, BC, 1990.

O. Shmueli, S. Tsur, and C. Zaniolo. Compilation of set terms in the logic data language (LDL). *Journal of Logic Programming*, 12(1&2):89–119, 1992.

J. H. Siekmann. Unification theory. *Journal of Symbolic Computation*, 7(1):207–274, 1989.

F. Stolzenburg. An algorithm for general set unification and its complexity. In E. G. Omodeo and G. Rossi, editors, *Proceedings of the Workshop on Logic Programming with Sets, in Conjunction with the 10th International Conference on Logic Programming*, pages 17–22, Budapest, Hungary, June 1993.

F. Stolzenburg. Logic programming with sets by membership-constraints. In N. E. Fuchs and G. Gottlob, editors, *Proceedings of the 10th Logic Programming Workshop*, Universität Zürich, 1994. Institut für Informatik. Technical Report ifi 94.10.

F. Stolzenburg. Membership-constraints and some applications. Fachberichte Informatik 5/94, Universität Koblenz-Landau, Koblenz, May 1994.

P. Van Hentenryck. *Constraint Satisfaction in Logic Programming*. MIT Press, Cambridge, MA, London, England, 1989.

AGOSTINO DOVIER, ALBERTO POLICRITI, AND GIANFRANCO ROSSI

INTEGRATING LISTS, MULTISETS, AND SETS IN A LOGIC PROGRAMMING FRAMEWORK

Abstract. The first order theories of lists, bags, compact-lists (i.e., lists where the number of contiguous occurrences of each element is immaterial), and sets are introduced via axioms. Such axiomatizations are shown to be especially suitable for the integration with free functor symbols governed by the classical Clark's axioms in the context of Constraint Logic Programming. Adaptations of the extensionality principle to the various theories taken into account is then exploited in the design of unification algorithms for the considered data structures. All the theories presented can be combined providing frameworks to deal with several of the proposed data structures simoultaneously. The unification algorithms proposed can be combined (merged) as well to produce engines for such combination theories.

1. Introduction

Lists are a fundamental data structure in programming languages. In lists, the order of elements and the number of occurrences of each element are meaningful information. Applications can advantageously exploit these properties when dealing with lists. However, there are a number of cases where these properties turn out to be too restrictive: different data abstractions may fit more naturally the problem requirements. In particular, it seems convenient to consider the following data abstractions:

- *sets*, in which the order of elements and the number of occurrences of each element do not matter;
- *multi-sets* (or bags), in which the number of occurrences of each element is important, whereas the order of elements does not matter;
- *compact-lists*, in which the order of elements is important whereas the number of contiguous occurrences of each element is not.

F. Baader and K.U. Schulz (eds.), Frontiers of Combining Systems, 303–319.
© 1996 *Kluwer Academic Publishers.*

Athough only relatively few programming languages support sets and multi-sets as primitive features of the language, they are well-known data structures, whose usefulness as a powerful data abstraction mechanism is widely recognized.

In contrast, compact-lists have been scarcely studied. To give an intuitive semantics of compact-lists, consider the following real-life problem: assume a laser printer can print on sheets of different sizes, but it has a unique paper feeder. Any time the size of the paper changes, the printer must switch paper feeder. The information necessary for the scheduler in order to correctly perform such switches is conveniently modeled by a compact-list (cf. § 2.3 for the formal definition of compact-lists).

The framework in which we assume the above mentioned data structures are to be incorporated is a (constraint) logic programming one, where a careful choice of suitable data structures is crucial for the control of the execution. All the considered data structures will be represented by terms of a logic language, moreover, we assume that the language provides, a number (possibly zero or infinite) of free functional symbols in addition to the data structure constructors. Thus, we will consider so-called *hybrid theories*, in which the basic objects are Herbrand terms. Therefore, our data objects will be standard terms, lists, bags, compact-lists, and sets of standard terms, as well as any possible combination of them.

The paper is organized as follows. In § 2 we give an axiomatic characterization of the considered data structures, showing the close connections among the relevant theories. Then, in § 3, the unification algorithms somehow suggested by the axioms expressing equality in the different contexts considered in the previous section are presented.

2. An axiomatic view of lists, compact-lists, multi-sets, and sets

In what follows we will use standard Prolog syntactic conventions and notations. In particular, $[\cdot\,|\,\cdot]$ will be used as the list constructor and the constant nil as the empty list. Thus, for instance, $[a\,|\,\text{nil}]$ and $[a\,|\,[b\,|\,X]]$, where X is a variable, are two lists, which can also be denoted simply as $[a]$ and $[a,b\,|\,X]$, respectively.

In addition, the following functional symbols are introduced to denote multi-sets, compact-lists, and sets (empty multi-sets, compact-lists, and sets are all denoted by nil):

- $\{\!\!\{\cdot\,|\,\cdot\}\!\!\}$ (of arity 2) for multi-sets,
- $[\![\cdot\,|\,\cdot]\!]$ (of arity 2) for compact-lists,
- $\{\cdot\,|\,\cdot\}$ (of arity 2) for sets.

Notational conventions similar to those used for lists will be freely exploited also for multi-sets, compact-lists, and sets. For example, $\{\,a, b \,|\, X\}$ is used to denote a partially specified set with two elements a and b and a variable part X.

2.1. LISTS

Consider the first order theory with equality '$\doteq$', and membership '$\in$' predicate symbols, and consisting of the two axioms

$$(N) \qquad \exists z \forall x\, (x \notin z)$$
$$(W) \qquad \forall y\, v \exists w \forall x\, (x \in w \leftrightarrow x \in v \lor x \doteq y).$$

Skolemizing (N) and (W), we introduce the two functional symbols $\texttt{nil}$ and $[\,\cdot\,|\,\cdot\,]$, and we can rewrite (N) and (W) as

$$(N^l) \qquad \forall x\, (x \notin \texttt{nil}), \text{ and}$$
$$(W^l) \qquad \forall y\, v\, x\, (x \in [\,y\,|\,v\,] \leftrightarrow x \in v \lor x \doteq y).$$

The language of the theory consists of the signatures $\Pi = \{\doteq, \in\}$ for the predicate symbols and $\Sigma = \{\texttt{nil}, [\,\cdot\,|\,\cdot\,], \cdots\}$ for the functional symbols. It can be proved (see (Dovier, 1996)) that any model of NW (the theory consisting of the axioms (N) and (W)) must necessarily be infinite and that this theory is not complete (nor model-complete). Moreover, it has been shown (e.g. in (Vaught, 1962; Parlamento & Policriti, 1988; Bellé & Parlamento, 1994)) that NW is not decidable. Nevertheless, this theory can be strengthened, by adding new axioms, in order to obtain a complete and decidable theory for suitable classes of sentences.

The following three axiom shemata (called freeness axioms, or Clark's equality axioms—see (Clark, 1978)) are usually introduced in logic programming and will play an important role in our axiomatization:

$$(F_1) \qquad \forall x_1 \cdots x_n y_1 \cdots y_n \quad \left(\begin{array}{c} f(x_1, \ldots, x_n) \doteq f(y_1, \ldots, y_n) \\ \rightarrow x_1 \doteq y_1 \land \cdots \land x_n \doteq y_n \end{array} \right) \quad f \in \Sigma$$
$$(F_2) \qquad \forall x_1 \cdots x_m y_1 \cdots y_n \quad f(x_1, \ldots, x_m) \not\doteq g(y_1, \ldots, y_n) \qquad f \not\equiv g$$
$$(F_3) \qquad \qquad\qquad \forall x\, (x \not\doteq t[x])$$

where $t[x]$ denotes a Σ-term having x as a proper subterm.

Axiom (F_1) holds also for $[\,\cdot\,|\,\cdot\,] \in \Sigma$, and expresses the adaptation of the classical extensionality principle to lists.

Axiom (F_3) states that there exists no term which is also a substerm of itself. Removing this axiom would allow us to accept equalities of the form $x \doteq f(x)$ whose solution requires to extend the interpretation domain so as to take into account also the so-called rational terms. Such an extension is considered in (Dovier, 1996) and, with regards to sets and multi-sets only, in (Omodeo *et al.*, 1993). In this paper, on the contrary, we prefer for the sake of simplicity not considering this kind of extension.

In the following three subsections we will adapt the above axioms so as to fulfill the intended meaning of bags, compact-lists, and sets. Before proceeding, however, we need to consider the following problem: what are the elements of an object denoted by a term $f(t_1, \ldots, t_n)$, with f free? For example one can write a term t of the form $[t_1, \ldots, t_n \,|\, a]$, and (W) states that $t_1, \ldots, t_n$ are elements of t, whereas the remaining elements of t are those of a.

To keep the notion of list (as well as of bag, compact-list and set) as distinct as possible from the notion of 'free' term, we will strenghten axiom (N) so as to keep *empty* any term which is not a list (a bag, a compact-list, a set) of terms.

Therefore, we introduce the axiom schema

(K) $\forall x\, y_1 \cdots y_n\, (x \notin f(y_1, \ldots, y_n))$
for any $f \in \Sigma$, $ar(f) = n$, f distinct from $[\,\cdot\,|\,\cdot\,]$, $\{\!\!\{\,\cdot\,|\,\cdot\,\}\!\!\}$, $[\![\,\cdot\,|\,\cdot\,]\!]$, $\{\,\cdot\,|\,\cdot\,\}$

which generalizes (N) and will be used in its place hereinafter.[1]

We will call *Ur-Element* (cf. (Tarski & Givant, 1986)) any term of the form $f(t_1, \ldots, t_n)$, where $f \in \Sigma$, f distinct from $[\,\cdot\,|\,\cdot\,]$, $\{\!\!\{\,\cdot\,|\,\cdot\,\}\!\!\}$, $[\![\,\cdot\,|\,\cdot\,]\!]$, $\{\,\cdot\,|\,\cdot\,\}$, $ar(f) = n$. When an ur-element is ground, it will be called a *kernel*. Intuitively, lists (bags, compact-lists and sets) can be seen as built starting from a kernel (in particular, from nil) and then adding to the kernel the other elements composing the data structure.

Axioms (K), (W^l), (F_1), (F_2), and (F_3), along with standard equality axioms, identify the hybrid theory of lists over the alphabets Π and Σ.

Remark 1.1 *In the context of lists, as well as in the other axiomathic theories we will consider, objects denoted by <u>ground</u> terms are forced to have a finite number of elements. We do not consider the case of objects non-denoted by terms, which can easily be avoided by suitable forms of the domain closure axiom and are outside the scope of this paper.*

2.2. BAGS

The signature of an hybrid theory of bags must contain the binary functional symbol $\{\!\!\{\,\cdot\,|\,\cdot\,\}\!\!\}$ and nil. By skolemization of (W) we can rewrite (W) as follows

---

[1] Indeed, since nil belongs to Σ, (N^l) is an instance of (K). Note also that when one of the two symbols f and g in (F_2) is the interpreted functional symbol $[\,\cdot\,|\,\cdot\,]$ ($\{\!\!\{\,\cdot\,|\,\cdot\,\}\!\!\}$, $[\![\,\cdot\,|\,\cdot\,]\!]$, $\{\,\cdot\,|\,\cdot\,\}$), then (F_2) is a theorem of KW.

(W^m) $\forall y \, v \, x \, (x \in \{\!\!\{ \, y \, | \, v \, \}\!\!\} \leftrightarrow x \in v \lor x \doteq y).$

For the context of bags the symbol $[\cdot \, | \, \cdot]$ is replaced by $\{\!\!\{ \, \cdot \, | \, \cdot \, \}\!\!\}$ in Σ. The behavior of the interpreted functional symbol $\{\!\!\{ \, \cdot \, | \, \cdot \, \}\!\!\}$ is regulated by the following axiom (permutativity axiom):

(E_1^m) $\forall xyz \, \{\!\!\{ \, x, y \, | \, z \, \}\!\!\} \doteq \{\!\!\{ \, y, x \, | \, z \, \}\!\!\}$

which, intuitively, states that the order of elements in a bag is immaterial. This means, for example, that

$$\{\!\!\{ \, a \, | \, \{\!\!\{ \, b \, \}\!\!\} \, \}\!\!\} \, (i.e. \, \{\!\!\{ \, a, b \, \}\!\!\}) \quad \doteq \quad \{\!\!\{ \, b \, | \, \{\!\!\{ \, a \, \}\!\!\} \, \}\!\!\} \, (i.e. \, \{\!\!\{ \, b, a \, \}\!\!\})$$

even if a is distinct from b. More generally:

Lemma 1.2 *Let* $\pi : \{1, \ldots, n\} \longrightarrow \{1, \ldots, n\}$ *be a permutation, then*

$$KW^m E_1^m \; \vdash \; \{\!\!\{ \, s_1, \ldots, s_n \, | \, t \, \}\!\!\} \doteq \{\!\!\{ \, s_{\pi_1}, \ldots, s_{\pi_n} \, | \, t \, \}\!\!\}$$

for any tuple of Σ-*terms* $s_1, \ldots, s_n, t$.

As opposed to what happens for lists, in the context of bags, axiom schema (F_1) does not hold when f is istantiated to $\{\!\!\{ \, \cdot \, | \, \cdot \, \}\!\!\}$, as it ensues from the above example. Since it will turn out that axiom schema (F_1) does not hold also in the cases of compact-lists and sets, we will modify it as follows:

(F_1') $\forall x_1 \cdots x_n y_1 \cdots y_n$ $\left(\begin{array}{c} f(x_1, \ldots, x_n) \doteq f(y_1, \ldots, y_n) \\ \rightarrow x_1 \doteq y_1 \land \cdots \land x_n \doteq y_n \end{array} \right)$
for any $f \in \Sigma$, f *distinct from* $\{\!\!\{ \, \cdot \, | \, \cdot \, \}\!\!\}$, $[\![\, \cdot \, | \, \cdot \,]\!]$, $\{\cdot \, | \, \cdot\}$, $ar(f) = n$.

A by-product of the fact that (F_1) does not hold is that in $KW^m E_1^m$ we lack in a principle for establishing equality between objects. On the other hand, any such equality principle must be consistent with the following simple result relating '$\in$' and '$\doteq$':

Lemma 1.3 *For all* $n \in \omega$ *and for all* $x, y_1, \ldots, y_n$

$$KW^m E_1^m \vdash x \in \{\!\!\{ \, y_1, \ldots, y_n \, \}\!\!\} \; iff \; KW^m E_1^m \vdash \exists z \, (\{\!\!\{ \, y_1, \ldots, y_n \, \}\!\!\} \doteq \{\!\!\{ \, x \, | \, z \, \}\!\!\}) \, .$$

We can introduce the following extensionality axiom for bag comparison: *bag-extensionality* with kernels

(E_k^m) $\forall y_1 y_2 v_1 v_2$ $\left(\begin{array}{c} \{\!\!\{ \, y_1 \, | \, v_1 \, \}\!\!\} \doteq \{\!\!\{ \, y_2 \, | \, v_2 \, \}\!\!\} \leftrightarrow \\ (y_1 \doteq y_2 \land v_1 \doteq v_2) \lor \\ \exists z \, (v_1 \doteq \{\!\!\{ \, y_2 \, | \, z \, \}\!\!\} \land v_2 \doteq \{\!\!\{ \, y_1 \, | \, z \, \}\!\!\}) \end{array} \right)$

Axiom (E_k^m) states that two non-empty bags are equal if and only if they are based on the same *kernel* (in particular `nil`) and they have the same number of occurrences of each element, regardless their order.

Axioms (K), (W^m), (E_k^m), (F_1'), (F_2), and (F_3), along with standard equality axioms, identify the hybrid theory of multi-sets over the alphabets Π and Σ.

2.3. COMPACT-LISTS

Let Σ be $\{\texttt{nil}, [\![\cdot \,|\, \cdot]\!], \ldots\}$, relative to the new version of (W) for compact-lists

$$(W^c) \qquad \forall y\, v\, x\, (x \in [\![\, y \,|\, v\,]\!] \leftrightarrow x \in v \vee x \doteq y).$$

The fundamental property of the compact-list constructor $[\![\cdot \,|\, \cdot]\!]$ is the *absorption property*, described by the following axiom

$$(E_2^c) \qquad \forall xy\, [\![\, x, x \,|\, y\,]\!] \doteq [\![\, x \,|\, y\,]\!]$$

which, intuitively, states that contiguous duplicates in a compact-list are immaterial.

Similarly to bags, also for compact-lists axiom schema (F_1) does not correctly interpret the behavior of the functional symbol $[\![\cdot \,|\, \cdot]\!]$, as it ensues from the following example:

$$[\![\, a \,|\, [\![\, a\,]\!]\,]\!]\, (i.e.\ [\![\, a, a\,]\!]) \quad \doteq \quad [\![\, a \,|\, \texttt{nil}\,]\!]\, (i.e.\ [\![\, a\,]\!]).$$

Moreover, also the freeness axiom (F_3) must be modified so as to agree with the intended meaning of $[\![\cdot \,|\, \cdot]\!]$. Indeed, one can observe, in particular, that an equation such as $x \doteq [\![\, a \,|\, x\,]\!]$ admits a finite tree solution, namely a solution that binds x to the term $[\![\, a \,|\, t\,]\!]$, where t is any term.

We first modify axiom (F_1) by introducing the extensionality principle for compact-lists: *compact-lists extensionality* with kernels

$$(E_k^c) \qquad \forall y_1 y_2 v_1 v_2 \left(\begin{array}{l} [\![\, y_1 \,|\, v_1\,]\!] \doteq [\![\, y_2 \,|\, v_2\,]\!] \leftrightarrow \\ (y_1 \doteq y_2 \wedge v_1 \doteq v_2) \vee \\ (y_1 \doteq y_2 \wedge v_1 \doteq [\![\, y_2 \,|\, v_2\,]\!]) \vee \\ (y_1 \doteq y_2 \wedge [\![\, y_1 \,|\, v_1\,]\!] \doteq v_2) \end{array}\right)$$

The first disjunct takes care of the simplest case where no duplicates occur in the two compact-lists. The second and third cases, instead, are relative to the case in which there are duplicates in the right-hand side term and in the left-hand side term, respectively.

As far as axiom (F_3) is concerned, notice that the following lemma holds:

Lemma 1.4 *In any model of $KW^c E_2^c$ such that every element has a finite number of members, the following holds:*

$$\exists x\, (x \doteq [\![\, y_1, \ldots, y_n \,|\, x\,]\!]) \quad \leftrightarrow \quad (y_1 \doteq y_2 \doteq \cdots \doteq y_n).$$

Notice that the finiteness requirement is necessary for Lemma 1.4 since, if we would accept infinite solutions, then, for any $y_1, \ldots, y_n$, the equation $x \doteq [\![y_1, \ldots, y_n \,|\, x]\!]$ would admit always the infinite (rational) solution

$$x = [\![y_1, \ldots, y_n, y_1, \ldots, y_n, y_1, \ldots, y_n, \ldots]\!] .$$

Axiom F_3 is therefore replaced by

(F_3^c) $\forall x \; (x \neq t[x])$
unless t has the form $[\![t_1, \ldots, t_n \,|\, x]\!]$, x not occurring in $t_1, \ldots, t_n$, and $t_1 \doteq \cdots \doteq t_n$.

The hybrid theory of compact-lists over Π and Σ is identified by axioms (K), (W^c), (E_k^c), (F_1'), (F_2), and (F_3^c), along with standard equality axioms.

2.4. SETS

The last theory we consider is a simple theory of sets. Σ is now required to contain $\texttt{nil}$ and $\{\cdot \,|\, \cdot\}$, and (W) becomes

(W^s) $\forall y \, v \, x \, (x \in \{y \,|\, v\} \leftrightarrow x \in v \lor x \doteq y).$

Sets have both the *permutativity* and the *absorption* properties which, in the case of the set constructor $\{\cdot \,|\, \cdot\}$, can be rewritten as follows:

(E_1^s) $\forall xyz \, \{x, y \,|\, z\} \doteq \{y, x \,|\, z\}$

(E_2^s) $\forall xy \, \{x, x \,|\, y\} \doteq \{x \,|\, y\}.$

Similarly to what has been done in the previous two subsections, we define an extensionality criterion for testing equality between two sets

(E_k^s) $\forall y_1 y_2 v_1 v_2$
$$\left(
\begin{array}{l}
\{y_1 \,|\, v_1\} \doteq \{y_2 \,|\, v_2\} \;\leftrightarrow \\
\quad (y_1 \doteq y_2 \land v_1 \doteq v_2) \lor \\
\quad (y_1 \doteq y_2 \land v_1 \doteq \{y_2 \,|\, v_2\}) \lor \\
\quad (y_1 \doteq y_2 \land \{y_1 \,|\, v_1\} \doteq v_2) \lor \\
\quad \exists k \, (v_1 \doteq \{y_2 \,|\, k\} \land v_2 \doteq \{y_1 \,|\, k\})
\end{array}
\right)$$

Such principle combines the results proved separately for (E_k^m) and (E_k^c), so that both duplicates and ordering of elements in sets are immaterial.

The modification of axiom (F_3) for sets simplifies the one used for compact-lists:

(F_3^s) $\forall x \; (x \neq t[x])$
unless t has the form $\{t_1, \ldots, t_n \,|\, x\}$, x not occurring in $t_1, \ldots, t_n$.

The hybrid theory of sets (with $\Pi = \{\doteq, \in\}$, and $\Sigma = \{\texttt{nil}, \{\cdot \,|\, \cdot\}, \ldots\}$) is therefore identified by axioms (K), (W^s), (E_k^s), (F_1'), (F_2), and (F_3^s), along with standard equality axioms.

Remark 1.5 *Notice that it is easily seen that in every model of each of the considered theories in which all elements are denoted by terms (hence finite—cf. remark at the end of § 1.1) the membership cannot form either cycles or infinite descending chains. To this extent the above theories can be considered well-founded.*

Remark 1.6 *The axiomatizations presented can easily be combined in order to obtain axiomatic theories capable to deal with any subset of the collection of proposed data structures. Moreover, the unification algorithms presented in the next section can easily be merged to solve the unification problem relative to such "combined" context.*

3. Unification of bags, compact-lists and sets

While unification for (hybrid) lists can be performed in linear time (cf. (Paterson & Wegman, 1976; Martelli & Montanari, 1982)), the unification problems for all the other data structures analyzed in this paper are NP-complete. NP-hardness can easily be proved for example via reduction of the 3-SAT famous problem to the unification problem for bags, compact-lists and sets (Dovier, 1996; Dovier *et al.*, 1993). For the NP-completeness, see (Kapur & Narendran, 1986; Omodeo & Policriti, 1994; Dovier, 1996).

We start recalling the unification algorithm for hybrid lists, that is basically the standard unification algorithm à la Robinson. This will allow us to introduce the style and the notation we will also employ for the subsequent algorithms. Moreover, all actions of the standard algorithm remain almost unchanged in all the other cases.

Unify_lists terminates on any input system of equations $\mathcal{E}$, returning an equivalent system in solved form[2] from which it is immediate to obtain the *unique most general unifier* for the given unification problem. Moreover, correctness and completeness of the algorithm can be proved w.r.t. the corresponding theory presented in § 2.1 (cf., for instance, (Lloyd, 1987; Lassez *et al.*, 1986)).

Action ($l4$) performs the so-called *occur-check*: it checks the well-foundedness of the unique most general solution of the system.

In the algorithms for bags, compact-lists, and sets, action ($l6$), when f is the corresponding interpreted functional symbol, will be replaced by non-deterministic actions which reflect the axioms (E_k^b), (E_k^c), and (E_k^s), respectively.

---

[2]Remember that, a system of equations $\mathcal{E}$ is said to be in *solved form* if it has the form $\{X_1 \doteq t_1, \ldots, X_n \doteq t_n\}$ and the X_is are distinct variables which do not occur in r.h.s. terms of any equation of $\mathcal{E}$.

function Unify_lists($\mathcal{E}$):

($l1$) $\qquad\qquad\qquad\qquad X \doteq X \wedge \mathcal{E} \quad \mapsto \quad \mathcal{E}$

($l2$) $\qquad\qquad\left.\begin{array}{c} t \doteq X \wedge \mathcal{E} \\ t \text{ is not a variable} \end{array}\right\} \mapsto X \doteq t \wedge \mathcal{E}$

($l3$) $\qquad\left.\begin{array}{c} X \doteq t \wedge \mathcal{E} \\ X \text{ does not occur in } t \\ X \text{ occurs in } \mathcal{E} \end{array}\right\} \mapsto \mathcal{E}[X/t] \wedge X \doteq t$

($l4$) $\qquad\qquad\left.\begin{array}{c} X \doteq t \wedge \mathcal{E} \\ X \text{ occurs in } t \end{array}\right\} \mapsto \texttt{fail}$

($l5$) $\left.\begin{array}{c} f(s_1, \ldots, s_m) \doteq g(t_1, \ldots, t_n) \wedge \mathcal{E} \\ f \text{ different from } g \end{array}\right\} \mapsto \texttt{fail}$

($l6$) $\qquad f(s_1, \ldots, s_m) \doteq f(t_1, \ldots, t_m) \wedge \mathcal{E} \quad \mapsto$
$$s_1 \doteq t_1 \wedge \ldots \wedge s_m \doteq t_m \wedge \mathcal{E}.$$

3.1. UNIFICATION OF HYBRID BAGS

The first five actions are exactly the same as in the algorithm for hybrid lists. In addition, we need to restrict applicability of action ($l6$) to non-bag terms only (action ($m6$)), and to introduce a new action to deal with bag-bag equations (action ($m7$)). This case introduces a source of don't know non-determinism: both alternatives (i) and (ii) must be exploited in order to guarantee completeness.

($m1$)—($m5$) $\qquad\qquad$ as ($l1$)—($l5$) in Unify_lists

($m6$) $\qquad\left.\begin{array}{c} f(s_1, \ldots, s_m) \doteq f(t_1, \ldots, t_m) \wedge \mathcal{E} \\ f \in \Sigma \setminus \{\!\{\cdot \mid \cdot\}\!\} \end{array}\right\} \mapsto$
$$s_1 \doteq t_1 \wedge \ldots \wedge s_m \doteq t_m \wedge \mathcal{E}$$

($m7$) $\qquad\qquad\qquad \{\!\{t \mid s\}\!\} \doteq \{\!\{t' \mid s'\}\!\} \wedge \mathcal{E} \quad \mapsto$
$\qquad\qquad$ (i) $\quad t \doteq t' \wedge s \doteq s' \wedge \mathcal{E}$
$\qquad\qquad$ (ii) $\quad s \doteq \{\!\{t' \mid N\}\!\} \wedge \{\!\{t \mid N\}\!\} \doteq s' \wedge \mathcal{E}.$

Although sound and complete the above algorithm does not always terminate. As an example, consider the following input system:

$\{\!\{T \mid S\}\!\} \doteq \{\!\{T' \mid S\}\!\} \overset{m7(ii)}{\mapsto}$
$S \doteq \{\!\{T' \mid N\}\!\} \wedge \{\!\{T \mid N\}\!\} \doteq S \overset{m3}{\mapsto}$
$S \doteq \{\!\{T' \mid N\}\!\} \wedge \{\!\{T \mid N\}\!\} \doteq \{\!\{T' \mid N\}\!\}.$

The last system is a system at least as difficult as the starting one.

More generally, for any situation of the form

$$
\begin{aligned}
\{\!\!\{\cdots \mid S_1\}\!\!\} &\doteq \{\!\!\{\cdots \mid S_2\}\!\!\} \quad \wedge \\
\{\!\!\{\cdots \mid S_2\}\!\!\} &\doteq \{\!\!\{\cdots \mid S_3\}\!\!\} \quad \wedge \\
&\;\;\vdots \qquad\qquad\qquad \wedge \\
\{\!\!\{\cdots \mid S_n\}\!\!\} &\doteq \{\!\!\{\cdots \mid S_1\}\!\!\},
\end{aligned}
$$

it is easy to find a non-deterministic sequence of actions leading to non-termination of the bag unification algorithm. An intuitive reason for non-termination is that the unification algorithm looks for all possible substitutions for S such that $\{\!\!\{T \mid S\}\!\!\} \doteq \{\!\!\{T' \mid S\}\!\!\}$ holds. In this process an infinite number of possibilities, corresponding to subsequently incrementations of S is generated. Clearly the solution in which S is unbounded is the most general one.

The situation described in the first example above (the base case) can be easily handled as special. Let **tail** and **de_tail** be the following functions:

$$
\begin{aligned}
\mathsf{tail}(\mathtt{nil}) &= \mathtt{nil} \\
\mathsf{tail}(X) &= X \\
\mathsf{tail}(\{\!\!\{t \mid s\}\!\!\}) &= \mathsf{tail}(s)
\end{aligned}
\qquad
\begin{aligned}
\mathsf{de\_tail}(X) &= \mathtt{nil} \\
\mathsf{de\_tail}(\{\!\!\{t \mid s\}\!\!\}) &= \{\!\!\{t \mid \mathsf{de\_tail}(s)\}\!\!\}
\end{aligned}
$$

Action $(m7)$ can be splitted into two sub-actions. If $\mathsf{tail}(s)$ and $\mathsf{tail}(s')$ are not the same variable then perform action $(m7)$. Otherwise replace $\{\!\!\{t \mid s\}\!\!\} \doteq \{\!\!\{t' \mid s'\}\!\!\}$ with $\mathsf{de\_tail}(\{\!\!\{t \mid s\}\!\!\}) \doteq \mathsf{de\_tail}(\{\!\!\{t' \mid s'\}\!\!\})$.

However, to solve the problem also in the more general case we find it convenient to split the system $\mathcal{E}$ into two parts, $\mathcal{E}_1$ and $\mathcal{E}_2$, the second of which is dealt with as a stack. In this way, we can ensure enough determinism in the algorithm to guarantee that when an equation $\{\!\!\{s_1, \ldots, s_m \mid S\}\!\!\} \doteq \{\!\!\{t_1, \ldots, t_n \mid S'\}\!\!\}$ is encountered, the sequence of actions executed is such that the algorithm returns a conjunction of equations of the form $s_{i_1} \doteq t_{j_1}, \ldots, s_{i_k} \doteq t_{j_k}$, along with either:

- two equations of the form $S \doteq \{\!\!\{t_{j_{k+1}}, \ldots, t_{j_n} \mid N\}\!\!\}$,
  $S' \doteq \{\!\!\{s_{i_{k+1}}, \ldots, s_{i_m} \mid N\}\!\!\}$, or
- one equation of the form $S \doteq \{\!\!\{t_{j_{k+1}}, \ldots, t_{j_n} \mid S'\}\!\!\}$, or
- one of the form $S' \doteq \{\!\!\{s_{i_{k+1}}, \ldots, s_{i_m} \mid S\}\!\!\}$.

After that, by applying the substitutions for the variables S and S', even if a new variable N is introduced into the system, S and/or S' become eliminable, that is they occur in $\mathcal{E}$ only as l.h.s. of one equation, so that, in a sense, they disappear from the system.

The bag unification algorithm is shown in Figure 1.

function Unify_bags($\mathcal{E}$);
$\mathcal{E}_2 := \emptyset$;
repeat
    repeat
        move the first equation e of $\mathcal{E}_2$ (if any) to $\mathcal{E}_1$;
        if e is not in solved form w.r.t. $\mathcal{E}_1$ and $\mathcal{E}_2$
        then Unify_bags_actions($\mathcal{E}$, e)
    until $\mathcal{E}_2 = \emptyset$;
    select arbitrarily from $\mathcal{E}_1$ an equation e <u>not</u> in solved form;
    Unify_bags_actions($\mathcal{E}$, e)
until $\mathcal{E}_1$ is in solved form and $\mathcal{E}_2 = \emptyset$.

function Unify_bags_actions($\mathcal{E},e$);
let $\mathcal{E}'$ be $\mathcal{E}_1 \setminus \{e\}$;
  case e of

$(m1)$ $\qquad\qquad\qquad\qquad X \doteq X \quad \mapsto \quad \mathcal{E}_1 := \mathcal{E}'$

$(m2)$ $\qquad\qquad\left.\begin{array}{c} t \doteq X \\ t \text{ is not a variable} \end{array}\right\} \mapsto \mathcal{E}_1 := X \doteq t \wedge \mathcal{E}'$

$(m3)$ $\qquad\left.\begin{array}{c} X \doteq t \\ X \text{ does not occur in } t \\ X \text{ occurs in } \mathcal{E}' \end{array}\right\} \mapsto$

$\qquad\qquad\qquad \mathcal{E}_1 := \mathcal{E}'[X/t] \wedge X \doteq t; \mathcal{E}_2 := \mathcal{E}_2[X/t]$

$(m4)$ $\qquad\qquad\left.\begin{array}{c} X \doteq t \\ X \text{ occurs in } t \end{array}\right\} \mapsto \texttt{fail}$

$(m5)$ $\qquad f(s_1,\ldots,s_m) \doteq g(t_1,\ldots,t_m) \quad \mapsto \quad \texttt{fail}$

$(m6)$ $\qquad\left.\begin{array}{c} f(s_1,\ldots,s_m) \doteq f(t_1,\ldots,t_m) \\ f \in \Sigma \setminus \{\!\{\!\{\cdot\,|\,\cdot\}\!\}\!\} \end{array}\right\} \mapsto$

$\qquad\qquad\qquad \mathcal{E}_1 := s_1 \doteq t_1 \wedge \ldots \wedge s_m \doteq t_m \wedge \mathcal{E}'$

$(m7)$ $\qquad\left.\begin{array}{c} \{\!\{t\,|\,s\}\!\} \doteq \{\!\{t'\,|\,s'\}\!\} \\ \text{tail}(s) \text{ and tail}(s') \text{ are} \\ \text{not the same variable} \end{array}\right\} \mapsto$

$\qquad$ (i) $\quad \mathcal{E}_1 := t \doteq t' \wedge \wedge\mathcal{E}'; \text{add } s \doteq s' \text{ to } \mathcal{E}_2$
$\qquad$ (ii) $\quad \mathcal{E}_1 := \mathcal{E}'; \text{add } s \doteq \{\!\{t'\,|\,N\}\!\} \text{ and } \{\!\{t\,|\,N\}\!\} \doteq s' \text{ to } \mathcal{E}_2$

$(m8)$ $\qquad\left.\begin{array}{c} \{\!\{t\,|\,s\}\!\} \doteq \{\!\{t'\,|\,s'\}\!\} \\ \text{tail}(s) \text{ and tail}(s') \text{ are} \\ \text{the same variable} \end{array}\right\} \mapsto$

$\qquad \mathcal{E}_1 := \mathcal{E}'; \text{add de\_tail}(\{\!\{t\,|\,s\}\!\}) \doteq \text{de\_tail}(\{\!\{t'\,|\,s'\}\!\}) \text{ to } \mathcal{E}_2\,.$

Figure 1. Bag unification algorithm

As an example, we can see that one of the critical situations pointed out above, can be dealt with correctly by this algorithm:

$$\{\!| T_1 | S_1 |\!\} \doteq \{\!| T_2 | S_2 |\!\} \wedge \{\!| T_3 | S_2 |\!\} \doteq \{\!| T_4 | S_1 |\!\} \overset{m7(ii)}{\mapsto}$$
$$S_1 \doteq \{\!| T_2 | N_1 |\!\} \wedge \{\!| T_1 | N_1 |\!\} \doteq S_2 \wedge \{\!| T_3 | S_2 |\!\} \doteq \{\!| T_4 | S_1 |\!\} \overset{m2-m3-m3}{\mapsto}$$
$$S_1 \doteq \{\!| T_2 | N_1 |\!\} \wedge S_2 \doteq \{\!| T_1 | N_1 |\!\} \wedge \{\!| T_3, T_1 | N_1 |\!\} \doteq \{\!| T_4, T_2 | N_1 |\!\}.$$

The last equation will be replaced by $\{\!| T_3, T_1 |\!\} \doteq \{\!| T_4, T_2 |\!\}$, avoiding the
loop.

More generally, it can be proved that

Theorem 1.7 (Termination) Unify_bags *always terminates, for any input
system* $\mathcal{E}$.

Moreover, soundness and completeness of this algorithm are assured by
the following

Theorem 1.8 *Let* $e \wedge \mathcal{E}$ *be an equation system, e an equation not in solved
form, and $\mathcal{E}_1, \dots, \mathcal{E}_h$ be the equation systems non-deterministically resulting
from the application of the action of* Unify_bags *fired by e. Let $N_1, \dots, N_k$
be the variables occurring in $\mathcal{E}_1, \dots, \mathcal{E}_h$ but not in $e \wedge \mathcal{E}$; then $T \vdash e \wedge \mathcal{E} \leftrightarrow
\exists N_1, \dots, N_k \bigvee_{i=1}^{h} \mathcal{E}_i$, where T is the theory of hybrid bags presented in* § 2.2.

The proof of Theorem 1.8 can be performed by case analysis. In particular one can observe that, since two bags are equal if and only if they
contain the same number of occurrences of each element, then the problem
$\{\!| s_1, \dots, s_m | X |\!\} \doteq \{\!| t_1, \dots, t_n | X |\!\}$ is perfectly equivalent to the problem
$\{\!| s_1, \dots, s_m |\!\} \doteq \{\!| t_1, \dots, t_n |\!\}$ (action $(m8)$). Moreover, soundness and completeness of action $(m7)$ follows from the strict analogy of this action with
axiom (E_k^m). (For the complete proofs of Theorems 1.7 and 1.8 see (Dovier,
1996).)

Before concluding this subsection, we want to point out an open problem
for the bag unification problem presented above, namely: is there a unification algorithm for bags that incrementally computes a *minimal* complete
set of unifiers?

Indeed, Unify_bags is not minimal in general, since it may compute
through non-determinism repeated equivalent solutions, as well as less general solutions. For example, Unify_bags returns three repeated solutions
for $\{\!| A, A | S |\!\} \doteq \{\!| A | S' |\!\}$, whereas the complete set of unifiers can be de-
scribed by the unique solution $S' = \{\!| A | S |\!\}$. On the other hand, there are
problems for which Unify_bags can be proved to compute exactly the mi-
nimal complete set of unifiers (e.g., $\{\!| A_1, \dots, A_m | S |\!\} \doteq \{\!| B_1, \dots, B_n | S' |\!\}$,
A_is, B_js, S and S' pairwise distinct variables, is such a problem).

The corresponding minimality problems for compact-lists and sets is
also open. To obtain minimality in general, the unification algorithms should

function Unify_clists($\mathcal{E}$);

$(c1)$—$(c3)$ as $(l1)$—$(l3)$ in Unify_lists

$(c4)$ $\left.\begin{array}{c} X \doteq t \\ t \text{ is not a compact-list} \\ \text{and } X \text{ occurs in } t \end{array}\right\} \mapsto \texttt{fail}$

$(c5)$ $\left.\begin{array}{c} X \doteq [\![t_0, \ldots, t_n \,|\, t]\!] \wedge \mathcal{E} \\ t \text{ is a variable or } [\![\,]\!] \\ \text{and } X \text{ occurs in } t_i,\ 0 \leq i \leq n \end{array}\right\} \mapsto \texttt{fail}$

$(c6)$ $\left.\begin{array}{c} X \doteq [\![t_0, \ldots, t_n \,|\, X]\!] \wedge \mathcal{E} \\ X \text{ does not occur in } t_0, \ldots, t_n \end{array}\right\} \mapsto$

$(Y \doteq t_0 \wedge \ldots \wedge Y \doteq t_n \wedge \mathcal{E})[X / [\![Y \,|\, Z]\!]] \wedge X \doteq [\![Y \,|\, Z]\!]$

$(c7)$ $\left.\begin{array}{c} f(s_1, \ldots, s_m) \doteq g(t_1, \ldots, t_n) \wedge \mathcal{E} \\ f \text{ different from } g \end{array}\right\} \mapsto \texttt{fail}$

$(c8)$ $\left.\begin{array}{c} f(s_1, \ldots, s_m) \doteq f(t_1, \ldots, t_m) \wedge \mathcal{E} \\ f \in \Sigma \setminus \{[\![\cdot \,|\, \cdot]\!]\} \end{array}\right\} \mapsto$

$\qquad\qquad\qquad\qquad s_1 \doteq t_1 \wedge \ldots \wedge s_m \doteq t_m \wedge \mathcal{E}$

$(c9)$ $\qquad\qquad [\![t \,|\, s]\!] \doteq [\![t' \,|\, s']\!] \wedge \mathcal{E} \quad \mapsto$

$\qquad\qquad\qquad (i) \quad t \doteq t' \wedge s \doteq s' \wedge \mathcal{E}$

$\qquad\qquad\qquad (ii) \quad t \doteq t' \wedge s \doteq [\![t' \,|\, s']\!] \wedge \mathcal{E}$

$\qquad\qquad\qquad (iii) \quad t \doteq t' \wedge [\![t \,|\, s]\!] \doteq s' \wedge \mathcal{E}.$

Figure 2.   Compact-list unification algorithm

be modified so as to deal with various significant special cases. This type of improvement, very important from a "practical" point of view, is not considered here for space limits. A first contribution for a solution—limited to the set unification problem but easily adaptable to other contexts—is presented in (Arenas & Dovier, 1995). Such optimizations, however, are independent from the non-deterministic complexity analysis of the algorithms, which is another interesting line of research.

3.2. UNIFICATION OF HYBRID COMPACT-LISTS

As shown in § 2.3, axiom (F_3) should be modified accordingly to the semantics of the compact-list constructor $[\![\cdot \,|\, \cdot]\!]$. This is reflected into the occur-checks performed by the compact-list unification algorithm shown in Figure 2 (actions $(c4)$—$(c6)$):

The other important difference w.r.t. standard list unification is the addition of action $(c9)$ which reflects the extensionality principle for compact-lists expressed by axiom (E_k^c).

Termination, soundness and completeness of the unification algorithm for compact-lists shown in Figure 2 has been proved in (Dovier, 1996).

3.3. UNIFICATION OF HYBRID SETS

The main (deterministic) part of the unification algorithm for hybrid sets is exactly the same as that of the unification algorithm for bags of Figure 1 (clearly calls to Unify_bag_actions are replaced by calls to Unify_sets_actions) and is not repeated here. The remaining part of the algorithm (function Unify_sets_actions—see Figure 3) is in a sense a combination of the unification algorithms for bags and for compact-lists shown in the previous two subsections.

As done with bags, we find convenient to split $\mathcal{E}$ into two parts, $\mathcal{E}_1$ and $\mathcal{E}_2$, where $\mathcal{E}_2$ is dealt with as a stack, and to add a little deterministic control to the algorithm in order to assure its termination.

The most important difference with the algorithms presented so far are actions $(s9)$ and $(s10)$, whose aim is the reduction of set-set equations. In particular, cases (ii) and (iii) take care of duplicates in the left-hand side term and in the right-hand side term, respectively. Case (iv), instead, takes care of permutativity of the set constructor $\{\cdot \,|\, \cdot\}$.

Equations of the form $\{t_0, \ldots, t_m \,|\, X\} \doteq \{t'_0, \ldots, t'_n \,|\, X\}$, where the two sides are set terms with the same variable tail element, are handled as a special case by action $(s10)$. The problem is the same singled out in § 3.1. However, in this case, the given equation can not be simply replaced by the equation $\{t_0, \ldots, t_m\} \doteq \{t'_0, \ldots, t'_n\}$, since there are other possible solutions not covered by this new equation. For instance, the equation $\{a, b \,|\, X\} \doteq \{b \,|\, X\}$, which has no solution as a bag unification problem, has the two distinct solutions, $X = \{a \,|\, N\}$ and $X = \{a, b \,|\, N\}$, as a set unification problem. To preserve completeness, therefore, the algorithm is forced to consider non-deterministically each element of one of the two sets involved in the set-set equation, so as to explore all possible combinations. Clearly, this solution opens a big (though finite) number of alternatives, possibly leading to redundant solutions. In (Arenas & Dovier, 1995) it is shown how to improve the algorithm from this point of view.

The algorithm of Figure 3 is the very same algorithm used in the language {log}, a logic programming language enriched with finite sets (Dovier et al., 1991; Dovier et al., 1996). In (Dovier et al., 1996) it is proved to terminate, and to be sound and complete.

{log} has been also reconsidered as an instance of the general CLP scheme (Dovier & Rossi, 1993). Also in this case, the algorithm of Figure 3 is used as a fundamental component (the one dealing with equality constraints) of the constraint satisfiability procedure.

function Unify_sets_actions($\mathcal{E}$, e);
  let $\mathcal{E}'$ be $\mathcal{E}_1 \setminus \{e\}$;
  case e **of**

($s1$)—($s3$) as ($l1$)—($l3$) in Unify_bags

($s4$) $\left.\begin{array}{c} X \doteq t \\ t \text{ is not a set term and } X \text{ occurs in } t \end{array}\right\} \mapsto \texttt{fail}$

($s5$) $\left.\begin{array}{c} X \doteq \{t_0, \ldots, t_n \,|\, t\} \\ t \text{ is a member-less and } X \text{ occurs in } t \\ \text{or } X \in vars(t_0) \cup \cdots \cup vars(t_n) \end{array}\right\} \mapsto \texttt{fail}$

($s6$) $\left.\begin{array}{c} X \doteq \{t_0, \ldots, t_n \,|\, X\} \\ X \text{ does not occur in } t_0, \ldots, t_n \end{array}\right\} \mapsto$

                 $\mathcal{E}_1 := \mathcal{E}'; \text{ add } X \doteq \{t_0, \ldots, t_n \,|\, N\} \text{ to } \mathcal{E}_2$

($s7$) $\left.\begin{array}{c} f(s_1, \ldots, s_m) \doteq g(t_1, \ldots, t_n) \\ f \text{ different from } g \end{array}\right\} \mapsto \texttt{fail}$

($s8$) $\left.\begin{array}{c} f(s_1, \ldots, s_m) \doteq f(t_1, \ldots, t_m) \\ f \in \Sigma\{\{\cdot \,|\, \cdot\}\} \end{array}\right\} \mapsto$

                 $\mathcal{E}_1 := s_1 \doteq t_1 \wedge \ldots \wedge s_m \doteq t_m \wedge \mathcal{E}'$

($s9$) $\left.\begin{array}{c} \{t \,|\, s\} \doteq \{t' \,|\, s'\} \\ \text{tail}(s) \text{ and tail}(s') \text{ are} \\ \text{not the same variable} \end{array}\right\} \mapsto$

      (i) $\mathcal{E}_1 := t \doteq t' \wedge \mathcal{E}'; \text{ add } s \doteq s' \text{ to } \mathcal{E}_2$
      (ii) $\mathcal{E}_1 := t \doteq t' \wedge \mathcal{E}'; \text{ add } \{t \,|\, s\} \doteq s' \text{ to } \mathcal{E}_2$
      (iii) $\mathcal{E}_1 := t \doteq t' \wedge \mathcal{E}'; \text{ add } s \doteq \{t' \,|\, s'\} \text{ to } \mathcal{E}_2$
      (iv) $\mathcal{E}_1 := \mathcal{E}'; \text{ add } s \doteq \{t' \,|\, N\} \text{ and}$
             $\{t \,|\, N\} \doteq s' \text{ to } \mathcal{E}_2$

($s10$) $\left.\begin{array}{c} \{t_0, \ldots, t_m \,|\, X\} \doteq \{t'_0, \ldots, t'_n \,|\, X\} \\ X \text{ variable} \end{array}\right\} \mapsto$

  select arbitrarily i in $\{0, \ldots, m\}$; choose one of the following actions:
    (i) $\mathcal{E}_1 := t_0 \doteq t'_i \wedge \mathcal{E}';$
        add $\{t_1, \ldots, t_m \,|\, X\} \doteq \{t'_0, \ldots, t'_{i-1}, t'_{i+1}, \ldots, t'_n \,|\, X\}$ to $\mathcal{E}_2$
    (ii) $\mathcal{E}_1 := t_0 \doteq t'_i \wedge \mathcal{E}';$
        add $\{t_0, \ldots, t_m \,|\, X\} \doteq \{t'_0, \ldots, t'_{i-1}, t'_{i+1}, \ldots, t'_n \,|\, X\}$ to $\mathcal{E}_2$
    (iii) $\mathcal{E}_1 := t_0 \doteq t'_i \wedge \mathcal{E}';$
        add $\{t_1, \ldots, t_m \,|\, X\} \doteq \{t'_0, \ldots, t'_n \,|\, X\}$ to $\mathcal{E}_2$
    (iv) $\mathcal{E}_1 := \mathcal{E}'; \text{ add } X \doteq \{t_0 \,|\, N\} \text{ and}$
        $\{t_1, \ldots, t_m \,|\, N\} \doteq \{t'_0, \ldots, t'_n \,|\, N\}$ to $\mathcal{E}_2$.

Figure 3. Set unification algorithm

4. Conclusions and further researches

In the first part of the paper we have presented axiomatically specified first order theories of lists, bags, compact-lists, and sets. Such axiomatizations have been shown to be especially suitable for the integration with free functor symbols governed by the classical Clark's axioms in the context of Constraint Logic Programming. Moreover, the adaptations of the extensionality principle to the various theories taken into account have been exploited in the design of unification algorithms presented in the second part of the paper.

All the theories presented can be combined providing frameworks to deal with several of the proposed data structures simoultaneously. The unification algorithms proposed can be combined (merged) as well to produce engines for such combination theories.

The open problem of devising minimal (irredundant) unification algorithms for lists, bags, compact-lists, and sets, has been briefly discussed and represents one of the most important (at least from a practical point of view) next step in this research.

Other two important lines for further research are the following: the adaptation of the results presented to the case of non-well-founded sets and/or rational terms; the extension of the unification algorithms into general purpose constraint solvers capable of dealing with negative information. Both these problems have been tackled, in the case of sets, in (Omodeo & Policriti, 1994; Dovier & Rossi, 1993).

Acknowledgements

We are grateful to Eugenio G. Omodeo for his precious suggestions and comments in many parts of the paper.

AGOSTINO DOVIER
Dipartimento di Informatica
Università di Pisa, Italia

ALBERTO POLICRITI
Dipartimento di Matematica e Informatica
Università di Udine, Italia

AND

GIANFRANCO ROSSI
Dipartimento di Matematica
Università di Parma, Italia

References

Arenas-Sánchez, P., and Dovier, A. Minimal Set Unification. In *Proc. Seventh Int'l Symp. on Programming Language Implementation and Logic Programming* (1995), M. Hermenegildo and S. D. Swierstra, Eds., vol. 982 of *Lecture Notes in Computer Science*, Springer-Verlag, Berlin, pp. 397–414.

Bellé, D., and Parlamento, F. Undecidability of Weak Membership Theories. In *Proceedings of the International Conference on Logic and Algebra (in memory of R. Magari)* (1994). Siena.

Clark, K. L. Negation as Failure. In *Logic and Databases*, H. Gallaire and J. Minker, Eds. Plenum Press, 1978, pp. 293–321.

Dovier, A. *Computable Set Theory and Logic Programming*. PhD thesis, Università degli Studi di Pisa, 1996. In preparation.

Dovier, A., Omodeo, E. G., and Policriti, A. Hyperset constraint handling. Rr 21/94, Dipartimento di Matematica ed Informatica, Univ. di Udine, December 1994.

Dovier, A., Omodeo, E. G., Pontelli, E., and Rossi., G. {log}: A Logic Programming Language with Finite Sets. In *Proc. Eighth Int'l Conf. on Logic Programming* (1991), K. Furukawa, Ed., The MIT Press, Cambridge, Mass., pp. 111–124.

Dovier, A., Omodeo, E. G., Pontelli, E., and Rossi, G. Embedding Finite Sets in a Logic Programming Language. In *Selected papers from 3rd Int'l Workshop on Extension of Logic Programming* (1993), E. Lamma and P. Mello, Eds., vol. 660 of *Lecture Notes in Artificial Intelligence*, Springer-Verlag, Berlin, pp. 150–167.

Dovier, A., Omodeo, E. G., Pontelli, E., and Rossi, G. {log}: A Language for Programming in Logic with Finite Sets. To appear in the Journal of Logic Programming, 1996.

Dovier, A., and Rossi, G. Embedding Extensional Finite Sets in CLP. In *Proc. of Int'l Logic Programming Symposium, ILPS'93* (1993), D. Miller, Ed., The MIT Press, Cambridge, Mass., pp. 540–556.

Kapur, D., and Narendran, P. NP-completeness of the set unification and matching problems. In *8th International Conference on Automated Deduction* (1986), J. H. Siekmann, Ed., vol. 230 of *Lecture Notes in Computer Science*, Springer-Verlag, Berlin, pp. 489–495.

Lassez, J. L., Maher, M. J., and Marriot, K. Unification revisited. In *Lecture Notes in Computer Science* (1986), vol. 306.

Lloyd, J. W. *Foundations of Logic Programming*. Springer-Verlag, Berlin, 1987. Second edition.

Martelli, A., and Montanari, U. An efficient unification algorithm. *ACM Transactions on Programming Languages and Systems 4* (1982), pp. 258–282.

Omodeo, E. G., Policriti, A., and Rossi., G. Che genere di insiemi/multi-insiemi/iperinsiemi incorporare nella programmazione logica? In D. Saccà, Ed., *Proc. Eigth Italian Conference on Logic Programming* (1993), pp. 55–70.

Parlamento, F., and Policriti, A. Decision Procedures for Elementary Sublanguages of Set Theory IX. Unsolvability of the Decision Problem for a Restricted Subclass of δ_0-Formulas in Set Theory. *Communications of Pure and Applied Mathematics 41* (1988), pp. 221–251.

Paterson, M. S., and Wegman, M. N. Linear unification. Tech. rep., IBM Thomas J. Watson Research Center, Yorktown Heights, 1976.

Tarski, A., and Givant, S. *A Formalization of Set Theory without Variables*, vol. 41 of *Colloquium Publications*. American Mathematical Society, 1986.

Vaught, R. L. On a Theorem of Cobham Concerning Undecidable Theories. In *Proceedings of the 1960 International Concress* (1962), E. Nagel, P. Suppes, and A. Tarski, Eds., Stanford University Press, Stanford, pp. 14–25.

FRED MESNARD, SÉBASTIEN HORAU, AND ALEXANDRA MAILLARD

CLP(χ) FOR PROVING PROGRAM PROPERTIES

Abstract. Various proof methods have been proposed to solve the implication problem, i.e. proving that properties of the form : $\forall(P \rightarrow Q)$ - where P and Q denote conjunctions of atoms - are logical consequences of logic programs. Nonetheless, it is a commonplace to say that it is still quite a difficult problem. Besides, the advent of the constraint logic programming scheme constitutes not only a major step towards the achievement of efficient declarative logic programming systems but also a new field to explore. By recasting and simplifying the implication problem in the constraint logic programming framework, we define a generic proof method for the implication problem, which we prove sound from the algebraic point of view. We present four examples using CLP($\mathbb{N}$), CLP($\mathcal{RT}$), CLP(Σ^*) and RISC-CLP($\mathcal{REAL}$). The logical point of view of the constraint logic programming scheme enables the automation of the proof method. At last, we prove the unsolvability of the implication problem, we point out the origins of the incompleteness of the proposed proof method and we identify two classes of programs for which we give a decision procedure for the implication problem.

1. Introduction

Various proof methods, e.g. (Clark, 1979), (Kanamori & Fujita, 1986), (Lever, 1991), (Colussi & Marchiori, 1991) and (Deransart, 1993), have been proposed to solve the implication problem, i.e. proving that properties of the form: $\forall(P \rightarrow Q)$ - where P and Q denote conjunction of atoms - are logical consequences of logic programs (Lloyd, 1987). Nonetheless, it is a commonplace to say that it is still quite a difficult problem.

Besides, the advent of the constraint logic programming (CLP) scheme (cf. (Jaffar & Lassez, 1986), (Jaffar & Maher, 1994)) whose main instances are described in (Aiba et al., 1988), (Colmerauer, 1990), (Dincbas et al., 1988) and (Jaffar et al., 1987), constitutes not only a major step towards the

F. Baader and K.U. Schulz (eds.), Frontiers of Combining Systems, 321–338.
© 1996 *Kluwer Academic Publishers.*

achievement of efficient declarative logic programming systems, but also a
new field to explore.

By simplifying and recasting the implication problem in the CLP frame-
work, i.e. proving that properties of the form $\forall (d_1 \wedge p \rightarrow d_2)$, where p is an
atom and d_1 and d_2 disjunctions of constraints, are logical consequences of
definite constraint logic programs, we extend our previous work (Mesnard
et al., 1992) and propose a *generic* proof procedure for the implication problem which may be easily implemented.

Let us give an intuitive idea of the implementation. The standard goal of a
constraint solver is to check the solvability of a conjunction of constraints.
A non-standard use is to test whether a constraint is a logical consequence
of a set of constraints. It relies on the following trick: an atomic constraint
c is a logical consequence of a conjunction of constraints C_s iff $C_s \cup \neg c$ is
unsolvable. So, in order to prove the property Prop: $\forall (p \rightarrow c)$, we first prove
that Prop is true for the non-recursive clauses defining p. Then, assuming
it holds for p in the body of a recursive clause, we thus prove that Prop is
true for the head of the clause.

The paper is organized as follows: section 2 recalls basic notations, definitions and results of the CLP scheme. Section 3 presents the proof method
from the algebraic point of view. We propose four examples of its use in
section 4. In section 5, we show that the logical point of view of the CLP
framework enables the automation of the proof method. We prove the unsolvability of the implication problem and we discuss the completeness of
the method in section 6. We conclude by comparing our work with other
techniques and sketching possible extensions of the proposed proof method.

2. Preliminaries

Let us *briefly* recall some basic concepts of the CLP framework (see (Jaffar
& Lassez, 1986) and (Jaffar & Maher, 1994) for more details). In order to
be concise, we only consider the mono-sorted case.

Let V be a denumerable set of variables, F a finite set of function symbols, C a finite set of constraint symbols and P a finite set of predicate
symbols. A finite arity is assigned to each function, constraint and predicate symbol. We assume that C contains the binary symbol $=$. The set of
terms, *atomic constraints* and *atoms*, defined as usually done, are denoted
by $T(F, V), A(C, F, V)$ and $A(P, F, V)$. $var(o)$ denotes the set of variables
of the syntactic object o, and we write $o(\tilde{x})$ as a shorthand for o where
$var(o) \subseteq \{\tilde{x}\}$.

Definition 2.1 *A constraint is a (possibly empty) conjunction of atomic
constraints.*

Definition 2.2 *A generalized constraint is a (possibly empty) disjunction of constraints.*

Note that at this point of the statement we do not impose any finiteness condition on a constraint or a generalized constraint.

When we switch from syntax to semantics, the key notion of the CLP(χ) scheme lies in the introduction of a *structure* which embodies the meaning of the specific intended constraint domain χ. More precisely, a structure χ over $< F, C >$ consists in a non-empty domain D_χ, an assignment to each n-ary function symbol of a function $(D_\chi)^n \to D_\chi$ and an assignment to each n-ary constraint symbol of a subset of $(D_\chi)^n$ (apart from the binary constraint symbol $=$, which is interpreted as equality). A χ-*valuation* is a mapping $V \to D_\chi$, extended in the obvious way to formulae. An atomic constraint c is χ-*solvable* iff there is a χ-valuation θ such that χ models $c\theta$ (which means that $c\theta$ evaluates to true wrt χ); θ is a χ-*solution* to c and we write: $\models_\chi c\theta$. Otherwise c is χ- *unsolvable*. χ-solvability and χ-unsolvability extend naturally to constraints (the empty conjunction of atomic constraints is trivially χ-solvable) and disjunctions of constraints (the empty disjunction of constraints is trivially χ-unsolvable). Let $\neg c$ denote the complement[1] of the solution space of c. This notation extends naturally to constraints and generalized constraints.

The structure χ has to be *solution-compact* wrt $< F, C >$, i.e. it should satisfy:

(SC$_1$)   every element of D_χ can be defined by a constraint (i.e. $\forall d \in D_\chi$, there exists a constraint c with $\{x\} \subseteq var(c)$ s.t. for every solution θ of c we have: $x\theta = d$)

and

(SC$_2$)   the complement of each atomic constraint can be described by a generalized constraint (i.e. for every atomic constraint c, there exists a generalized constraint c' s.t. $\models_\chi \forall(\neg c \leftrightarrow c')$).

Let *Pgm* be a definite CLP(χ) program, P be the set of predicate symbols in *Pgm* and D_χ^* be the set of finite sequences of elements of D_χ. The χ-*base* is the cross product $P \times D_\chi^*$ respecting the arities of the predicate symbols. A χ-*interpretation* is any subset of the χ-base. A χ-*model* of *Pgm* is a χ-interpretation in which all the clauses of *Pgm* are true. We have a mapping $T_{Pgm,\chi}$ from and into the χ-base:

---

[1]If we consider the structure $< \mathbf{N}, \{0, 1, +\}, \{=\} >$ and the constraint $c(x, y) \equiv x = y + 1$ then $\neg c(x, y)$ denotes for instance the following generalized constraint: $x = y \lor x = y + 2 \lor x = y + 3 \lor \ldots \lor x + 1 = y \lor x + 2 = y \lor \ldots$.
But $\neg c(x, y)$ denotes the generalized constraint $x \geq y + 2 \lor y \geq x$ if the structure is $< \mathbf{N}, \{0, 1, +\}, \{=, \geq\} >$.

$T_{Pgm,\chi}(S) = \{\ d \in \chi$-base: there is a clause in Pgm: $A \leftarrow c \diamond Body$ where A is an atom, $Body$ a conjunction of atoms, c a finite constraint and a χ-valuation θ such that $\models_\chi c\theta$, $\models_\chi (A\theta = d)$ and $\{Body\theta\} \subseteq S\}$

whose powers are defined as usually done:

$$\begin{cases} T_{Pgm,\chi} \uparrow 0 = \emptyset\ ; \\ T_{Pgm,\chi} \uparrow n+1 = T_{Pgm,\chi}(T_{Pgm,\chi} \uparrow n)\ ; \\ T_{Pgm,\chi} \uparrow \omega = \bigcup_{n \in \mathbf{N}}(T_{Pgm,\chi} \uparrow n). \end{cases}$$

At last, the following fundamental property establishes the equivalence between the algebraic and the fixpoint semantics:

Theorem 2.3 *(semantics of constraint logic programs* (Jaffar & Lassez, 1986*))*
The least χ-model of Pgm is the least fixpoint of $T_{Pgm,\chi}$ i.e. we have $M_{\chi,Pgm} = T_{Pgm,\chi} \uparrow \omega$

3. The proof method: the algebraic point of view

Let B_1 and B_2 be two generalized constraints. We write $B_1 \models_\chi B_2$ iff every χ-solution of B_1 is a χ-solution of B_2. We have the following property:

Observation 3.1 $B_1 \wedge \neg B_2$ *is* $\chi - unsolvable$ *iff* $B_1 \models_\chi B_2$.

We now give the format of the properties we want to prove.

Definition 3.2 *(system of implications)*
Let $\tilde{x}$ be a vector of distinct variables, $P = \{p_1, \cdots, p_n\}$ a set of predicate symbols, and $B_1, \cdots, B_n$ n finite generalized constraints. We call a system of implications the following conjunction of implications:

$$\bigwedge_{i=1}^{n} \left[\forall \tilde{x}(p_i(\tilde{x}) \rightarrow B_i(\tilde{x})) \right]$$

Before formulating the main result of this section, we give a syntax of constraint logic procedures. Let Pgm be a definite CLP(χ) program and P the set of all predicate symbols that appear in at least one clause of Pgm. If $p \in P$, let n_p be the total number of clauses from Pgm which define p. The ith clause defining p is denoted $Cl_{p,i}$ and can be written as:

$$(Cl_{p,i}) \qquad p(\tilde{h}_i) \leftarrow C_{p,i} \diamond p_{i,1}(\tilde{t}_{i,1}) \dots p_{i,k_i}(\tilde{t}_{i,k_i})$$

where $p_{i,1}, \dots, p_{i,k_i}$ are predicates from P, C's denote finite constraints, $\tilde{h}$'s and $\tilde{t}$'s are vectors of terms from $T(F, V)$, well formed wrt the corresponding predicate arity. Notice that we can have $k_i = 0$, which means $Cl_{p,i}$ is a unitary clause and, that in the body of a clause of a predicate p, we can have $p_{i,j} = p$, which means $Cl_{p,i}$ is a recursive clause.

Now, let (SI) be the following system of implications:

$$\bigwedge_{p \in P} \left[\forall \tilde{x}(p(\tilde{x}) \to B_p(\tilde{x})) \right]$$

Let τ_{SI} be the mapping which turns each clause $Cl_{p,i}$ into the generalized constraint[2]:

$$\tau_{SI}(Cl_{p,i}) \equiv \left(\bigwedge_{j=1}^{k_i} B_{p_{i,j}}(\tilde{t}_{i,j}) \right) \wedge C_{p,i} \wedge (\tilde{h}_i = \tilde{x}) \wedge \neg B_p(\tilde{x})$$

We have the following result:

Theorem 3.3 *(the proof method and its correctness)*
If

$$\left[\bigvee_{\substack{p \in P \\ 1 \le i \le n_p}} \tau_{SI}(Cl_{p,i}) \right] \text{ is } \chi - \text{unsolvable} \qquad (H)$$

then the least χ-model of Pgm is a χ-model of (SI).

Proof: Let I be the set $\{p(\tilde{s}) \mid p \in P, \models_\chi B_p(\tilde{s})\}$. I is a χ-interpretation which is a χ-model of (SI) (by construction). We first prove that I is a χ-model of Pgm.

Let $Cl_{p,i}$ be a clause of Pgm, say $p(\tilde{h}_i) \leftarrow C_{p,i} \circ p_{i,1}(\tilde{t}_{i,1}) \ldots p_{i,k_i}(\tilde{t}_{i,k_i})$ and let θ be a χ-valuation verifying the following properties:

$$\begin{cases} p_{i,1}(\tilde{t}_{i,1}\theta) \in I, \ldots p_{i,k_i}(\tilde{t}_{i,k_i}\theta) \in I & (1) \\ \models_\chi C_{p,i}\theta & (2) \end{cases}$$

We must prove that $p(\tilde{h}_i\theta) \in I$. By hypothesis (H) we know that:

$$\left[\bigvee_{\substack{p \in P \\ 1 \le i \le n_p}} \tau_{SI}(Cl_{p,i}) \right] \text{ is } \chi - \text{unsolvable}$$

and so, we know that for the predicate p and the considered clause:

$$\left[(\bigwedge_{j=1}^{k_i} B_{p_{i,j}}(\tilde{t}_{i,j})) \wedge C_{p,i} \wedge (\tilde{h}_i = \tilde{x}) \wedge \neg B_p(\tilde{x}) \right] \text{ is } \chi - \text{unsolvable}$$

---

[2]One may write $\neg B_p(\tilde{h}_i)$ instead of $(\tilde{h}_i = \tilde{x}) \wedge \neg B_p(\tilde{x})$. But this simplification hides a crucial point of the proof, namely that $\tilde{x}$ is a vecteur of *fresh distinct* variables.

The previous formula is universally quantified and hence, if we apply θ:

$$\left[(\bigwedge_{j=1}^{k_i} B_{p_{i,j}}(\tilde{t}_{i,j}\theta)) \wedge C_{p,i}\theta \wedge (\tilde{h}_i\theta = \tilde{x}\theta) \wedge \neg B_p(\tilde{x}\theta) \right] \ is \ \chi - unsolvable$$

But by (1) we have: $\models_\chi \bigwedge_{j=1}^{k_i} B_{p_{i,j}}(\tilde{t}_{i,j}\theta)$. So by (2) and because $\tilde{h}_i\theta = \tilde{x}\theta$ is always solvable, we conclude that $\neg B_p(\tilde{h}_i\theta)$ is χ-unsolvable i.e. $\models_\chi B_p(\tilde{h}_i\theta)$ (by observation 3.1). Consequently, $p(\tilde{h}_i\theta) \in I$ which implies that I is a χ-model of $Cl_{p,i}$, hence a χ-model of Pgm.

To conclude the proof, it remains to show that the least χ-model $M_{\chi,Pgm}$ of Pgm is a χ-model of (SI). Let $p \in P$ and $Imp_p : \forall \tilde{x} \ (p(\tilde{x}) \rightarrow B_p(\tilde{x}))$ be the implication corresponding to p in (SI). If there is no $p(\tilde{s}) \in M_{\chi,Pgm}$ then Imp_p is trivially true. Else, as $M_{\chi,Pgm} \subseteq I$ (by definition of the least χ-model) and I is a χ-model of (SI), we have $\models_\chi B_p(\tilde{s})$. $\square$

Corollary 3.4 *The proof method can be applied to the following system of extended implications:*

$$\bigwedge_{p \in P} \left[\forall \tilde{x} \ (A_p(\tilde{x}) \wedge p(\tilde{x})) \rightarrow B_p(\tilde{x}) \right]$$

where A_p and B_p are generalized constraints.

Proof: This result is due to two reasons. We have the logical equivalence:

$$\forall \tilde{x} \ (A_p(\tilde{x}) \wedge p(\tilde{x})) \rightarrow B_p(\tilde{x}) \ \equiv \ \forall \tilde{x} \ p(\tilde{x}) \rightarrow (\neg A_p(\tilde{x}) \vee B_p(\tilde{x}))$$

and we notice that the negation of a generalized constraint can be rewritten (SC_2) as a generalized constraint. $\square$

4. Examples

Let us give four applications of theorem 3.3. We would like to point out that the simplification of constraints relies on algebraic reasoning in the involved structures.

Example 4.1 This example emphasizes the use of a generalized constraint in an implication. Let Pgm be the following CLP($\mathbb{N}$) definite program (McCarthy's 91 function), where $\mathbb{N}$ denotes the set of natural numbers, with $F = \{0, 1, +, -\}$ and $C = \{=, \neq, >, \leq\}$ (Contejean, 1993):

$$\begin{aligned}
(Cl_{p,1}) \qquad & p(u, u-10) \quad \leftarrow \quad u > 100 \diamond \\
(Cl_{p,2}) \qquad & p(u, w) \qquad \leftarrow \quad u \leq 100 \diamond p(u+11, v), \ p(v, w)
\end{aligned}$$

and the implication:

$$(SI) \qquad \forall(x, y) \ p(x, y) \rightarrow [(x > 100 \wedge y = x - 10) \vee (x \leq 100 \wedge y = 91)]$$

We apply the mapping τ_{SI}:

$\tau_{SI}(Cl_{p,1}) \equiv [u > 100] \land [(u, u - 10) = (x, y)] \land \neg[(x > 100 \land y = x - 10) \lor (x \le 100 \land y = 91)]$

$\tau_{SI}(Cl_{p,2}) \equiv [(u + 11 > 100 \land v = u + 11 - 10) \lor (u + 11 \le 100 \land v = 91)] \land [(v > 100 \land w = v - 10) \lor (v \le 100 \land w = 91)] \land [u \le 100] \land [(u, w) = (x, y)] \land \neg[(x > 100 \land y = x - 10) \lor (x \le 100 \land y = 91)]$

We must prove $(\tau_{SI}(Cl_{p,1}) \lor \tau_{SI}(Cl_{p,2}))$ is $\mathbb{N}$-unsolvable.

- $\tau_{SI}(Cl_{p,1}) \rightarrow [u > 100] \land \neg[(u > 100 \land u - 10 = u - 10)] \land \neg[(u \le 100 \land u - 10 = 91)]$

$\rightarrow [u > 100] \land [(u \le 100 \lor u - 10 \ne u - 10)] \land [(u > 100 \lor u - 10 \ne 91)]$

$\rightarrow [(u > 100) \land (u \le 100)] \land [(u > 100 \lor u - 10 \ne 91)]$

$\tau_{SI}(Cl_{p,1})$ is $\mathbb{N}$-unsolvable.

- $\tau_{SI}(Cl_{p,2}) \rightarrow [(u > 89 \land v = u + 1) \lor (u \le 89 \land v = 91)] \land [(v > 100 \land w = v - 10) \lor (v \le 100 \land w = 91)] \land [u \le 100] \land [u \le 100 \lor w \ne u - 10] \land [u > 100 \lor w \ne 91]$

By simplifications and the application of ($\lor/\land$)-distributivity we obtain:

$\tau_{SI}(Cl_{p,2}) \rightarrow [(w = 91) \land (w \ne 91) \land [\ldots]] \lor [(v = 91) \land (v > 100) \land [\ldots]] \lor [(w = 91) \land (w \ne 91) \land [\ldots]] \lor [(w = 91) \land (w \ne 91) \land [\ldots]]$

Once again, $\tau_{SI}(Cl_{p,2})$ is $\mathbb{N}$-unsolvable. So we conclude to the unsolvability of $(\tau_{SI}(Cl_{p,1}) \lor \tau_{SI}(Cl_{p,2}))$, i.e. the least $\mathbb{N}$-model of Pgm is a $\mathbb{N}$-model of (SI). Notice that the converse:

$$\forall(x, y) \quad [(x > 100 \land y = x - 10) \lor (x \le 100 \land y = 91)] \rightarrow p(x, y)$$

is true and can be proved directly by any system including CLP($\mathbb{Q}$). ◁

Example 4.2 This example shows that it might be necessary to prove a 'strong enough' system of implications (see section 6). Let Pgm be the following CLP($\mathcal{RT}$) definite program, where $\mathcal{RT}$ denotes the set of rational trees, with the two constraints = and $\ne$ (Colmerauer, 1984).

$$
\begin{array}{llll}
(Cl_{q,1}) & q(u) & \leftarrow & u = g(v), \ v = h(w), \ w = f(u) \diamond \\
(Cl_{q,2}) & q(u) & \leftarrow & u = g(v) \diamond p(f(u), v) \\
(Cl_{p,1}) & p(u, v) & \leftarrow & v = h(u) \diamond q(g(v))
\end{array}
$$

And let (SI) be the system of implications we want to prove (we write $fgh(x)$ as a shorthand for $f(g(h(x)))$):

$$
\begin{cases}
\forall x \quad q(x) \rightarrow x = ghf(x) & (3) \\
\forall(x, y) \quad p(x, y) \rightarrow x = fgh(x) \land y = hfg(y) & (4)
\end{cases}
$$

We apply the mapping τ_{SI}:

$\tau_{SI}(Cl_{q,1}) \equiv [(u = g(v)) \wedge (v = h(w)) \wedge (w = f(u)) \wedge (u = x) \wedge (x \neq ghf(x))]$

$\tau_{SI}(Cl_{q,2}) \equiv [(f(u) = fghf(u)) \wedge (v = hfg(v)) \wedge (u = x) \wedge (x \neq ghf(x))]$

$\tau_{SI}(Cl_{p,1}) \equiv [(g(v) = ghfg(v)) \wedge (v = h(u)) \wedge ((u,v) = (x,y)) \wedge \neg(x = fgh(x) \wedge y = hfg(y))]$

We must prove $(\tau_{SI}(Cl_{q,1}) \vee \tau_{SI}(Cl_{q,2}) \vee \tau_{SI}(Cl_{p,1}))$ is $(\mathcal{RT})$-unsolvable.

- $\tau_{SI}(Cl_{q,1}) \rightarrow (u = ghf(u)) \wedge (u = x) \wedge (x \neq ghf(x))$

$\rightarrow (x = ghf(x)) \wedge (x \neq ghf(x))$ and is $(\mathcal{RT})$-unsolvable

- $\tau_{SI}(Cl_{q,2}) \rightarrow (u = ghf(u)) \wedge (v = hfg(v)) \wedge (u = g(v)) \wedge (u \neq ghf(u))$, is $(\mathcal{RT})$-unsolvable

Note that the implication (4) is used to achieve the demonstration of the $(\mathcal{RT})$-unsolvability of $\tau_{SI}(Cl_{q,2})$.

Similarly, we can prove that $\tau_{SI}(Cl_{p,1})$ is $(\mathcal{RT})$-unsolvable. Hence the least $(\mathcal{RT})$-model of Pgm is a $(\mathcal{RT})$-model of (SI). ◁

Example 4.3 We study a formal system using CLP(Σ^*). Consider the set S of words (or strings) over $\Sigma = \{\mathbf{m,i,u}\}$ defined in (Hofstadter, 1979) as follows: S is the least set containing $\mathbf{mi}$ and verifying the four rules (α and β are any words over Σ):

	if		then	
	if	$\alpha \mathbf{i} \in S$	then	$\alpha \mathbf{iu} \in S$;
	if	$\mathbf{m}\alpha \in S$	then	$\mathbf{m}\alpha\alpha \in S$;
	if	$\alpha \mathbf{iii}\beta \in S$	then	$\alpha \mathbf{u}\beta \in S$;
	if	$\alpha \mathbf{uu}\beta \in S$	then	$\alpha\beta \in S$.

We now briefly propose a possible definition of CLP(Σ^*) (see also (Walinsky, 1989)). Let $F_0 = \{\mathbf{m,i,u,.}\}$ where $\mathbf{m}$, $\mathbf{i}$ and $\mathbf{u}$ are constant symbols and . is a binary function symbol, $C = \{=, \in\}$, where $=$ and $\in$ are two binary constraint symbols. Let t and s be two elements of $T(F_0, V)$, $F = F_0 \cup \{\emptyset, \Lambda, *, +\}$, and R be an element of $T(F, V)$, which is a regular expression over $\{\mathbf{m,i,u,.}\}$. An atomic constraint is either of the form $t = s$ or $t \in R$. The associated domain D_{Σ^*} is $\{\mathbf{m,i,u}\}^*$, the constant symbols $\mathbf{m}$, $\mathbf{i}$ and $\mathbf{u}$ are interpreted as the strings $\mathbf{m}$, $\mathbf{i}$ and $\mathbf{u}$, and . as concatenation of strings. The solution space of the constraint $x \in R$ is the set of words over $\{\mathbf{m,i,u}\}$ which the regular expression R denotes. This structure is clearly solution-compact wrt $< F, C >$. Moreover, it is well known that there is a regular expression R' such that $\neg(x \in R) \equiv x \in R'$; we write $x \notin R$ as a shorthand for $\neg(x \in R)$.

We can now easily construct a CLP(Σ^*) program Pgm (with a slight abuse of notation) that exactly recognizes the words of S:

$$
\begin{aligned}
dh(\mathbf{mi}) &\leftarrow \diamond \\
dh(x.\mathbf{iu}) &\leftarrow \diamond \; dh(x.\mathbf{i}) \\
dh(\mathbf{m}.x.x) &\leftarrow \diamond \; dh(\mathbf{m}.x) \\
dh(x.\mathbf{u}.y) &\leftarrow \diamond \; dh(x.\mathbf{iii}.y) \\
dh(x.y) &\leftarrow \diamond \; dh(x.\mathbf{uu}.y)
\end{aligned}
$$

Let us prove that each element of S begins with $\mathbf{m}$ and, after the $\mathbf{m}$, contains only $\mathbf{i}$'s and $\mathbf{u}$'s. Formally, we have:

$$\forall x \quad dh(x) \to x \in \mathbf{m}.(\mathbf{i}+\mathbf{u})^* \tag{5}$$

We try to apply theorem 3.3:
- $x = \mathbf{mi} \wedge x \notin \mathbf{m}.(\mathbf{i}+\mathbf{u})^*$ is clearly Σ^*-unsolvable;
- $x.\mathbf{i} \in \mathbf{m}.(\mathbf{i}+\mathbf{u})^*$ implies $x.\mathbf{iu} \in \mathbf{m}.(\mathbf{i}+\mathbf{u})^*$ hence $x.\mathbf{i} \in \mathbf{m}.(\mathbf{i}+\mathbf{u})^* \wedge$ $x.\mathbf{iu} \notin \mathbf{m}.(\mathbf{i}+\mathbf{u})^*$ is Σ^*-unsolvable;
- likewise, $\mathbf{m}.x \in \mathbf{m}.(\mathbf{i}+\mathbf{u})^* \wedge \mathbf{m}.x.x \notin \mathbf{m}.(\mathbf{i}+\mathbf{u})^*$ is Σ^*-unsolvable;
- if $x.\mathbf{iii}.y \in \mathbf{m}.(\mathbf{i}+\mathbf{u})^*$ then $x = \mathbf{m}.x'$ and $x'.\mathbf{iii}.y \in (\mathbf{i}+\mathbf{u})^*$ so $x'.\mathbf{u}.y \in (\mathbf{i}+\mathbf{u})^*$ therefore $\mathbf{m}.x'.\mathbf{u}.y \in \mathbf{m}.(\mathbf{i}+\mathbf{u})^*$. Consequently, $x.\mathbf{iii}.y \in \mathbf{m}.(\mathbf{i}+\mathbf{u})^* \wedge x.\mathbf{u}.y \notin \mathbf{m}.(\mathbf{i}+\mathbf{u})^*$ is Σ^*-unsolvable;
- All the same, $x.\mathbf{uu}.y \in \mathbf{m}.(\mathbf{i}+\mathbf{u})^* \wedge x.y \notin \mathbf{m}.(\mathbf{i}+\mathbf{u})^*$ is Σ^*-unsolvable.
From theorem 3.3, we conclude that (5) is true in the least Σ^*-model of Pgm.
In the same way, we could prove:

$$\forall x \quad dh(x) \to x \in \mathbf{m}.(\mathbf{u}^*.\mathbf{i}.\mathbf{u}^* + \mathbf{u}^*.\mathbf{i}.\mathbf{u}^*.\mathbf{i}.\mathbf{u}^*).(\mathbf{i}.\mathbf{u}^*.\mathbf{i}.\mathbf{u}^*.\mathbf{i}.\mathbf{u}^*)^* \tag{6}$$

which shows that the number n of $\mathbf{i}$'s in a word x of S verifies: $n \equiv 1 \; (mod \; 3)$ or $n \equiv 2 \; (mod \; 3)$. Consequently, (6) gives a negative answer to the main question asked by D. Hofstadter (Hofstadter, 1979) concerning S: 'does $\mathbf{mu}$ belong to S ?' ◁

Example 4.4 Let us consider the homographic series defined as follows:

$$
\begin{cases}
u_0 \neq -\dfrac{d}{c} \\
u_{n+1} = \dfrac{au_n + b}{cu_n + d}, \; n \geq 0
\end{cases}
$$

where a, b, c and d are real numbers with $c \neq 0$. We want to prove that the series is constant equal to $\frac{a}{c}$ from $n = 1$ if $ad - bc = 0$.

First, we define the following definite program on RISC-CLP($\mathcal{REAL}$), where the domain is the set of real numbers, $F = \{+, -, *\}$ and $C = \{=, \neq,$

$\geq, >\}$ (see (Hong, 1992) for more details):

(Cl_1) $s(0, u_0, a, b, c, d)$ $\leftarrow$ $-cu_0 \neq d, \ c \neq 0 \ \diamond$

(Cl_2) $s(n+1, x, a, b, c, d)$ $\leftarrow$ $-cy \neq d, \ x(cy+d) = ay + b, \ c \neq 0$
                                     $\diamond \ s(n, y, a, b, c, d)$

So, we want to prove the following implication:

$$\forall(n, x, a, b, c, d) \in \mathbb{R}^6, [(n > 0 \ \wedge \ ad - bc = 0) \wedge s(n, x, a, b, c, d)] \rightarrow cx = a$$

We recognize the format of extended implications described in corollary 3.4. So, we must prove (SI):

$$\forall(n, x, a, b, c, d) \in \mathbb{R}^6, s(n, x, a, b, c, d) \rightarrow (0 \geq n \ \vee \ ad - bc \neq 0 \ \vee \ cx = a)$$

$\tau_{SI}(Cl_1) \equiv n = 0 \ \wedge \ -cu_0 \neq d \ \wedge \ c \neq 0 \ \wedge \ n > 0 \ \wedge \ ad - bc = 0 \ \wedge \ cu_0 \neq a$,
           is clearly $\mathcal{REAL}$-unsolvable.

$\tau_{SI}(Cl_2) \equiv (0 \geq n \ \vee \ ad - bc \neq 0 \ \vee \ cy = a) \ \wedge \ -cy \neq d \ \wedge \ c \neq 0 \ \wedge x(cy+d) = ay + b \ \wedge \ n > 0 \ \wedge \ ad - bc = 0 \ \wedge \ cx \neq a$

If $0 \geq n$ then $\tau_{SI}(Cl_2)$ is trivially $\mathcal{REAL}$-unsolvable. And the same if $ad - bc \neq 0$. If $cy = a$ then:

$$
\begin{aligned}
x(cy + d) = ay + b \ &\equiv \ cx(cy + d) = acy + bc \ \text{(multiply by } c \neq 0) \\
&\equiv \ cx(a + d) = a^2 + bc \ \text{(because } cy = a) \\
&\equiv \ cx(a + d) = a^2 + ad \ \text{(because } ad - bc = 0) \\
&\equiv \ cx(a + d) = a(a + d) \quad\quad\quad\quad\quad\quad (7)
\end{aligned}
$$

If $a + d = 0$ then $(cy = a) \equiv (-cy = d)$ and then $\tau_{SI}(Cl_2)$ is $\mathcal{REAL}$-unsolvable. Else, we simplify (7) by $a + d$ which gives $cx = a$ and $\tau_{SI}(Cl_2)$ is $\mathcal{REAL}$-unsolvable. So the property holds. $\triangleleft$

5. Automating the proof method: the logical point of view

First, in order to automate the proof method described in section 3, we take advantage of the complete correspondence between the algebraic and the logical semantics for definite constraint logic programs, by considering a *satisfaction-complete* theory Th which *corresponds* to χ:

- χ is a model of Th, i.e. $\models_\chi Th$
- for every constraint c, $\models_\chi \exists c$ iff $Th \models \exists c$
- for every constraint c, either $Th \models \exists c$ or $Th \models \neg \exists c$

Next, we impose that the negation of the generalized constraints appearing in a system of (extended) implications are *finite* generalized constraints themselves. A sufficient condition for this requirement is that for each atomic constraint c of each generalized constraint, there are n atomic constraints $c_1, \ldots, c_n$ such that:

$$\neg c \equiv c_1 \vee \ldots \vee c_n$$

So, by replacing χ-unsolvability by Th-unsatisfiability, theorem 3.3 directly provides us with a correct procedure to prove that a system SI of (extended) implications is true in the least model of a program Pgm and Th. The finiteness condition concerning SI, as stated above, ensures its termination. However, its computational complexity heavily depends on the complexity of the constraint solver, for which Th gives a theoretical lower bound (Grigorieff, 1989).

Example 5.1 We explain why one might use CLP($\mathbb{Q}$) for proving properties of CLP($\mathbb{N}$) programs. Let $F = \{0, 1, +\}$, where 0 and 1 are constant symbols, and $+$ is a binary function symbol. Let $C = \{=, \geq\}$, where $=$ and $\geq$ are two binary constraint symbols. If t is a term from $T(F, V)$ and n a natural number, we note nt the term defined by $0t = 0$ and $(n+1)t = nt + t$. Furthermore, we write n to abbreviate $n1$. One can easily show that $\mathbb{N}$ and $\mathbb{Q}$ - wrt the obvious interpretation of F and C - are two solution-compact structures for $< F, C >$. We define the '$\mathbb{N} - complement$' of $t = s$ and $t \geq s$ as:

$$\neg(t = s) \equiv (t \geq s + 1) \vee (s \geq t + 1) \quad and \quad \neg(t \geq s) \equiv (s \geq t + 1)$$

We have the following proposition:

Observation 5.2 $(c \wedge \neg c')$ is $\mathbb{Q} - unsolvable$ implies $c \models_\mathbb{N} c'$

that enables us to use a symbolic simplex-like algorithm as a constraint solver to implement theorem 3.3, and provides us with a correct procedure for proving properties of CLP($\mathbb{N}$) programs. The complexity of a solver for CLP($\mathbb{N}$) (Contejean, 1993) and observation 5.2 justify the switch from $\mathbb{N}$ to $\mathbb{Q}$. ◁

An implementation of theorem 3.3 based on observation 5.2 written in Prolog III is available from the authors. The idea is that for all predicate p of a program and for each clause defining p, we use the Prolog III interpreter to check the unsolvability of the formulae: $(\bigwedge_{j=1}^{k_i} B_{p_{i,j}}(\tilde{t}_{i,j})) \wedge C_{p,i} \wedge (\tilde{h}_i = \tilde{x}) \wedge \neg B_p(\tilde{x})$. If we succeed for all clauses of p then we conclude that the hypothesis of theorem 3.3 is true and so we have the implication for p.

Example 5.3 Consider the following program which defines the Ackermann's function:

$$
\begin{aligned}
ack(0, y, y + 1) &\leftarrow y \geq 0 \diamond \\
ack(x + 1, 0, z) &\leftarrow x \geq 0, z \geq 0 \diamond ack(x, 1, z) \\
ack(x + 1, y + 1, z) &\leftarrow x \geq 0, y \geq 0, z \geq 0, t \geq 0 \diamond \\
&\quad ack(x + 1, y, t), ack(x, t, z)
\end{aligned}
$$

Using our implementation we can show that $\forall (x, y, z)$ $[ack(x, y, z)$ $\rightarrow$ $z \geq y + 1]$, and thus specialize the third clause:

$$ack(x + 1, y + 1, z) \leftarrow x \geq 0,\ y \geq 0,\ t \geq y + 1,\ z \geq t + 1 \diamond$$
$$ack(x + 1, y, t), ack(x, t, z)$$

in order to be sure to find all the solutions to the goal:

$$\begin{aligned}
&> \ z \leq 61 \ \diamond \ ack(3, y, z) \\
&\{y = 0\ ,\ z = 5\} \\
&\{y = 1\ ,\ z = 13\} \\
&\{y = 2\ ,\ z = 29\} \\
&\{y = 3\ ,\ z = 61\} \\
&> \ \triangleleft
\end{aligned}$$

6. Inherent incompleteness of the proof method

First of all, let us point out how difficult is the problem we address:

Theorem 6.1 *The implication problem, i.e. proving that properties of the form $\forall (d_1 \wedge p \rightarrow d_2)$ - where p is an atom and d_1 and d_2 disjunctions of constraints - are logical consequences of definite constraint logic programs, is unsolvable.*

Proof: The proof we present relies on CLP($\mathbb{N}$) and can be generalized to most domains under reasonable assumptions.

A 2-register machine (see (Minsky, 1961), (Johnstone, 1986) and (Shepherdson, 1991)) has a pair of registers R_1 and R_2 which may hold an arbitrary natural number. A program for the machine is defined by specifying a finite number of states $S_0, (S_1, \ldots, S_n)$, together with, for each i, $1 \leq i \leq n$, an instruction to be carried out whenever the machine is in the state S_i. S_1 is the initial state and S_0 is the terminal state. Suppose we are in state S_i $(1 \leq i \leq n)$, there are the two kinds of instruction:

(1) add 1 to register R_j ($j = 1$ or 2) and move to state S_k, $(0 \leq k \leq n)$
(2) test if R_j holds 0: if it does, move to state S_l, otherwise, subtract 1 from it and move to state S_k, $(0 \leq k, l \leq n)$.

We can describe such a machine by a program in CLP($\mathbb{N}$). These following clauses represent the instructions (1) and (2):

$$\begin{aligned}
p(R_1, R_2, i) &\leftarrow p(R_1 + 1, R_2, k) \\
p(R_1, R_2, i) &\leftarrow R_2 = 0 \ \diamond \ p(R_1, R_2, l) \\
p(R_1, R_2, i) &\leftarrow R_2 \geq 1 \ \diamond \ p(R_1, R_2 - 1, k)
\end{aligned}$$

Consider a 2-register machine and *Pgm* its CLP($\mathbb{N}$) associated program. The proposition "*Pgm* $\models_{\mathbf{N}} p(R_1, R_2, q)$", where R_1, R_2 and q are integers, means that starting from the state S_q with registers R_1 and R_2, the machine halts in state S_0.

Example 6.2 Here is a program which computes $n_1 - n_2$ if $n_1 \geq n_2$ and fails to terminate if $n_1 < n_2$.

$$
\begin{aligned}
S_0 &\leftrightarrow \text{Stop} \\
S_1 &\leftrightarrow \text{If } R_2 = 0 \text{ then move to } S_0 \\
&\qquad \text{else subtract 1 from } R_2 \text{ and move to } S_2 \\
S_2 &\leftrightarrow \text{If } R_1 = 0 \text{ then move to } S_3 \\
&\qquad \text{else subtract 1 from } R_1 \text{ and move to } S_1 \\
S_3 &\leftrightarrow \text{add 1 to } R_1 \text{and move to } S_3.
\end{aligned}
$$

And its associated CLP($\mathbb{N}$) program:

$$
\begin{aligned}
p(R_1, R_2, 0) &\leftarrow \diamond \\
p(R_1, 0, 1) &\leftarrow \diamond\, p(R_1, 0, 0) \\
p(R_1, R_2, 1) &\leftarrow R_2 \geq 1 \diamond\, p(R_1, R_2 - 1, 2) \\
p(0, R_2, 2) &\leftarrow \diamond\, p(0, R_2, 3) \\
p(R_1, R_2, 2) &\leftarrow R_1 \geq 1 \diamond\, p(R_1 - 1, R_2, 1) \\
p(R_1, R_2, 3) &\leftarrow \diamond\, p(R_1 + 1, R_2, 3)
\end{aligned}
$$

◁

2-register machines are useful for coding any unary recursive function. Let f be an unary recursive function, *Dom f* be the definition domain of f and *Pgm* the program in CLP($\mathbb{N}$) associated with the 2-register machine coding f. We have the equivalence:

$$
n \in Dom\, f \iff Pgm \models_{\mathbf{N}} p(n, R_2, 1) \tag{8}
$$

Now, consider the following proposition:

$$
(I) \qquad \forall R_1, R_2, q \in \mathbb{N} \quad p(R_1, R_2, q) \implies q \neq 1
$$

If *Pgm* $\models_{\mathbf{N}} I$ then $p(R_1, R_2, 1)$ is not in the $\mathbb{N}$-model of *Pgm*. By (8), we conclude that *Dom f* $= \emptyset$. Conversely, if *Pgm* $\not\models_{\mathbf{N}} I$ then $\exists R_1, R_2, q \in \mathbb{N}$ such as $p(R_1, R_2, 1)$. Thus by (8) we have $R_1 \in Dom\, f$. Finally (*Pgm* $\models_{\mathbf{N}}$ *I*) $\iff$ (*Dom f* $= \emptyset$).

Since for recursive functions, the problem "*Dom f* $= \emptyset$" is unsolvable and as *I* is a particular case of the implications we consider, we conclude to the unsolvability of the implication problem. $\square$

Now, let us try to identify three weakness of the proof method we propose.

The first one is a drawback of the implementation (but not of theorem 3.3) we present in example 5.1 lies in the fact that we make use of $CLP(\mathbb{Q})$ to prove $CLP(\mathbb{N})$ properties: clearly, there is a lack of precision. Consider for instance the following program:

$$
\begin{aligned}
p(0) &\leftarrow \diamond \\
p(x) &\leftarrow 2x = 1 \diamond
\end{aligned}
$$

The implication $\forall x,\ p(x) \to x = 0$ is true in $CLP(\mathbb{N})$ but false in $CLP(\mathbb{Q})$.

The second is a disadvantage of the proof method lies in the fact that we must have a 'strong enough' system of implications. For instance, in example 4.2, the proof method cannot work if we consider the following system of implications:

$$
\begin{cases}
\forall x & q(x) & \to x = x \\
\forall (x,y) & p(x,y) & \to x = fgh(x) \ \land \ y = hfg(y)
\end{cases}
$$

Yet this system is clearly true. It seems difficult to define what is exactly a 'strong enough' system of implications.

The third and main drawback of our proof method lies deeper. If the clauses defining a logic procedure p are unitary clauses, the method is obviously complete. However, as soon as the definition of p contains one recursive clause, problems may arise. The χ-solvability of a $\tau_{SI}(Cl_{p,i})$ produces a set S of solutions. Roughly speaking, the question is:

$$(Q) \quad M_{\chi,p} \cap S = \emptyset \ ?$$

If the intersection is empty, then the property is true wrt the ith clause of p. If the intersection is not empty, we have at the same time a set of counterexamples together with a clause that does not satisfy the expected property. *However, we do not dispose of any finite means to answer the question Q.* Here stands the heart of the incompleteness of the proof method rejoining the theorem 6.1.

This last remark leads us to recall a well-known technique for proving implications such as I: $\forall \tilde{x}[p(\tilde{x}) \to c(\tilde{x})]$, which resides in giving the goal $p(\tilde{x}) \land \neg c(\tilde{x})$ to the corresponding CLP system. If the answer is negative, then I is true; if the system computes some values for $\tilde{x}$, then I is false. However, it may also loop: this is the case for most examples in this article. This fact constitutes the main difference with our proof procedure which always terminates.

Finally, for some classes of programs, there is at least a decision procedure [3] for proving systems of implications. Let us present two classes of programs and a possible proof method.

Definition 6.3 *(Lloyd, 1987) A level mapping of a program is a mapping from its set of predicate symbols to the non-negative integers. We refer to the value of a predicate symbol p under this mapping as the level of that predicate symbol, denoted by level(p).*

Definition 6.4 *(Lloyd, 1987) A program is hierarchical if it has a level mapping such that, in every program statement $p(\tilde{h}) \leftarrow Body$, the level of every predicate symbol in Body is less than the level of p.*

Proposition 6.5 *If a program Pgm is hierarchical, then, for each predicate p of Pgm, there is a finite generalized constraint A_p which characterize p i.e.: $\forall \tilde{x}\ (p(\tilde{x}) \leftrightarrow A_p)$.*

Proof: By induction on $level(p)$. □

Definition 6.6 *(Fribourg & Veloso Peixoto, 1994) A program is 3-recursive if it is a definite CLP($\mathcal{Z}$) program such that all its predicates are defined as follows:*

$$
\begin{aligned}
p(\tilde{x}) &\leftarrow \xi(\tilde{x}) \diamond \\
p(\tilde{x} + \tilde{a}) &\leftarrow \Phi_1(\tilde{x}) \diamond p(\tilde{x}) \\
p(\tilde{x} + \tilde{b}) &\leftarrow \Phi_2(\tilde{x}) \diamond p(\tilde{x}) \\
p(\tilde{x} + \tilde{c}) &\leftarrow \Phi_3(\tilde{x}) \diamond p(\tilde{x})
\end{aligned}
$$

where $\tilde{a}$, $\tilde{b}$ and $\tilde{c}$ are vectors of integers, $\xi(\tilde{x}), \Phi_1(\tilde{x}), \Phi_2(\tilde{x})$ and $\Phi_3(\tilde{x})$ are finite linear arithmetic constraints.

Theorem 6.7 *(Fribourg & Veloso Peixoto, 1994) If a program Pgm is 3-recursive then each predicate of Pgm can be characterized by a finite generalized arithmetic constraint A_p.*

Let *Pgm* be a definite program. Consider the following system of implications:

$$
\bigwedge_{p \in P} \left[\forall \tilde{x}(p(\tilde{x}) \rightarrow B_p(\tilde{x})) \right].
$$

Proposition 6.8 *If Pgm is hierarchical or 3-recursive, then the following proof method is complete:*
  for each p of Pgm:
  - compute the finite generalized constraint A_p characterizing p,
  - check if $A_p \models_\chi B_p$.

---

[3] i.e. a correct, complete and terminating procedure.

Proof: Obvious by propositions 3.1, 6.5 and theorem 6.7. □

Automation of the above proof method follows the lines of section 5. In contrast, the proof method we propose in section 3 does not constitute a decision procedure but our procedure can be applied to every definite program.

7. Conclusion

Let us first compare our technique with other works. The proof method we propose can be seen as an instance of computational induction (Manna, 1974) (see also (Lloyd, 1987), chapter 2) specially adapted to constraint logic programming. Extended execution (Kanamori & Fujita, 1986), (Renault, 1994) may prove a larger class of systems of implications. The main difference is that our technique can be completely automated. Bottom-up answer analysis (Giacobazzi *et al.*, 1992) may solve a similar problem, with the advantage that such techniques will infer and prove systems of implications. However, example 4.4 shows that our method can prove accurate properties which bottom-up answer analysis cannot take into account, because of its generality.

We now give some research directions in order to pursue this work. An obvious extension of the proof technique we present is the ability to process many-sorted structures. Negation inside the body of the clauses could be considered. At last, the introduction of existential quantifiers in the consequent part of an implication seems to raise interesting problems. This is work in progress.

Finally let us summarize our work. We have presented a correct method for proving definite constraint logic program properties which can quite easily be implemented. Unfortunately any proof method is incomplete, as explained in section 6, but we would like to emphasize on the *generality* of our work: theorem 3.3 relies on properties of the structure χ which all domains that satisfy the framework described in (Jaffar & Lassez, 1986) possess. The only requirement (which strengthens SC_2) concerns the atomic constraints c's appearing in a system of implications, namely: $\neg\ c \equiv c_1 \lor \ldots \lor c_n$. To our knowledge, for all constraint logic programming languages currently available, it does not seem to be too severe a restriction.

FRED MESNARD, SÉBASTIEN HOARAU AND ALEXANDRA MAILLARD

Iremia, Université de La Réunion
Saint-Denis, France

Acknowledgments

We would like to thank Philippe Devienne who gave us the sketch of the proof of theorem 6.1 and anonymous referees for their helpful comments.

References

A. Aiba, A. Sakai, Y. Sato, D.J. Hawley, and R. Hasegawa. Constraint logic programming language CAL. *Proc. of FGCS'88*, pages 263–276, 1988.

K.L. Clark. Predicate logic as a computational formalism. Technical Report Doc 79/59, Logic Programming Group, Imperial College, London, 1979.

A. Colmerauer. Equations and inequations on finite and infinite trees. *Proc. of FGCS'84*, pages 85–99, 1984.

A. Colmerauer. An introduction to Prolog III. *CACM*, 33 (7):70–90, July 1990.

L. Colussi and E. Marchiori. Proving correctness of logic programs using axiomatic semantics. In *Logic Programming - Proceedings of 8th International Conference*, pages 629–642. MIT press, 1991.

E. Contejean. Solving linear diophantine constraints incrementally. *Proc. of ICLP'93*, 1993.

P. Deransart. Proof methods of declarative properties of definite programs. *Theoretical Computer Science*, pages 99–166, 1993.

M. Dincbas, P. Van Hentenrick, H. Simonis, A. Aggoun, T. Graf, and F. Berthier. The constraint logic programming language CHIP. *Proc. of FGCS'88*, pages 693–702, 1988.

L. Fribourg and M. Veloso Peixoto. Bottom-up evaluation of datalog programs with arithmetic constraints: The case of 3 recursive rules. Technical report, L.I.E.N.S, France, 1994.

R. Giacobazzi, S.K. Debray, and G. Levi. A generalized semantics for constraint logic programs. *Proc. of FGCS'92*, pages 581–591, 1992.

S. Grigorieff. Décidabilité et complexité des théories logiques. *Logique et Informatique : une introduction, Collection Didactique, INRIA*, pages 7–97, 1989.

D. Hofstadter. *Gödel, Escher, Bach : an Eternal Golden Braid*. Basic Books, Inc., 1979.

H. Hong. Non-linear real constraints in constraint logic programming. In *LNCS 632*, pages 201–212. Springer Verlag, 1992.

J. Jaffar and J.L. Lassez. Constraint logic programming. Technical Report 74, Monach University, Australia, 1986.

J. Jaffar and M.J. Maher. Constraint logic programming: a survey. *J. Logic Programming*, pages 503–581, 1994.

J. Jaffar, S. Michaylov, P.J. Stuckey, and R.H.C. Yap. The CLP($\mathbf{R}$) language and system. *Proc. of the 4th ICLP*, 1987.

P.T. Johnstone. *Notes on logic and set theory*. cambridge mathematical textbooks. Cambridge University Press, 1986.

T. Kanamori and H. Fujita. Formulation of induction formulas in verification of prolog programs. *Proc. of the 8th CADE*, pages 281–299, 1986.

J.M. Lever. Proving program properties by means of SLS-resolution. *Proc. of the 8th ICLP*, pages 614–628, 1991.

J.W. Lloyd. *Foundations of Logic Programming*. Springer-Verlag, 1987.

Z. Manna. *Mathematical theory of computation*. McGraw-Hill, 1974.

F. Mesnard et al. CLP($\mathbb{Q}$) for proving interargument relations. In *LNCS 649*, pages 308–320. Springer Verlag, 1992.

M.L. Minsky. Recursive unsolvability of post's problem of 'tag' and other topics in the theory of Turing machines. *Ann. of Math.*, 74:437–455, 1961.

S. Renault. Generalized extended execution for normal programs. In *LNCS 883*. Springer Verlag, 1994.

J.C. Shepherdson. Unsolvable problems for SLDNF resolution. *J. Logic Programming*, pages 19–22, 1991.

C. Walinsky. CLP(Σ^*): Constraint logic programming with regular sets. *Proc. of the 6th ICLP*, pages 181–196, 1989.

FIRST-ORDER CONSTRAINED LAMBDA CALCULUS

Abstract. In (Crossley *et al.*, 1993a; Crossley *et al.*, 1993b; Mandel, 1995) a calculus which extends the traditional lambda calculus by the addition of constraints was presented. The constraints can be used passively for restricting the range of variables and actively for computing solutions of goals. Here we present an extension of that calculus obtained by adding existential quantifiers and new rules for handling the new terms. This avoids the problem of shared variables problem and gives a very smoothly working language. We present a proof of the Church-Rosser property and the denotational semantics of the whole calculus.

Introduction

The constraint programming paradigm is based on the idea of using a special mechanism – constraint solving – to enhance a conventional programming language by making use of the particular – albeit partial – information provided by the specialized constraint solver. The first approach which uses iterative approximation techniques to solve constraint programs was Sketchpad (Sutherland, 1963) in '63. In '87 Jaffar and Lassez introduced Constraint Logic Programming (CLP). This approach gives a general setting for incorporating constraints in a logic programming language. This has been implemented in the language CLP(IR) (Jaffar *et al.*, 1992) which contains a simplex algorithm for solving constraints over the real numbers.

The idea of combining constraints with functional programming was introduced by Darlington et al. in FALCON (Guo & Pull, 1991) which is a functional-logic programming language with constraints. This language extends functional and logic programming to general constraint programming, and permits the use of constraints for programming as well as for queries.

In previous work (Crossley *et al.*, 1993a; Crossley *et al.*, 1993b; Mandel, 1995) we presented a constraint enhanced version of lambda calculus.

F. Baader and K.U. Schulz (eds.), Frontiers of Combining Systems, 339–356.

For this language we added, in particular, a new formation rule for terms whereby if M is a term and C a constraint then $\{C\}M$ is a term. The constraints are used in two ways. On the one hand, passively, the constraints restrict the universe of validity of a term, and on the other, actively, where inferences that are made from constraints cause the replacing of subterms in the target term. In (Mandel, 1993) the semantics were given and the correctness of the reduction rules demonstrated. A work in related direction can be found in (Henz *et al.*, 1993).

In the work described here we add a new term constructor, the existential quantifier ∃. This gives us first-order constrained lambda calculus. The addition of the quantifier gives added strength to the constraint solving. It also solves a problem caused by the binding of variables by both constraints and λ-abstraction (see (Mandel, 1995)). Along with the new term formation rules reduction rules are introduced.

Indeed, the global nature of the constraints of the propositional approach can be limited by the ∃-quantifier. This new binder is inspired by the so-called Henkin theories. It defines "local" variables, in the same way as the λ-abstractor. The value of the variable attached to a λ-abstractor is calculated by the functional theory (i.e. β-reduction), the value of the variable attached to a ∃-quantifier is calculated by the constraint theory. In this way, the active use of constriants is explicitly stimulated and both theories are orthogonally combined.

Outline. In section 1 we present the notation of the constraint language, the term formation rules and some examples. In section 2 we present examples showing how constraints and functional programming are amalgamated in the present approach. In section 3 the denotational semantics of the calculus is given and it is demonstrated that the semantic function is well defined and that the reduction rules are correct, that is, interconvertible terms have the same semantics. In section 4 we show that the new calculus satisfies the Church-Rosser property. This means that if a term has a normal form, then it is unique. Finally we mention an implementation and indicate future research directions.

1. First-Order Constrained Lambda Calculus

We begin by defining the language of constrained lambda terms, and then we list the associated reduction rules. In the same way as in the propositional approach, we present a formulation independent of the particular choice of constraint domain and theory. Given a constraint theory with equality $\Pi = \langle \pounds, \mathcal{A}, \mathcal{R}, \vdash_\pi \rangle$ in a language $\pounds = \langle \mathcal{K}, \mathcal{V}, \mathcal{F}, \mathcal{P} \rangle$, where $\mathcal{K}$, $\mathcal{V}$, $\mathcal{F}$, $\mathcal{P}$ are the countable sets of individual constants, variables, function letters, and predicate letters, respectively, $\mathcal{A}$ is a set of $\pounds$-wffs, $\mathcal{R}$ is a set of inference rules over $\mathcal{L}$-wffs, and $\vdash_\pi$ is an inference relation with axioms $\mathcal{A}$ and inference rules $\mathcal{R}$; then a *constrained lambda term* uses in addition the λ-abstractor, the ∃-quantifier, and the auxiliary symbols {, } and (,) and the constant **fail**.

The term formation rules extend the set of rules of the λ_c-calculus defined in (Crossley *et al.*, 1993a; Crossley *et al.*, 1993b) by adding the ∃ term constructor. The extended set of terms is denoted by $\Lambda_c^\exists$, and we write $\Lambda_c^\exists = \Lambda_c + \exists$. The set of constraints is denoted by **Constraint**.

The central rules are of the form: If M is a term and C a constraint, then $\lambda\,x\,.\,M$, $\{C\}M$ and $\exists\,x\,.\,M$ are terms. Formally we have:

Definition 1.1 (Constrained Lambda Term, Constraint)   The set $\Lambda_c^\exists$ of constrained lambda terms and the set **Constraint** of constraints over lambda terms are defined by mutual induction as follows:

(I)
- $\mathsf{fail} \in \Lambda_c^\exists$;   (Fail)
- $k \in \mathcal{K} \Rightarrow k \in \Lambda_c^\exists$;   (Constant)
- $x \in \mathcal{V} \Rightarrow x \in \Lambda_c^\exists$;   (Var)
- $f^n \in \mathcal{F}, \mathrm{M}_1, \ldots, \mathrm{M}_n \in \Lambda_c^\exists \Rightarrow f^n(\mathrm{M}_1, \ldots, \mathrm{M}_n) \in \Lambda_c^\exists$;   (Functors)
- $x \in \mathcal{V}, \mathrm{M} \in \Lambda_c^\exists \Rightarrow (\lambda\,x\,.\,\mathrm{M}) \in \Lambda_c^\exists$;   (Abstr)
- $\mathrm{M}, \mathrm{N} \in \Lambda_c^\exists \Rightarrow (\mathrm{M}\ \mathrm{N}) \in \Lambda_c^\exists$;   (Applic)
- $C \in \mathsf{Constraint}, \mathrm{M} \in \Lambda_c^\exists \Rightarrow (\{C\}\mathrm{M}) \in \Lambda_c^\exists$;   (Constr)
- $\mathrm{M} \in \Lambda_c^\exists \Rightarrow (\exists\,x\,.\,\mathrm{M}) \in \Lambda_c^\exists$.   (Exist)

(II)
- $P^m \in \mathcal{P}, \mathrm{M}_1 \cdots \mathrm{M}_m \in \Lambda_c^\exists \Rightarrow \{P^m(\mathrm{M}_1, \cdots, \mathrm{M}_m)\} \in \mathsf{Constraint}$;
- $\mathrm{M}_1, \mathrm{M}_2 \in \Lambda_c^\exists \Rightarrow \{\mathrm{M}_1 = \mathrm{M}_2\} \in \mathsf{Constraint}$;
- $\{C\}, \{D\} \in \mathsf{Constraint} \Rightarrow \{C, D\} \in \mathsf{Constraint}$.   □

As usual $\exists\,x_1\,.\,(\exists\,x_2\,.\,\mathrm{M})$ will be written as $\exists\,x_1\,x_2\,.\,\mathrm{M}$. This equivalence is made at syntactic level, as well as α-conversion. Moreover, we make the *variable convention* that the free and bound variables in a term have different names; see (Barendregt, 1984).

REDUCTION RULES

As in (Crossley *et al.*, 1993a; Crossley *et al.*, 1993b), we completely reduce constraints which have unique solutions. The key rule which introduces reductions implied by the constraint theory is as follows:

if $\mathcal{A} \vdash (\exists!\,x\,.\,C) \wedge (\forall\,x\,.\,C\ \to\ x = \mathrm{N})^1$
and N is in normal form (see 1.2 below) then
$\exists\,x\,.\,\{C\}\mathrm{M} \longrightarrow_\exists (\{C\}\mathrm{M})[x := \mathrm{N}]$

---

[1]The expression $\exists!\,x\,.\,\varphi$ denotes "there is a unique x s.t. φ," and the operator $\to$ denotes implication. We are definitely sloppy here, since we use a quantifier ∃ both in the λ-theory and in the constraint theory. Normally, however, no confusion can arise since the contexts of use are rather distinguishable.

where $\mathcal{A}$ is the set of axioms of the inference relation of the underlying constraint theory. (Note that the variables in N and M must satisfy the variable convention mentioned above.) This rule is inspired by the so-called Henkin theories, i.e. theories $\mathcal{T}$ for which, given any sentence $(\exists x \, . \, \varphi)$, there is a constant k in the language such that: $\mathcal{T} \models (\exists x \, . \, \varphi) \supset (\varphi[x := k])$.

This reduction rule presents similarities with the β-rule as shown by the reductions in figure 1. For β-reduction, the substitution is induced by the presence of a formal parameter in the abstraction. In the case of the $\exists$-reduction the value N for x is calculated by the constraint solver as the only value valid for x in the constraint theory under the conditions in $\{C\}$.

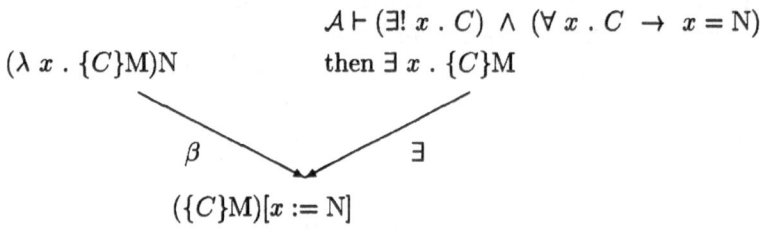

Figure 1. $\exists$ and β rules

Remark 1.2 There may be two syntactically different terms N_1 and N_2 such that

$$\mathcal{A} \vdash (\exists! x \, . \, C) \wedge (\forall x \, . \, C \rightarrow x = N_1) \text{ and}$$
$$\mathcal{A} \vdash (\exists! x \, . \, C) \wedge (\forall x \, . \, C \rightarrow x = N_2)$$

For instance, let the set $\mathbb{N}$ of natural numbers be the underlying constraint theory with its usual axiomization. Then

$$\mathbb{N} \vdash (\exists! x \, . \, x = 2 + 2) \wedge (\forall x \, . \, x = 2 + 2 \rightarrow x = 2 + 2)$$
$$\mathbb{N} \vdash (\exists! x \, . \, x = 2 + 2) \wedge (\forall x \, . \, x = 2 + 2 \Rightarrow x = 4).$$

This would give rise to two different reductions:

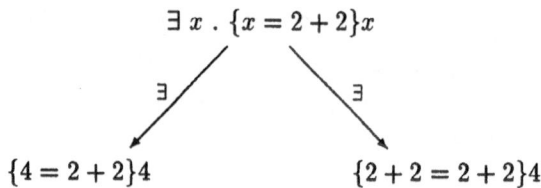

Now, of course, $\mathbb{N} \vdash 2 + 2 = 4$ and, in general we shall have $\mathcal{A} \vdash N_1 = N_2$. In this case, we require the constraint solver to calculate a normal form for the term computed.

We require that for all closed terms M such that $\mathcal{A} \vdash M = M'$ there should be a distinguished closed term N such that $\mathcal{A} \vdash M = N$. Such an N is called the *normal form* of M. For example, for positive rational numbers a standard normal form would be m/n where m, n are mutually prime. □

The rules of the $\lambda_c^\exists$-calculus supplement those of λ-calculus by adding (i) reductions for $\exists x . \{C\}M$ where the constraint C has a unique solution for x, and (ii) rules which move existential quantifiers to the left. Formally we have:

Definition 1.3 (Reduction Rules)

(β) $(\lambda x . M) N \longrightarrow_\beta M[x := N]$;

$(\exists)$ $\exists x . \{C\}M \longrightarrow_\exists (\{C\}M)[x := N]$

  if $\mathcal{A} \vdash (\exists! x . C) \land (\forall x . C \rightarrow x = N)$;

(A_c) $\{C\}(\{D\}M) \longrightarrow_{A_c} \{C, D\}M$;

(P_c) $\{D, P(\ldots, \{C_i\}M_i, \ldots)\}N \longrightarrow_{P_c} \{D, C_i, P(\ldots, M_i, \ldots)\}N$;

(F_c) $f(\ldots, \{C_i\}M_i, \ldots) \longrightarrow_{F_c} \{C_i\}f(\ldots, M_i, \ldots)$;

$(F_\exists)$ $f(\ldots, \exists x_i . M_i, \ldots) \longrightarrow_{F_\exists} \exists x_i . f(\ldots, M_i, \ldots)$;

$(P_\exists)$ $\{D, P(\ldots, \exists x_i . M_i, \ldots)\}N \longrightarrow_{P_\exists} \exists x_i . \{D, P(\ldots, M_i, \ldots)\}N$;

$(C_\exists)$ $\{C\}(\exists x . M) \longrightarrow_{C_\exists} \exists x . (\{C\}M)$;

(fail)   The fail-rule is the union of the following five rules:

  (f_1) (fail M) $\longrightarrow_{f_1}$ fail;

  (f_2) $(\lambda x . \text{fail}) \longrightarrow_{f_2}$ fail;

  (f_3) $\{C\}M$ $\longrightarrow_{f_3}$ fail if $\{C\}$ is *inconsistent*;

  (f_4) $\{C\}\text{fail}$ $\longrightarrow_{f_4}$ fail;

  (f_5) $\exists x . \text{fail} \longrightarrow_{f_5}$ fail.   □

The constraint theory extended with λ_c-terms is required to be strict on the **fail** element. That is, $f(\ldots \text{fail} \ldots) \longrightarrow \text{fail}$ and $P(\ldots \text{fail} \ldots)$ is inconsistent. This requirement allows the Church-Rosser property to hold.

2. Examples

In this section we present the reader two examples.

Example 2.1 (The Plane) *A plane is travelling horizontally at 40 metres per second at a height of 100 metres above the ground and drops a package. How far away, horizontally, does the package strike the ground relative to the point at which it was released?*

Let t be the time required to drop, V_{x_0}, V_{y_0} the initial horizontal and vertical speeds. Then the horizontal and vertical coordinates satisfy the equations

$$x = V_{x_0} \cdot t = 40 \cdot t, \qquad\qquad y = -\tfrac{1}{2} \cdot G \cdot t^2 = -\tfrac{1}{2} \cdot (9.81) \cdot t^2 = -100.$$

Therefore, we can build the following $\lambda_c^{\exists}$-term:

$$((\lambda\, V_{x_0}\, y\,.\,(\exists\, t\,.\,\{y = -\tfrac{1}{2} \cdot 9.81 \cdot t^2, t > 0\} V_{x_0} \cdot t))40) - 100$$
$$\longrightarrow_\beta \exists\, t\,.\,\{-100 = -\tfrac{1}{2} \cdot 9.81 \cdot t^2, t > 0\} 40 \cdot t \equiv M$$

Assuming the constraint solver can solve the system of equations in the variable t, so that

$$\{-100 = -\tfrac{1}{2} \cdot 9.81 \cdot t^2, t > 0\} \vdash_\pi t = 4.52.$$

Then, applying the $\exists$-rule we get

$$M \longrightarrow_\exists \{-100 = -\tfrac{1}{2} \cdot 9.81 \cdot 20.39, 4.5152 > 0\}\ 180.61.$$

And from this term we have that the distance relative to the point at which the package was released is 180.61 metres. □

Next consider a triangle of sides a, b, c where we are given $a = 3$, $b = 5$ but we only have the constraint $c > 3$. The triangle must satisfy the triangle constraint, that is: $a > b+c$, etc. We want to calculate the area of the given triangle. Without loss of generality we suppose that the side a is the base of the triangle, i.e. it is the largest side. We divide the given triangle into two right-angled triangles as in the figure 2. The area of the given triangle is the sum of the areas of both right-angled triangles. For this example we show only the most important reduction steps.

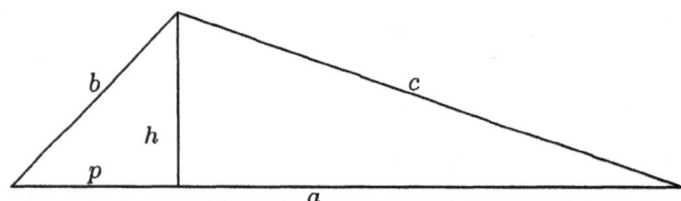

Figure 2.   Triangle example

Example 2.2 (Area of a Triangle)

$$area\text{-}right\text{-}angled\text{-}triangle \equiv (\lambda\, l_1\, l_2\,.\,\{l_1 > 0,\ l_2 > 0\} \tfrac{l_1 \times l_2}{2})$$
$$area\text{-}triangle \equiv (\lambda\, a\, b\, c\,.\,\exists\, p\, h\,.\,\{a < b+c,\ b < c+a,\ c < b+a,$$
$$p^2 + h^2 = b^2,\ (a-p)^2 + h^2 = c^2,\ h > 0\}$$
$$(area\text{-}right\text{-}angled\text{-}triangle\ p\ h) +$$
$$(area\text{-}right\text{-}angled\text{-}triangle\ (a-p)\ h)).$$

We now calculate the area of a triangle where the side c is unknown but restricted to satisfy the constraint $\{c > 3\}$. We can take advantage of the possibility of computing functions with incomplete information as follows:

$(\lambda\ a\ b\ c\ .\ \exists\ p\ h\ .\ \{a < b + c,\ b < c + a,\ c < b + a,$
$\qquad p^2 + h^2 = b^2,\ (a - p)^2 + h^2 = c^2,\ h > 0\}$
$\qquad (area\text{-}right\text{-}angled\text{-}triangle\ p\ h) +$
$\qquad (area\text{-}right\text{-}angled\text{-}triangle\ (a - p)\ h))\ 5\ 3\ (\{c > 3\}c).$

By β-reduction we get:

$\exists\ p\ h\ .\ \{5 < 3 + (\{c > 3\}c),\ 3 < (\{c > 3\}c) + 5,\ (\{c > 3\}c) < 3 + 5,$
$\qquad p^2 + h^2 = 3^2,\ (5 - p)^2 + h^2 = (\{c > 3\}c)^2,\ h > 0 \qquad\qquad \}$
$\qquad (area\text{-}right\text{-}angled\text{-}triangle\ p\ h) +$
$\qquad (area\text{-}right\text{-}angled\text{-}triangle\ (5 - p)\ h).$

Applying the β-rule twice we get:

$\exists\ p\ h\ .\ \{5 < 3 + (\{c > 3\}c),\ 3 < (\{c > 3\}c) + 5,\ (\{c > 3\}c) < 3 + 5,$
$\qquad p^2 + h^2 = 3^2,\ (5 - p)^2 + h^2 = (\{c > 3\}c)^2,\ h > 0 \qquad\qquad \}$
$\qquad (\{p > 0, h > 0\}\ \frac{p \times h}{2}) +$
$\qquad (\{(5 - p) > 0, h > 0\}\ \frac{(5 - p) \times h}{2}).$

Using the F_c-rule twice we have:

$\exists\ p\ h\ .\ \{5 < 3 + (\{c > 3\}c),\ 3 < (\{c > 3\}c) + 5,\ (\{c > 3\}c) < 3 + 5,$
$\qquad p^2 + h^2 = 3^2,\ (5 - p)^2 + h^2 = (\{c > 3\}c)^2,\ h > 0,$
$\qquad p > 0,\ (5 - p) > 0 \qquad\qquad\qquad\qquad\qquad\qquad \}$
$\qquad (\frac{p \times h}{2}) + (\frac{(5 - p) \times h}{2}).$

By use of the P_c-rule we have:

$\exists\ p\ h\ .\ \{5 < 3 + c,\ 3 < c + 5,\ c < 3 + 5,$
$\qquad p^2 + h^2 = 3^2,\ (5 - p)^2 + h^2 = c^2,\ h > 0,$
$\qquad p > 0,\ (5 - p) > 0,\ c > 3 \qquad\qquad \}\ \frac{5 \times h}{2}.$

We assume that the constraint solver can solve[2] the system on p and h returning values which are terms of the constraint language. From the constraint solver we then get:

$\{5 < 3 + c,\ 3 < c + 5,\ c < 3 + 5,$
$\quad p^2 + h^2 = 3^2,\ (5 - p)^2 + h^2 = c^2,\ h > 0,$
$\quad p > 0,\ (5 - p) > 0,\ c > 3 \qquad\qquad \} \vdash h = \frac{\sqrt{-c^4 + 68c^2 - 256}}{10},\ p = \frac{34 - c^2}{10}$

Then, by $\exists$-rule in two steps we get:

$\{\ 5 < c + 3,\ c < 8,\ 3 < c + 5,\ c > 3,$
$\quad p^2 + \frac{-c^4 + 68c^2 - 256}{100} = 9,$
$\quad (5 - p)^2 + \frac{-c^4 + 68c^2 - 256}{100} = c^2,$
$\quad \sqrt{-c^4 + 68c^2 - 256} > 0 \qquad\qquad \}\ \frac{\sqrt{-c^4 + 68c^2 - 256}}{4}$

$\{5 < 3 + c,\ 3 < c + 5,\ c < 8,$
$\quad (\frac{34 - c^2}{10})^2 + (\frac{\sqrt{-c^4 + 68c^2 - 256}}{10})^2 = 9,$
$\quad (5 - \frac{34 - c^2}{10})^2 + (\frac{\sqrt{-c^4 + 68c^2 - 256}}{10})^2 = c^2,$
$\quad \frac{\sqrt{-c^4 + 68c^2 - 256}}{10} > 0,$
$\quad \frac{34 - c^2}{10} > 0,\ (5 - \frac{34 - c^2}{10}) > 0,\ c > 3\quad \}\ \frac{5 \times \frac{\sqrt{-c^4 + 68c^2 - 256}}{10}}{2}.$

---

[2] It is important to note that variables under the scope of an $\exists$-quantifier are computed using the underlying constraint theory whereas variables under the scope of the λ-abstractor are computed using the functional theory.

Writing

$$\{C\} = \{5 < 3+c, \; 3 < c+5, \; \frac{(34-c^2)^2}{100} + \frac{-c^4+68c^2-256}{100} = 9,$$
$$(5 - \frac{34-c^2}{10})^2 + \frac{-c^4+68c^2-256}{100} = c^2, \; \sqrt{-c^4+68c^2-256} > 0,$$
$$34 - c^2 > 0, \; 50 - 34 + c^2 > 0 \qquad\qquad\qquad \}$$

the set of those atomic constraints that do not affect the constraint, together with the set of subsumed atomic constraints, we can abbreviate the whole term to:

$$\boxed{\{C, c > 3, \; c < 8\}^{\frac{\sqrt{-c^4+68c^2-256}}{4}}}$$

□

3. Semantics

Here we give the denotational semantics of the language. The semantics for a general scheme for first-order constraint functional logic programming was presented in (López-Fraguas, 1992). Also in (Panangaden *et al.*, 1991) a semantics for concurrent constraint programming languages is given.

The semantic function is, as usual, defined in a recursive fashion. In order to prove its well-definedness, the semantics of the constraint logic was defined to be a 3-valued one. These values may be explained to be *false* (denoted by $\top_C$), *true* (denoted by **Comp** for compatible), and *I (still) don't know* (denoted by $\bot_C$). Moreover, these values are ordered by $\bot_C \sqsubseteq \textbf{Comp} \sqsubseteq \top_C$, and this domain is denoted by **C**. Motivation for this ordering is the fact that the more information we get the bigger is the risk of contradiction being introduced.

In the same spirit, the semantics of terms can be any of three possibilities. On the one hand we have the meaningful values. On the other, the semantics of a term can be $\top_V$ if contradiction was introduced, or $\bot_V$ if there are variables or, in general, subterms whose values cannot be calculated in the context given. These values are also ordered. Typically, $\mathbf{V} = \mathbf{C} \oplus \textbf{Num} \oplus \textbf{Char} \oplus \textbf{String}$. Sometimes we call $\top_V$ the overspecified, $\bot_V$ the underspecified value. **F** denotes the domain of functions, and **Env** the one of environment functions.

The semantic functions will have, besides the target term, not only an environment with values for variables but also a global constraint as argument. Indeed, consider the term

$$f(\{x <= 0\}g(x), \{x >= 0\}h(x)) \xrightarrow{\ \ }_{A_c} \{x <= 0, x >= 0\}f(g(x), h(x))$$

$$\begin{array}{ccc}
 & f & \\
 \swarrow & & \searrow \\
\{x <= 0\} & & \{x >= 0\} \\
\downarrow & & \downarrow \\
g & & h \\
\downarrow & & \downarrow \\
x & & x
\end{array}$$

The information contained in the constraints is global and affects the whole term, in the previous case the value 0 is inferred for the variable x. This not necessarily implies that a contradictory constraint will provoke the whole term to fail; see the semantic function defined in detail below. We will use complete lattices as domains, where for each domain $\bot$ and $\top$ are the least, resp. greatest, elements. **Env** denotes the domain of environment functions, **V** the domain of values (typically $\mathbf{V} = \mathbf{C} \oplus \mathbf{Num} \oplus \mathbf{Char} \oplus \mathbf{String}$), **F** the one of functions and **C** is the domain of *compatible* values consisting of three points ordered by $\bot_{\mathbf{C}} \sqsubseteq \mathbf{Comp} \sqsubseteq \top_{\mathbf{C}}$.

$\mathcal{CLE} : \Lambda_c^{\exists} \to (\mathbf{Env} \times \mathsf{Constraint}) \to \mathbf{V} \oplus \mathbf{F}$

$\mathcal{CON} : \mathsf{Constraint} \to (\mathbf{Env} \times \mathsf{Constraint}) \to \mathbf{C}$

$\mathcal{AC} : \mathsf{AtomicConstraint} \to (\mathbf{Env} \times \mathsf{Constraint}) \to \mathbf{C}$

$\mathbf{F} \quad \cong \quad \mathbf{V} \oplus \mathbf{F} \to \mathbf{V} \oplus \mathbf{F}$

$\mathbf{Env} \quad = \quad \mathcal{V} \to \mathbf{V} \oplus \mathbf{F}$

$\mathbf{V} \quad = \quad \mathcal{U}_Q \cup \{\top_{\mathbf{V}}, \bot_{\mathbf{V}}\}$ ordered by $\bot_{\mathbf{V}} \sqsubseteq v \sqsubseteq \top_{\mathbf{V}}$ for all v in $\mathcal{U}_Q$,

where $\mathcal{U}_Q$ is the underlying constraint domain.

Whenever we want to define the semantics of a term M, we have to collect the information of the constraints contained in M. This is done by the **collect** function and will allow the manipulation of the side effects a constraint has over its surrounding terms. For instance, consider the term $M \equiv ((\lambda\ x\ .\ \{y = 0\}\ x + y)\ y)$. Intuitively, M has the same meaning as $N \equiv ((\lambda\ x\ .\ x + 0)\ 0) \longrightarrow_\beta 0 + 0$. In order to keep track of bound variables, we use a duplicate set of variables. These are indicated by superscripts. Moreover, given that a variable may be bound more than once in a term (e.g. $\lambda\ x\ .\ ((\lambda\ x\ .\ x)x)$), every duplicated variable will have an index, that will be updated using a **shift** function. These functions are inductively defined as follows:

Definition 3.1 shift : $\mathcal{V} \times (\Lambda_c^{\exists} \cup \mathsf{Constraint}) \to \Lambda_c^{\exists} \cup \mathsf{Constraint}$ is recursively defined by:

- **shift**$(x, \mathsf{fail}) = \mathsf{fail}$;
- **shift**$(x, k) = k$;
- **shift**$(x, y) = \begin{cases} x^1 & \text{if } x \equiv y, \\ x^{n+1} & \text{if } \exists\ n : y \equiv x^n, \\ y & \text{otherwise}; \end{cases}$
- **shift**$(x, f(M_1, \ldots, M_n)) = f(\mathbf{shift}(x, M_1), \ldots, \mathbf{shift}(x, M_n))$;
- **shift**$(x, \lambda\ y\ .\ M) = \begin{cases} \lambda\ x\ .\ M & \text{if } x \equiv \mathsf{var}, \\ \lambda\ y\ .\ \mathbf{shift}(x, M) & \text{otherwise}; \end{cases}$
- **shift**$(x, (M\ N)) = (\mathbf{shift}(x, M)\ \mathbf{shift}(x, N))$;
- **shift**$(x, \{C\}M) = \mathbf{shift}(x, \{C\})\mathbf{shift}(x, M)$;
- **shift**$(x, \{c_1, \ldots, c_k\}) = \{\mathbf{shift}(x, c_1), \ldots, \mathbf{shift}(x, c_k)\}$;
- **shift**$(x, P(M_1, \ldots, M_m)) = P(\mathbf{shift}(x, M_1), \ldots, \mathbf{shift}(x, M_m))$;
- **shift**$(x, M = N) = \mathbf{shift}(x, M) = \mathbf{shift}(x, N)$. $\qquad\qquad\square$

Definition 3.2 collect : $\Lambda_c^3 \to$ Constraint is recursively defined by:

- collect(fail) = {**False**};
- collect(k) = collect(x) = $\emptyset$;
- collect($f(M_1, \ldots, M_n)$) = $\bigcup_{i=1}^n$ collect(M_i);
- collect($\lambda\ x\ .\ M$) = shift(x, collect(M));
- collect(M N) = collect(M) $\cup$ collect(N);
- collect($\{C\}$M) = collect($\{C\}$) $\cup$ collect(M);
- collect($\{c_1, \ldots, c_n\}$) = $\bigcup_{i=1}^n$ collect(c_i);
- collect($\{M = N\}$) = $\{M = N\}$ $\cup$ collect(M) $\cup$ collect(N);
- collect($\{P(M_1, \ldots, M_m)\}$) =
  = $\{P(M_1, \ldots, M_m)\} \cup \bigcup_{i=1}^m$ collect(M_i). $\square$

The denotations are as follows.

- $\mathcal{CLE}[\![\text{fail}]\!] = \lambda\ e \in \mathbf{Env}, c \in$ Constraint . $\top_{\mathbf{V} \oplus \mathbf{F}}$
- $\mathcal{CLE}[\![k]\!] = \lambda\ e \in \mathbf{Env}, c \in$ Constraint . $\text{in}_1(\mathcal{I}_Q(k))$
  where $\mathcal{I}_Q$ is the interpretation function of the underlying constraint domain.
- $\mathcal{CLE}[\![x]\!] = \lambda\ e \in \mathbf{Env}, c \in$ Constraint .
  unique $\{v : e\{x := v\} \models \exists\ \vec{y}\ .\ c_x\}$
  where c_x is the union of all the atomic constraints dependent on the variable x, $\vec{y}$ are the free variables of c_x different from x, the expression $e\{x := v\} \models \exists\ \vec{y}\ .\ c_x$ abbreviates "there exist $v_i \in \mathbf{V} \oplus \mathbf{F}$ such that $e\{x := v\}\{y_i := v_i\} \models c_x$," the environment $e\{x := v\}$ is defined by

  - $e\{x := v\}(y) = e(y)$ if $x \not\equiv y$;
  - $e\{x := v\}(x) = \begin{cases} v & \text{if } e(x) = \bot_{\mathbf{V}} \text{ or } e(x) = v, \\ e(x) & \text{otherwise}, \end{cases}$

  and the function **unique** is defined as follows:
  $$\text{\textbf{unique}}\ A = \begin{cases} \top_{\mathbf{V}} & \text{if } A = \emptyset \\ a & \text{if } A = \{a\} \\ \bot_{\mathbf{V}} & \text{if } \exists\ a, b \in A \text{ such that } a \neq b. \end{cases}$$
  The **unique** operator returns the *unique* value v for x making the constraint c_x true. If there is no such a unique object, either there is no such v or there are more than one. In the first case, we let **unique** $\{v : e\{x := v\} \models \exists\ \vec{y}\ .\ c_x\}$ denote $\top_{\mathbf{V}}$, and in the second $\bot_{\mathbf{V}}$. The **unique** operator is monotonic, since if we are given more information (i.e. values assigned by the environment e and/or constraints in c), we can still ignore –in other words, know of more than one– or distinguish exact one or, because of contradiction, individualize no value for the variable x. In these cases, we speak of a underdefined, defined or overdefined variable, respectively; see theorem 3.4 and figure 3.

In the following we introduce denotations for terms which must *share* their values with other subterms.

- $\mathcal{CLE}[\![f(M_1,\ldots,M_n)]\!] = \lambda\, e \in \mathbf{Env}, c \in \text{Constraint}\,.$
  let $c' = c \cup \mathbf{collect}(f(M_1,\ldots,M_n))$
  in $\mathcal{FUNC}[\![f]\!]\,(\mathcal{CLE}[\![M_1]\!]\,e\,c',\ldots,\mathcal{CLE}[\![M_n]\!]\,e\,c')$
- $\mathcal{CLE}[\![(\lambda\, x\,.\,M)]\!] = \lambda\, e \in \mathbf{Env}, c \in \text{Constraint}\,.$
  $\mathbf{in}_2(\lambda\, v \in \mathbf{V}{\oplus}\mathbf{F}\,.\,\mathcal{CLE}[\![M]\!]\,e[x := v]\ \ c[x := x_{new}])$
  where x_{new} is new, that is $e(x_{new}) = \bot_{\mathbf{V}\oplus\mathbf{F}}$.
- $\mathcal{CLE}[\![(M\ N)]\!] = \lambda\, e \in \mathbf{Env}, c \in \text{Constraint}\,.$
  let $c_1 = \mathbf{collect}(M)$
     $c_2 = \mathbf{collect}(N)$
  in $[\mathbf{id}_{\mathbf{F}}, \lambda\, x \in \mathbf{V}\,.\,\bot_{\mathbf{F}}](\mathcal{CLE}[\![M]\!]\,e\,c \cup c_2)(\mathcal{CLE}[\![N]\!]\,e\,c \cup c_1)$
- $\mathcal{CLE}[\![(\{C\}\ M)]\!] = \lambda\, e \in \mathbf{Env}, c \in \text{Constraint}\,.$
  let $c_1 = \mathbf{collect}(\{C\})$
     $c_2 = \mathbf{collect}(\lambda_c\text{-term})$
  in if $\mathcal{CON}[\![c \cup c_1 \cup c_2]\!]\,e\ \emptyset$ then $\mathcal{CLE}[\![M]\!]\,e\,c \cup c_1$
- $\mathcal{CLE}[\![(\exists\, x\,.\,M)]\!]\,e\,c = \mathcal{CLE}[\![M[x := x_{new}]]\!]\,e\,c$

Now we have to define the semantics of a constraint taking into account that now $\lambda_c^{\exists}$-terms can appear *inside* a constraint. For example, an atomic constraint can test the equality of two λ-abstractions via the continuous equality $=_{\mathbf{V}\,\oplus\,\mathbf{F}}$. (In many cases the result of such a comparison will be $\bot_{\mathbf{C}}$, because such comparisson –in general– are not computable.)

- $\mathcal{CON}[\![\{C\}]\!] = \mathcal{CON}[\![\mathsf{ac}_1(M_1^1,\ldots,M_{n_1}^1),\ldots,\mathsf{ac}_k(M_1^k,\ldots,M_{n_k}^k)]\!] =$
  $\lambda\, e \in \mathbf{Env}, c \in \text{Constraint}\,.$
  let $c' = c \cup \mathbf{collect}(\{C\})$
  in and $(\mathcal{AC}[\![\mathsf{ac}_1(M_1^1,\ldots,M_{n_1}^1)]\!]\,e\,c'$
  $$\vdots$$
  $\mathcal{AC}[\![\mathsf{ac}_k(M_1^k,\ldots,M_{n_k}^k)]\!]\,e\,c')$
- $\mathcal{AC}[\![P(M_1,\ldots,M_m)]\!] = \lambda\, e \in \mathbf{Env}, c \in \text{Constraint}\,.$
  let $c' = c \cup \mathbf{collect}(\{P(M_1,\ldots,M_m)\})$
  in $\mathcal{PRED}[\![P]\!]\,(\mathcal{CLE}[\![M_1]\!]\,e\,c',\ \ldots,\ \mathcal{CLE}[\![M_m]\!]\,e\,c')$
- $\mathcal{AC}[\![M = N]\!] = \lambda\, e \in \mathbf{Env}, c \in \text{Constraint}\,.$
  let $c' = c \cup \mathbf{collect}(\{M = N\})$
  in $(\mathcal{CLE}[\![M]\!]\,e\,c' =_{\mathbf{V}\,\oplus\,\mathbf{F}} \mathcal{CLE}[\![N]\!]\,e\,c')$
- $\mathcal{PRED}[\![P]\!] = \mathcal{I}_{\mathbf{V}}(P) \in (\mathbf{V}{\oplus}\mathbf{F})^n \to \mathbf{C}$
  where $\mathcal{I}_{\mathbf{V}}$ is the underlying function $\mathcal{I}_Q$ of constraints interpretation augmented in order to handle least and greatest elements ($\bot_{\mathbf{V}\oplus\mathbf{F}}$ and $\top_{\mathbf{V}\oplus\mathbf{F}}$) in the obvious way.
- $\mathcal{FUNC}[\![f]\!] = \mathcal{I}_{\mathbf{V}}(f) \in (\mathbf{V}{\oplus}\mathbf{F})^m \to \mathbf{V}{\oplus}\mathbf{F}$
  (idem).

It is easy to see that the equalities added at syntactic level are correct, i.e. α-convertible terms have the same semantics as well as $(\exists\, x_1\,.\,(\exists\, x_2\,.\,M))$ and $(\exists\, x_2\,.\,(\exists\, x_1\,.\,M))$.

Theorem 3.3 The reduction rules preserve the semantics.　　　□

Proof: It is enough to see that if $M \to M'$ then $\mathcal{CLE}[\![M]\!] = \mathcal{CLE}[\![M']\!]$.
We analyze only the case of the $\exists$-rule. What we have to demonstrate
is if $\mathcal{A} \vdash (\exists! \, x \, . \, C) \wedge (\forall \, x \, . \, C \to x = N)$ then $\mathcal{CLE}[\![\exists \, x \, . \, \{C\}M]\!] \, e \, c = \mathcal{CLE}[\![(\{C\}M)[x := N]]\!] \, e \, c$ for any e, c.

First
$\mathcal{CLE}[\![\exists \, x \, . \, \{C\}M]\!] \, e \, c =$
$= \mathcal{CLE}[\![(\{C\}M)[x := y]]\!] \, e \, c$　　　　　　by definition where y is a
　　　　　　　　　　　　　　　　　　　　　　new variable, so $e(y) = \perp_{\mathbf{V} \oplus \mathbf{F}}$
$= \mathcal{CLE}[\![\{C[x := y]\}M[x := y]]\!] \, e \, c$　　　　　by definition

Secondly, if $\mathcal{A} \vdash (\exists! \, x \, . \, C) \wedge (\forall \, x \, . \, C \to x = N)$, then
$$\mathcal{A} \vdash (\exists! \, y \, . \, C[x := y]) \wedge (\forall \, y \, . \, C[x := y] \to y = N),$$
and thus
$\mathcal{CLE}[\![\exists \, x \, . \, \{C\}M]\!] \, e \, c =$
$= \mathcal{CLE}[\![\{C[x := y], \, y = N[x := y]\}M[x := y]]\!] \, e \, c$
　　　　　　　by correctness of the γ_1-rule (see (Mandel, 1993))
$= \mathcal{CLE}[\![(\{C, \, y = N\}M)[x := y]]\!] \, e \, c$
$= \mathcal{CLE}[\![(\{C, \, y = N\}M)[x := \{y = N\}y]]\!] \, e \, c$
　　　　　　　by correctness of the γ_2-rule (see (Mandel, 1993))
$= \mathcal{CLE}[\![\{C, \, y = N\}M]\!] \;\; e[x := \mathcal{CLE}[\![\{y = N\}y]\!] \, e \, c]$
　　　　　　　　　　　　$c[x := \mathcal{CLE}[\![\{y = N\}y]\!] \, e \, c]$
　　　　　　　by correctness of the β-rule (see (Barendregt, 1984))
$= \mathcal{CLE}[\![\{C, \, y = N\}M]\!] \;\; e[x := \mathcal{CLE}[\![\{y = N\}N]\!] \, e \, c]$
　　　　　　　　　　　　$c[x := \mathcal{CLE}[\![\{y = N\}N]\!] \, e \, c]$
　　　　　　　by correctness of the γ_3-rule (see (Mandel, 1993))
$= \mathcal{CLE}[\![(\{C, \, y = N\}M)[x := \{y = N\}N]]\!] \, e \, c$
　　　　　　　by correctness of the β-rule (see (Barendregt, 1984))
$= \mathcal{CLE}[\![(\{C, \, y = N\}M)[x := N]]\!] \, e \, c$　　　by correctness of γ_2-rule (see
(Mandel, 1993))
$= \mathcal{CLE}[\![(\{C\}M)[x := N]]\!] \, e \, c$ by correctness of γ_1-rule (see (Mandel, 1993))
Therefore, the thesis holds for the $\exists$-rule.　　　　　　　　　　　　□

Theorem 3.4 The semantic function $\mathcal{CLE}$ is monotonic.　　　□

Proof: It is enough to see that for any $\lambda_c^{\exists}$-term M, if $e \sqsubseteq e'$ and $c \sqsubseteq c'$, then
$\mathcal{CLE}[\![M]\!] \, e \, c \sqsubseteq \mathcal{CLE}[\![M]\!] \, e' \, c'$. The proof is done by induction on the structure
of M. For constants and **fail**, their semantics does not depend either on the
environment or on the global constraint, therefore in this cases the property
is trivially satisfied. If $M \equiv x$, the monotonicity is demonstrated graphically
in figure 3. The other cases are trivially satisfied because the semantics is
calculated by composition of monotonic functions, and by the induction
hypothesis.　　　　　　　　　　　　　　　　　　　　　　　　　　　　　□

As a corollary, the semantic function is well defined.

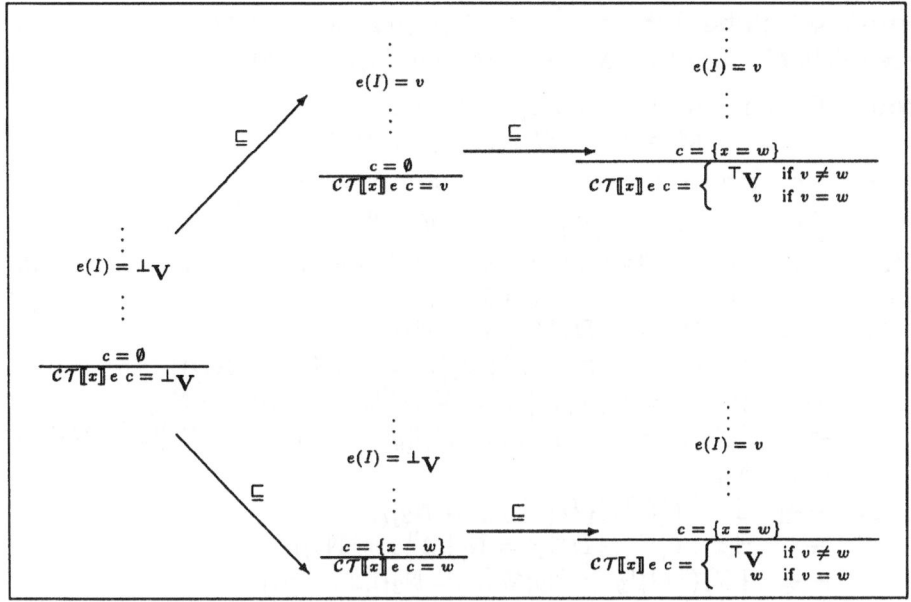

Figure 9.   Monotonicity

4. The Church-Rosser Property

In the present section we demonstrate that the calculus defined in this work satisfies the Church-Rosser (CR) property. The significance of this result is that the calculus is independent of any reduction strategy, and thus, given a starting term, any implementation will reach the same solution (if it reaches one). In other words, normal forms are unique.

The proof is organized as follows. First we demonstrate that the relation $\Phi = (\beta \cup \exists)$ satisfies the CR property. For this proof we demonstrate that the $\exists$-rule satisfies the CR property and commutes with the β-rule. As the β-rule satisfies the CR property then by the Hindley-Rosen lemma (see (Barendregt, 1984, page 64)) the Φ-relation is CR.

The satisfaction of the weak CR property of the relation defined as $\Upsilon = (A_c \cup \text{fail} \cup P_c \cup F_c \cup F_\exists \cup P_\exists \cup C_\exists)$ was demonstrated in (Crossley *et al.*, 1993a; Crossley *et al.*, 1993b). Basically all the overlapping cases were analyzed. We also prove that the Υ-relation terminates and then by (Newman, 1942) the Υ-relation is CR. Finally we demonstrate that the reflexive-transitive closure of the relations Φ and Υ commutes. Using the Hindley-Rosen lemma again the Γ-relation defined as $\Gamma = (\Phi \cup \Upsilon)$ satisfies the CR property. That is, the whole calculus satisfies the CR property.

Lemma 4.1 (The ∃-Rule is CR) If M $\longrightarrow\!\!\!\!\twoheadrightarrow_\exists$ M$_1$ and M $\longrightarrow\!\!\!\!\twoheadrightarrow_\exists$ M$_2$, then there exists M$_3$ such that M$_1$ $\longrightarrow\!\!\!\!\twoheadrightarrow_\exists$ M$_3$ and M$_2$ $\longrightarrow\!\!\!\!\twoheadrightarrow_\exists$ M$_3$.   □

Proof: The only overlap occurs in a term of the form
$$M \equiv \exists\, x\, .\, \{C\}(\ldots \exists\, y\, .\, \{D\}M' \ldots),$$
where $\mathcal{A} \vdash (\exists!\, x\, .\, C) \wedge (\forall\, x\, .\, C \;\to\; x = N_1)$ and
$$\mathcal{A} \vdash (\exists!\, y\, .\, D) \wedge (\forall\, y\, .\, D \;\to\; y = N_2).$$
Without loss of generality let M be $\exists\, x\, .\, \{C\}(\exists\, y\, .\, \{D\}M')$, and let N$_1$ and N$_2$ be normal forms as defined in 1.2.

$$
\begin{aligned}
\text{M} \quad &\longrightarrow_\exists \quad (\{C\}(\exists\, y\, .\, \{D\}M'))[x := N_1] \\
&\equiv \quad \{C\}[x := N_1](\exists\, y\, .\, (\{D\}[x := N_1]M'[x := N_1])) \\
&\longrightarrow_\exists \quad \{C\}[x := N_1](\{D\}[x := N_1]M'[x := N_1])[y := N_2] \\
&\equiv \quad \{C\}[x := N_1](\{D\}[x := N_1][y := N_2]M'[x := N_1][y := N_2]) \\
&\equiv_{\text{def}} \quad \text{M}_1 \\
\text{M} \quad &\longrightarrow_\exists \quad \exists\, x\, .\, (\{C\}((\{D\}M')[y := N_2])) \\
&\equiv \quad \exists\, x\, .\, (\{C\}(\{D\}[y := N_2]M'[y := N_2])) \\
&\longrightarrow_\exists \quad \{C\}(\{D\}[y := N_2]M'[y := N_2])[x := N_1] \\
&\equiv \quad \{C\}[x := N_1](\{D\}[y := N_2][x := N_1]M'[y := N_2][x := N_1]) \\
&\equiv_{\text{def}} \quad \text{M}_2
\end{aligned}
$$

The question is under what conditions M$_1$ ≡ M$_2$. Applying the substitution lemma (see (Barendregt, 1984, 2.1.16)) to M$_1$ we have the following.

$$
\begin{aligned}
\text{M}_1 \quad &\equiv \quad \{C\}[x := N_1](\{D\}[x := N_1][y := N_2]M'[x := N_1][y := N_2]) \\
&\equiv \quad \{C\}[x := N_1](\{D\}[y := N_2][x := N_1]M'[y := N_2][x := N_1]) \\
&\equiv \quad \text{M}_2
\end{aligned}
$$

The first step of the equivalence is correct due to the variable convention. Hence the substitution $[x := N_1[y := N_2]]$ is equivalent to $[x := N_1]$. Thus the property is satisfied and the ∃-rule satisfies the diamond property. By lemma (Barendregt, 1984, 3.2.2) the ∃-rule is CR.   □

Lemma 4.2 (The ∃-Rule and the β-Rule Commute) If M $\longrightarrow_\exists$ M$_1$ and M $\longrightarrow_\beta$ M$_2$, then there exists M$_3$ such that M$_1$ $\longrightarrow_\beta$ M$_3$ and M$_2$ $\longrightarrow_\exists$ M$_3$.   □

Proof: Taking $\mathcal{A} \vdash (\exists!\, y\, .\, C) \wedge (\forall\, y\, .\, C \;\to\; y = N_3)$, there are three overlapping cases:
- M ≡ $\exists\, y\, .\, \{C\}((\lambda\, x\, .\, N_1)N_2)$,
- M ≡ $(\lambda\, x\, .\, (\exists\, y\, .\, (\{C\}N_1)))N_2$ and
- M ≡ $(\lambda\, x\, .\, N_1)(\exists\, y\, .\, (\{C\}N_2))$.

Case 1:
$$
\begin{aligned}
\text{M} \quad &\longrightarrow_\exists \quad (\{C\}((\lambda\, x\, .\, N_1)N_2))[y := N_3] \\
&\equiv \quad \{C\}[y := N_3]((\lambda\, x\, .\, N_1[y := N_3])N_2[y := N_3]) \\
&\longrightarrow_\beta \quad \{C\}[y := N_3]N_1[y := N_3][x := N_2[y := N_3]] \\
&\equiv \quad \{C\}[y := N_3]N_1[x := N_2][y := N_3] \equiv \text{M}_3 \\
\text{M} \quad &\longrightarrow_\beta \quad \exists\, y\, .\, \{C\}N_1[x := N_2] \\
&\longrightarrow_\exists \quad \{C\}[y := N_3]N_1[x := N_2][y := N_3] \equiv \text{M}_3
\end{aligned}
$$

Case 2: $\mathrm{M} \longrightarrow_{\exists} (\lambda\ x\ .\ (\{C\}\mathrm{N}_1)[y := \mathrm{N}_3])\mathrm{N}_2$
$\qquad \longrightarrow_{\beta} (\{C\}\mathrm{N}_1)[y := \mathrm{N}_3][x := \mathrm{N}_2] \equiv \mathrm{M}_3$
$\qquad \mathrm{M} \longrightarrow_{\beta} (\exists\ y\ .\ (\{C\}\mathrm{N}_1))[x := \mathrm{N}_2]$
$\qquad \equiv\ \exists\ y\ .\ ((\{C\}\mathrm{N}_1)[x := \mathrm{N}_2])$
$\qquad \longrightarrow_{\exists} (\{C\}\mathrm{N}_1)[x := \mathrm{N}_2][y := \mathrm{N}_3]$
$\qquad \equiv\ (\{C\}\mathrm{N}_1)[y := \mathrm{N}_3][x := \mathrm{N}_2] \equiv \mathrm{M}_3$

Case 3: $\mathrm{M} \longrightarrow_{\exists} (\lambda\ x\ .\ \mathrm{N}_1)(\mathrm{N}_2[y := \mathrm{N}_3])$
$\qquad \longrightarrow_{\beta} \mathrm{N}_1[x := \mathrm{N}_2[y := \mathrm{N}_3]] \equiv \mathrm{M}_3$
$\qquad \mathrm{M} \longrightarrow_{\beta} \mathrm{N}_1[x := \exists\ y\ .\ (\{C\}\mathrm{N}_2)]$
$\qquad \longrightarrow_{\exists} \mathrm{N}_1[x := \mathrm{N}_2[y := \mathrm{N}_3]] \equiv \mathrm{M}_3$

In every case the reductions commute, and the property is satisfied. $\square$

Here we demonstrate that the Υ-relation terminates, i.e. that there is no infinite derivation $\mathrm{M}_1 \longrightarrow_{\Upsilon} \mathrm{M}_2 \longrightarrow_{\Upsilon} \cdots$. Following (Dershowitz & Jouannaud, 1984, page 270) we define a positive function $\mathcal{W}$ on the set of $\lambda_c^\exists$-terms. Termination is assured because the reduction Υ is $\mathcal{W}$-decreasing.

Definition 4.3 (Weight) The function $\mathcal{W} : \Lambda_c^\exists \cup \mathsf{Constraint} \to \mathbb{N}$ is defined as follows:

- $\mathcal{W}(\mathsf{fail}) = \mathcal{W}(k) = \mathcal{W}(x) = 3$;
- $\mathcal{W}(\lambda\ x\ .\ \mathrm{M}) = \mathcal{W}(\exists\ x\ .\ \mathrm{M}) = 1 + \mathcal{W}(\mathrm{M})$;
- $\mathcal{W}((\mathrm{M}\ \mathrm{N})) = \mathcal{W}(\mathrm{M}) + \mathcal{W}(\mathrm{N})$;
- $\mathcal{W}(f(\mathrm{M}_1, \ldots, \mathrm{M}_n)) = 2^{\Sigma\ W(M_i)}$;
- $\mathcal{W}(\{C\}\mathrm{M}) = 1 + \mathcal{W}(\{C\}) \cdot \mathcal{W}(\mathrm{M})$;
- $\mathcal{W}(\{c_1, \ldots, c_n\}) = \Sigma\ \mathcal{W}(c_i)$;
- $\mathcal{W}(P(\mathrm{M}_1, \ldots, \mathrm{M}_m)) = 2 \cdot \Sigma\ \mathcal{W}(\mathrm{M}_i)$. $\square$

Clearly $\mathcal{W}$ is an integer and positive function, i.e for each $\mathrm{M} \in \Lambda_c^\exists$, $\mathcal{W}(\mathrm{M}) > 0$. This function induces a reduction order which is well founded.

Lemma 4.4 (Υ-Reductions Terminate) The Υ-rule defined as $(A_c \cup \mathsf{fail} \cup P_c \cup F_c \cup F_\exists \cup P_\exists \cup C_\exists)$ is terminating. $\square$

Proof: Every Υ-*redex* is $\mathcal{W}$-greater than its Υ-*contractum*. Therefore, by (Dershowitz & Jouannaud, 1984, page 270) the Υ-relation is terminating. $\square$

Theorem 4.5 (The Υ-Relation is CR) If $\mathrm{M} \longrightarrow\!\!\!\!\rightarrow_{\Upsilon} \mathrm{M}_1$ and $\mathrm{M} \longrightarrow\!\!\!\!\rightarrow_{\Upsilon} \mathrm{M}_2$, then there exists M_3 such that $\mathrm{M}_1 \longrightarrow\!\!\!\!\rightarrow_{\Upsilon} \mathrm{M}_3$ and $\mathrm{M}_2 \longrightarrow\!\!\!\!\rightarrow_{\Upsilon} \mathrm{M}_3$. $\square$

Proof: By (Crossley *et al.*, 1993b) the Υ-relation is weak-CR. By 4.4 the Υ-relation terminates. Following the Newman's lemma (see (Newman, 1942)) the Υ-relation is CR. $\square$

Theorem 4.6 (The $\lambda_c^\exists$-Calculus is CR)   If M $\longrightarrow\!\!\!\!\twoheadrightarrow_\Gamma$ M$_1$ and M $\longrightarrow\!\!\!\!\twoheadrightarrow_\Gamma$ M$_2$, then there exists M$_3$ such that M$_1$ $\longrightarrow\!\!\!\!\twoheadrightarrow_\Gamma$ M$_3$ and M$_2$ $\longrightarrow\!\!\!\!\twoheadrightarrow_\Gamma$ M$_3$.   □

Proof: Both Φ and Υ are CR. Analyzing the overlapping cases we get that both relations commute in one step. By the Hindley-Rosen lemma, the whole calculus is CR.   □

5. Conclusions and Future Work

We have implemented an abstract machine for the system where the constraints are restricted to be linear equations and unequations. For this we have developed a version of the calculus with explicit substitutions. We specified the $\lambda_c^\exists$-calculus for the RAP and TIP (see (Fraus, 1994)) systems and we expect to prove its local confluence with their built-in Knuth-Bendix completion procedure.

The calculus presented is an elegant enhancement of the traditional λ-calculus. This allows us not only to deal with incomplete information but also to add some typically imperative features to the functional nature of λ-calculus.

We are working on an implementation of the first-order constrained lambda calculus presented, using de Bruijn indices. Some progress was reported in (Knapp & Mandel, 1994).

In future work we will remove the restriction to unique solutions. That is, we will allow solutions such as: $C \vdash x = N \vee x = M$. This fact only affects the $\exists$-rule which is extended in order to deal with more than one value. There are two major possibilities for this purpose. The first one consists in the addition of a restricted backtracking mechanism for each value obtained by the constraint solver. This approach is traditional in logic programming languages such as Prolog. The second is to use non-deterministic choice. We will study the second one following the non-deterministic λ-calculus considered by Paterson (see (Paterson, 1991)). An approach in this direction, restricted to finite domains, was presented in (Cengarle & Mandel, 1995). We must add a new term constructor to the term which defines non-deterministic choices among terms, and redefine the semantic function as a many-valued one.

Acknowledgements. We wish to thank the anonymous referees for their helpful comments on the paper. And also María Victoria Cengarle for her careful reading of previous versions of this work and insightful suggestions.

This research is supported by the DFG-project Constraints and partly by an IBM research agreement number 15760046.

JOHN N. CROSSLEY

*Department of Computer Science, Monash University
Clayton, Australia*

AND

LUIS MANDEL AND MARTIN WIRSING

*Institut für Informatik
Ludwig-Maximilians-Universität München, Germany*

References

H. P. Barendregt. *The Lambda Calculus. Its Syntax and Semantics.* North Holland, second edition, 1984.

M. V. Cengarle and L. Mandel. Finite Domains in the Constrained Lambda Calculus. In *Proceedings of the Post-Conference Workshop on Constraints, Databases and Logic Programming. International Logic Programming Symposium (ILPS)*, pages 75–89, December 8 1995. 15 pages.

J. N. Crossley, L. Mandel, and M. Wirsing. Una Extensión de Constraints al Cálculo Lambda. In *Proceedings of the Segundo Congreso de Programación Declarativa, ProDe '93*, Blanes, Girona Spain, September 29-30, October 1 1993. (In Spanish).

J. N. Crossley, L. Mandel, and M. Wirsing. Untyped Constrained Lambda Calculus. Institut für Informatik, Number 9318, Ludwig-Maximilians-Universität München, Leopoldstraße 11b, 80802 München, Germany, October 1993. 48 pages.

N. Dershowitz and J.-P. Jouannaud. Rewriting Systems. In J. V. Leeuwen, editor, *Handbook of Theoretical Computer Science*, volume B, chapter 6, pages 244–320. Elsevier Science Publishers, 1984.

U. Fraus. Inductive Theorem Proving for Algebraic Specifications – TIP System User's Manual. Technical Report MIP 9401, Universität Passau, 1994.

Y. Guo and H. Pull. FALCON: Functional and Logic Language with Constraints (Language Definition). Technical Report IC/FPG/Phoenix/15/3, Imperial College, March 1991.

M. Henz, G. Smolka, and J. Würz. Oz – a programming language for multi agent systems. In *Proceedings of the Int. Joint Conference on Artificial Intelligence*, 1993.

J. Jaffar, S. Michaylov, P. Stuckey, and R. Yap. The CLP(ℝ) Language and System. *ACM Transactions on Programming Languages and Systems*, pages 339–395, 1992.

A. Knapp and L. Mandel. A Rewriting System with Explicit Substitutions for the First Order Constrained Lambda Calculus. Interne Reports der FG Programmiersysteme FR I-Passau-1994-094, FORWISS (Universität Passau), 1994.

F. J. López-Fraguas. A General Scheme for Constraint Functional Logic Programming. In H. Kirchner and G. Levi, editors, *Proceedings of the Algebraic and Logic Programming Conference, ALP-1992*, pages 213–227, October 28–31 1992.

L. Mandel. The Semantics of the Untyped Constrained Lambda Calculus. Institut für Informatik, Number 9319, Ludwig-Maximilians-Universität München, Leopoldstraße 11b, 80802 München, Germany, October 1993.

L. Mandel. *Constrained Lambda Calculus.* PhD thesis, Ludwig-Maximilians-Universität München, Leopoldstraße 11b, 80802 München, Germany, December 1995. 213 pages.

M. H. A. Newman. On Theories with a Combinatorial Definition of "Equivalence". *Annals of Mathematics*, 43(2):223–243, 1942.

P. Panangaden, V. Saraswat, and M. Rinard. Semantic Foundations of Concurrent Constraint Programming. In *Proceedings of the Eighteenth Annual ACM Symposium on Principles of Programming Languages*, 1991.

R. Paterson. A Tiny Functional Language with Logical Features. In *Declarative Programming, Sasbachwalden*, 1991.

I. Sutherland. SKETCHPAD: A Man-Machine Graphical Communication System. In *IFIPS Proceedings of the Sprint Joint Computer Conference*, Jan. 1963. See also his Ph.D thesis from M.I.T.

NAJI HABRA AND BAUDOUIN LE CHARLIER

UNIFIED RELATIONAL FRAMEWORK FOR PROGRAMMING PARADIGM COMBINATION

1. Introduction

The goal of this work is to provide a clean formal framework to support a development methodology for programs integrating modules of different paradigms (e.g., Horn clauses modules, equational modules, functional modules, constraints based modules, ...). Modules can be developed each in its own paradigm, the framework presents a uniform and high-level view allowing to make reasoning about those modules, to build them using a uniform methodology and to combine them adequately in a realistic program.

The framework is based on a language whose syntax and semantics are sufficiently general so as the different modules of the different paradigms can be considered as particular (possibly incomplete) implementations of that language. Meanwhile the semantics is sufficiently close to operational features so as these implementations could be seen a as specializations of it. The language is thus at midway between a specification language and an implementation language. The key ideas of the language are:

- The syntax assumes simply that a program is a collection of relation definitions given in a typed form of first-order logic. Yet, variables are considered to denote objects of the corresponding domain, no more no less (no assumptions are made about the object representations).
- Besides the classical model-theoretic semantics, the language is given an original partial semantics based on a three-values modeling and expressed by a fixpoint operator.
- Negation is handled in a constructive way: positive and negative pieces of information are derived in parallel. This allows the different approaches used in logic programming (e.g., negation by failure,...) to be considered as particular implementations.
- A program is a collection of relation definitions. They are built on top

F. Baader and K.U. Schulz (eds.), Frontiers of Combining Systems, 357–375.
© 1996 *Kluwer Academic Publishers*.

of some relations which are considered as *primitives*. Primitives allow programs to be elaborated level-by-level in an incremental way. And program properties like correctness, completeness and termination will be conditioned by corresponding assumptions on the primitives used in that program.

A methodology based on that language and considering different programming paradigms as particular implementations of it, is under development.

The paper is organized as follows. Section 2 introduces the language syntax and semantics. Section 3 discusses different characteristics of the language, in particular, it outlines those features which make it suitable for combining paradigms. Section 4 shows how different implementation semantics (e.g., SLDNF-resolution, rewriting,...) can be viewed as particularizations of the semantics presented. This last section is less technical than the former ones since its contents is actually in an embryonic state.

2. Presentation of the Language

2.1. SYNTAX

Roughly speaking, well-formed formulae of the language are typed first-order logic formulae having a particular form of *relation definition*. The form corresponds to the completion of Prolog program introduced by Clark (Lloyd, 1987) and to the logical descriptions introduced by (Deville, 1990), but without explicit compound terms.

Definition 1
   1) Let S denotes a set of sorts. An S-sorted signature (Π, Σ) consists of a set of predicate names $\Pi = \{p, q, r,...\}$ and a set of function names $\Sigma = \{f, g, h,...\}$, where each predicate (resp. function) name is equipped with a profile, i.e., a n-tuple (resp. $n + 1$-tuple) of sorts $(n \geq 0)$.
   2) Let $X = \langle X_s \rangle_{s \in S}$ denotes an S-indexed family of variables, the S-sorted signature (Π, Σ, X) is an extension of (Π, Σ) made by considering each variables of each set X_s as a constant (a 0-ary function) of the corresponding sort s.
   3) Given an S-sorted signature (Π, Σ, X), a relational (resp. functional) atom has the form $p(x_1, \ldots, x_n)$ (resp. $f(x_1, \ldots, x_n) = x_{n+1}$) where p (resp. f) is a predicate (resp. function) name of the signature and where the x_i's are variables which respect the profile.
   4) A literal is an atom or the negation of an atom.
   5) A relation (resp. function) definition is an equivalence where the left-hand side is an atom and the right-hand side is a disjunction of conjunctions of literals. Each conjunction is prefixed by existential

quantification of variables so that the variables remaining free in the conjunction appear also in the left-hand side.

6) A program P is a set of definitions such that each relation/function is defined at most once. The relations/functions which are not defined in the program are considered as *primitives*.

∎

In the examples below, function s, constants 0, a and b are primitives.

even(X) ⇔	X=0	
	∨ ∃ Y even(Y) ∧ X=s(s(Y))	Example 1

even(X) ⇔	X=a	
	∨ ∃ Y ¬even(Y) ∧ X=s(Y)	Example 2

p(X) ⇔	∃ Y p(Y) ∧ q(Y) ∧ X=a	
	∨ ∃ Y p(Y) ∧ X=s(Y)	
q(X) ⇔	X=b	
	∨ ∃ Y q(Y) ∧ X=s(Y)	Example 3

2.2. MODEL-THEORETICAL SEMANTICS

A classical model-theoretic semantics can be defined in the usual way. Interpretations are algebraic structures; a program has as models the interpretations that satisfy all the definitions composing it:

Definition 2
1) A (Π, Σ)-interpretation I is
    1.a) A domain, that is an S-indexed family of domain sets $U = \bigcup_s U_s$.
    1.b) Assignments for relations and functions names, that is, for each relation(function) name p, a relation(function) I_p which respects the profile given in the signature.
2) A state (or a variable assignment) σ over a family of variables $\langle X_s \rangle$ in a domain $U = \bigcup_s U_s$ is a function that assigns to each variable x of sort s a value of the corresponding set U_s; this value is denoted by $\sigma(x)$.
3) An interpretation I and a state σ over the domain of I associate each formula *wff* with a *truth value*: true or false. We say that I with σ satisfy (resp. falsify) the formula *wff*, denoted by $I \models_\sigma wff$ (resp. $I \not\models_\sigma wff$). This valuation is defined inductively using the classical logical meaning of the connectors.

4) The interpretation I satisfies the definition $p(x_1, \ldots, x_n) \Leftrightarrow D$ (or is a model of it) iff for each well-typed state σ the interpretation I assigns the same truth value to both sides of the definition. This is denoted by $I \models p(x_1, \ldots, x_n) \Leftrightarrow D$,

5) The (Π, Σ)-interpretation I is a model of the program P iff it satisfies all the definitions in P.

∎

In an incremental program elaboration process, new relations are defined on top of predefined ones called primitives. Primitives are assigned a fixed meaning in I.

The semantics does not guarantee existence nor unicity of model. We take the point of view that such desirable properties should be considered as constraints on the elaboration of the theory(program). This should be contrasted with classical algebraic semantics where restrictions are imposed on the semantics itself to get a unique model, e.g., by choosing the initial or the final model.

Functions are arbitrary ones, not only constructors as in the usual model-theoretic semantics of logic programs. Terms are not a basic notion of our semantics. They can be domain values in some interpretation, or they can *represent* domain values in some *implementation*. Representation choices are discussed latter on (Section 4).

2.3. BIRELATIONAL SEMANTICS

The model-theoretic semantics is purely declarative in the sense that it determines which interpretations are satisfying, but does not refer to any deduction system. In practice, under an operational view, a theory (a program) is equipped with a *deduction system* which should be at least consistent with the declarative semantics (and ideally complete).

The birelational semantics developed in this section describes a kind of upper-bound of what can be deduced by any practical automated deduction system. And, since deduction can be incomplete in general, this semantics leads to a *partial modeling* in the sense that deduction can fail to assign a truth value to some closed formulae.

Birelation is the key concept used for this semantics. A birelation on a domain U is a couple of relations on U, $\langle Pos, Neg \rangle$, assigned to each defined predicate (with $Pos \cap Neg = \emptyset$). The relation Pos includes the tuples for which the predicate *is known* to hold, and symmetrically, the relation Neg includes those for which the defined predicate is known to not hold. In other words, the relation $Pre = Pos \cup Neg$ associated with a predicate represents a precondition to use that predicate correctly, tuples outside $Pos \cup Neg$ are those for which the deduction fail.

The definition of the birelational semantics uses classical primitives operators from the relational algebra of (Codd, 1970) themselves borrowed from the cylindric algebra of Tarski (Tarski & Thompson, 1952). Hereafter, we recall some definitions and we fix some notations:

Definition 3
1) A relation R over the domain $U_{s_1} \times \ldots \times U_{s_n}$ is a set of tuples $t \in U_{s_1} \times \ldots \times U_{s_n}$.
2) A *labeled* relation is the same as an ordinary relation except that explicit labels are associated with the different domain sets.

∎

Notation 4
1) Labels $L = \{l_1, \ldots, l_n\}$ assigned to a relation R are any distinguishing names; L is called the *relation scheme*. If the relation scheme is to be highlighted the relation is denoted by $R^{l_1, \ldots, l_n}$ or R^L, and tuples are denoted by $t^L = \{l_1 : a_1, \ldots, l_n : a_n\}$ with $a_i \in U_{s_i}$.
2) The scheme of a labeled relation R is denoted by $\alpha(R)$. [1]
3) The unlabeled version of a labeled relation R, obtained by forgetting labels and taking into account the elements order, is denoted by $unl(R)$.
4) If t is a tuple labeled by L and if L' is a subset of L, then $t[L']$ denotes the tuple corresponding to the restriction of the mapping t on L'.

∎

Definition 5
1) The binary operations $\cup$, $\cap$ and $-$ are defined only on relations of the same schemes; they correspond to the set theoretic operations having the same name applied on the corresponding two sets of tuples.
2) If R and S are relations labeled by X and Y, respectively, then the Natural Join $\bowtie$ and the Conjoint Union $\oplus$ (generalizations of $\cap$ and $\cup$) are defined by:
$$\alpha(R \bowtie S) = X \cup Y \qquad \alpha(R \oplus S) = X \cup Y$$
$$R \bowtie S = \{t : t[X] \in R \wedge t[Y] \in S\} \quad R \oplus S = \{t : t[X] \in R \vee t[Y] \in S\}$$
3) If R is a relation labeled by X, and if Y is a subset and Z a superset of its labels ($Y \subseteq X \subseteq Z$), then the projection $\Pi_Y R$ and the elevation $UP_Z R$ are defined by:
$$\alpha(\Pi_Y R) = Y \qquad \alpha(UP_Z R) = Z$$
$$\Pi_Y R = \{t[Y] : t \in R\} \quad UP_Z R = \{t : t[X] \in R\}$$
The labels $Z - X$ should have determined associated domain sets.
4) If R and S are relations labeled by X and Y, respectively, (with $Y \subset X$), then the quotient $R \div S$ is the relation defined by:

---

[1] Note that a labeled tuple $t^L \in U_{s_1} \times \ldots \times U_{s_n}$ can be viewed as a mapping from L to $U_{s_1} \cup \ldots \cup U_{s_n}$. In particular, if the labels correspond to variable names the tuples can be viewed as a state (an assignment) over those variables.

$$\alpha(R \div S) = X - Y$$
$$R \div S = \{t : \forall t' \in S \; \exists t'' \in R \; t''[X - Y] = t \wedge t''[Y] = t'\}$$
Usually the quotient is presented as a derived operator. In fact
$$R \div S = \Pi_{X-Y}R - \Pi_{X-Y}(((\Pi_{X-Y}R) \times S) - R)$$

∎

The set of birelations on a given domain can be assigned an ordering $\sqsubseteq$ defined by: $\langle Pos_2, Neg_2 \rangle \sqsubseteq \langle Pos_1, Neg_1 \rangle$ iff $Pos_2 \subseteq Pos_1$ and $Neg_2 \subseteq Neg_1$. The set together with the ordering constitute a complete partial order (cpo). On basis of this cpo a semantics can be very naturally described as the fixpoint of a transformation operator.

Thus, the birelational semantics (henceforth, B-semantics for short) associates with the program P a family of birelations $B = \{B_p, B_q, \ldots, B_f, \ldots\}$, one for each relation and each function name. This family is defined as the least fixpoint of an operator XB_P which transforms one family of birelations into another (XB stands for eXtend Birelations). More precisely:

Definition 6

1) Let P be a program composed of a collection of definitions $p(X) \Leftrightarrow D$, where D is a disjunction of *prefixed conjunctions* PC_j, each PC_j has the form $\exists C_j$, where C_j is a conjunction of literals L_i, each literal L_i is an atom A or the negation of an atom $\neg A$.
Let $B = \langle B_p, B_q \ldots B_f, B_g \ldots \rangle$ be a family of *birelations* associated with P. Using this family B, each sub-formula *wff* in each definition of P could be associated with a birelation labeled by X, the free variables of *wff*. This birelation, say $\hat{B}(wff)$, is defined inductively as follows:

   1.a) An atomic formula $p(y_1, \ldots, y_n)$ is assigned the labeled birelation determined by B_p and the labels $y_1, \ldots, y_n$, i.e., $B_p^{\{y_1, \ldots, y_n\}}$.

   1.b) If $\hat{B}(A) = \langle Pos, Neg \rangle$
   then $\hat{B}(\neg A) = \langle Neg, Pos \rangle$.

   1.c) If $\hat{B}(L_1) = \langle Pos_1, Neg_1 \rangle$ and $\hat{B}(L_2) = \langle Pos_2, Neg_2 \rangle$
   then $\hat{B}(L_1 \wedge L_2) = \langle Pos_1 \bowtie Pos_2, Neg_1 \oplus Neg_2 \rangle$

   1.d) If $\hat{B}(C) = \langle Pos^X, Neg^X \rangle$
   then $\hat{B}(\exists x : C) = \langle \Pi_{X-\{x\}} Pos^X, Neg^X \div U_x \rangle$
   where U_x is the domain of the variable x.

   1.e) If $\hat{B}(PC_1) = \langle Pos_1, Neg_1 \rangle$ and $\hat{B}(PC_2) = \langle Pos_2, Neg_2 \rangle$
   then $\hat{B}(PC_1 \vee PC_2) = \langle Pos_1 \oplus Pos_2, Neg_1 \bowtie Neg_2 \rangle$

It can be easily proved that the produced birelations $\hat{B}(wff)$ are indeed birelations, i.e., that $Pos \cap Neg = \emptyset$.

2) The operator XB_P is defined by means of $\hat{B}$ as follows:
   for each definition $p(X) \Leftrightarrow D$, the operator XB transforms the birelation B_p associated with p into the unlabeled version of $\hat{B}(D)$.

3) According to the B-semantics, the program P is assigned with the
   birelations family determined by *the least fixpoint of the operator*
   XB_P.

   ∎

One can check that the operator XB is monotonic. This follows from the
monotonicity of the operators composing it: the operators $\oplus$, $\bowtie$, $\div$, Π on
relations, the operators that label/unlabel a relation, the operator that
composes two relations to make a birelation, and the operator that swap
the two components of a birelation. Examples of successive applications of
XB are shown in Subsection 3.4

Note that nothing in the semantics guarantees the functional character
of a given symbol. We keep however a distinguished syntax for functions,
this in order to allow programmer to indicate the functional character of
a given predicate if he has sufficient confidence in that fact; e.g., he uses
known functional primitives.

As for the model-theoretic semantics definition, we can fix, once for all,
a typed domain and an interpretation (by birelations) for the primitives.

3. Discussion

3.1. CONSISTENCY W.R.T. THE MODEL-THEORETIC SEMANTICS

The model-theoretic semantics defines satisfying interpretations, while the
B-semantics provides an upper bound to what could be derived by means
of an effective computation method; practically, sets of positive and negative tuples. Obviously, we want the computed tuples to be consistent with
the model-theoretic semantics: the birelational semantics determines tuples which are *true in every interpretation* as well as tuples which are *false
in every interpretation*, with the additional requirement that both kind of
facts can be derived by a general computation mechanism.

More precisely, let P be a program; $I = \{I_{p_1}, I_{p_2}, \ldots\}$ a model of P and
$B = \{\langle Pos_{p_1}, Neg_{p_1} \rangle, \langle Pos_{p_2}, Neg_{p_2} \rangle, \ldots\}$ a B-interpretation of P according
to the B-semantics. B is said to be *consistent* with I iff for each predicate
p_i, for each tuple t: if $t \in Pos_{p_i}$ then $t \in I_{p_i}$, and if $t \in Neg_{p_i}$ then $t \notin I_{p_i}$.

In order to prove the consistency property defined above, we first prove
that the operator XB_P is *consistency preserving* in the sense that whenever a B-interpretation B is consistent with an interpretation I then the
B-interpretation $XB_P(B)$ is also consistent with I. This follows from the
definition of XB. As a consequence, the consistency of the B-semantics is

guaranteed because the initial (empty) birelations used to start the fixpoint computation are consistent. Practically, when a program is elaborated in an incremental way on top of some primitives, consistency depends on the use of consistent B-interpretations for those primitives.

3.2. COMPUTATIONAL ASPECTS

One can observe that the monotonic operator XB is not continuous in the general case. In fact, the operators composing XB are all continuous except $\div$. As a consequence, in the general case, the B-semantics is not constructive in the sense that the tuples belonging to *Pos* and *Neg* are not recursively enumerable.

Though the B-semantics is not constructive in the general case, it remains in accordance with our purpose. In fact, the idea of the language at this level is to provide a clean and elegant formal framework to describe composite programs and to reason about them independently of the representation and of the implementation strategy. Composite programs amalgamate different paradigms having each its own *implementation strategy*, i.e. a particularization of the language that makes the semantics constructive. The description above is too general and there are many *natural* ways to specialize the language and, in the same time, to make the semantics constructive.

For example, if we know that our domains are finite, we can choose an adequate *complete representation* and a continuous implementation for the operator $\div$. The sought fixpoint would then be reached after a finite number of steps and the semantics becomes thus computable directly through a general fixpoint algorithm, e.g., (Le Charlier & Van Hentenryck, 1993; Le Charlier & Van Hentenryck, 1994). Note that many practical languages, like relational database languages and constraints languages on finite domains, fit this case.

Now, if our domains are infinite, we have to deal with incomplete representations for the relations. Even in this case, some *restrictions* can be used to ensure that the relational operators ($\bowtie$, $\oplus$, $\div$, ...) have computable implementation in the chosen representation. Different restrictions on the use of negative literals are examples of this. For example, we can assume that there is no negation through recursion. Recall that non-continuity is induced by the operator $\div$ needed only for existentially quantified variables in the negative part. [2]

In addition, let us notice that, for practical reason, some computable versions of the B-semantics are incomplete (and even unsound) implemen-

---

[2]This kind of restriction is used in the language Toupie (Rauzy, 1995) on finite domains.

tations. For example, for logic programs, Prolog can be seen as a particular implementation of the B-semantics (of the Clark's completion of the program). In that case, the search rule can lead to incompleteness while the omission of occur check can lead to unsoundness (see (Deville, 1990)).

The idea underlying our work, is to take advantages of the different restrictions that lead to efficient and realistic implementations; it aims at dealing with these different restrictions through the use an adequate *construction methodology* which ensure that the produced program P satisfy the sought restriction. Such methodology has been developed for the case of Logic Programming in the FOLON project (Deville, 1990; Henrard & Le Charlier, 1992; De Boeck & Le Charlier, 1993; Le Charlier & Rossi, 1995), and the methodology we seek for is actually a generalization of this methodology so as to deal with the general language developed here and to take profit of a large library of implementation techniques.

3.3. COMPARISON WITH RELATED WORKS

One can observe that the B-semantics amounts to a three-valued logic where formulae are assigned truth values in $\{true, false, unknown\}$. Indeed, although the presentation is different, our semantics is close to the semantics of Fitting (Fitting, 1985), (based on Kleene three-valued logic) who uses a non continuous operator Φ like our XB. And, as pointed out by Fitting, such modeling allows for a natural consideration of negation and of the other connectives. We believe that such modeling is sufficiently powerful for our purpose: tuples that remain outside the birelation assigned to a given predicate correspond to those for which any reasonably practical deduction would fail.

On basis of Fitting's work, Kunen (Kunen, 1987) proposed a three-valued logic semantics for *Prolog-like* programs. Kunen also uses a monotonic discontinuous operator ext, but the recursion is cut off arbitrarily at the first ordinal ω even when the fixpoint is not reached. Although the operators are very close, the purpose of our approach is quite different. Kunen's purpose is to define a *declarative* semantics that models adequately the operational behavior of Prolog interpreters with its particular negation handling. Different works based on Clark's completion (Apt, 1992; Lloyd, 1987) have a similar purpose: they start with Prolog programs and *complete* them to give them a semantics. At the opposite, our approach starts with a general clean language for which Prolog represents one particular implementation.

3.4. EXAMPLES AND ANALOGIES

In Example 1 above, let us choose the interpretation universe to be N (represented as usual by $0, 1, 2, \ldots$) and the primitives to be :

$B_{X=0} = \langle \{X : X = 0\}, \{X : X \neq 0\} \rangle$

$B_{X=s(s(Y))} = \langle \{\langle X, Y \rangle : X = Y + 2\}, \{\langle X, Y \rangle : X \neq Y + 2\} \rangle$

The transformation operator XB_P is then defined by:

$Pos_{i+1,even} = unl \quad (\{X : X = 0\}$
$\qquad\qquad\qquad \oplus$
$\qquad\qquad\qquad \Pi_X(Pos^Y_{i,even} \bowtie \{\langle X, Y \rangle : X = Y + 2\}))$

$Neg_{i+1,even} = unl \quad (\{X : X \neq 0\}$
$\qquad\qquad\qquad \bowtie$
$\qquad\qquad\qquad (Neg^Y_{i,even} \oplus \{\langle X, Y \rangle : X \neq Y + 2\}) \div N^Y)$

So, starting with $Pos_{0,even} = Neg_{0,even} = \{\}$ we get:

$Pos_{1,even} = unl(\{X : X = 0\} \oplus \Pi_X(\{\}^Y \bowtie \{\langle X, Y \rangle : X = Y + 2\}))$
$\qquad\quad = unl(\{X : X = 0\} \oplus \{\})$
$\qquad\quad = unl(\{X : X = 0\})$
$\qquad\quad = \{0\}$

$Neg_{1,even} = unl(\{X : X \neq 0\} \bowtie (\{\}^Y \oplus \{\langle X, Y \rangle : X \neq Y + 2\}) \div N^Y)$
$\qquad\quad = unl(\{X : X \neq 0\} \bowtie (\{\langle X, Y \rangle : X \neq Y + 2\} \div N^Y))$
$\qquad\quad = unl(\{X : X \neq 0\} \bowtie \{X : X = 0 \vee X = 1\})$
$\qquad\quad = unl(\{X : X = 1\}$
$\qquad\quad = \{1\}$

$Pos_{2,even} = unl(\{X : X = 0\} \oplus \Pi_X(\{0\}^Y \bowtie \{\langle X, Y \rangle : X = Y + 2\}))$
$\qquad\quad = unl(\{X : X = 0\} \oplus \Pi_X\{Y : Y = 0\} \bowtie \{\langle X, Y \rangle : X = Y + 2\}))$
$\qquad\quad = unl(\{X : X = 0\} \oplus \Pi_X\{\langle X, Y \rangle : Y = 0 \wedge X = Y + 2\}))$
$\qquad\quad = unl(\{X : X = 0\} \oplus \{X = 2\})$
$\qquad\quad = \{0, 2\}$

$Neg_{2,even} = unl(\{X : X \neq 0\} \bowtie (\{1\}^Y \oplus \{\langle X, Y \rangle : X \neq Y + 2\}) \div N^Y)$
$\qquad\quad = unl(\{X : X \neq 0\} \bowtie (\{Y : Y = 1\}) \oplus \{\langle X, Y \rangle : X \neq Y + 2\}) \div N^Y)$
$\qquad\quad = unl(\{X : X \neq 0\} \bowtie (\{\langle X, Y \rangle : X \neq Y + 2 \wedge Y = 1\}) \div N^Y)$
$\qquad\quad = unl(\{X : X \neq 0\} \bowtie (\{X = 0 \vee X = 1 \vee X = 3\})$
$\qquad\quad = unl(\{X : X = 1 \vee X = 3\}$
$\qquad\quad = \{1, 3\}$

$Pos_{i,even} = \{0, 2, 4, \ldots, 2 * i - 2\}$
$Neg_{i,even} = \{1, 3, 5, \ldots, 2 * i - 1\}$

So, the fixpoint obtained at the first ordinal corresponds to the sets of even and odd naturals, i.e., *the intended meaning* of the program.

Example 2, which is taken from (Fitting, 1985) can be treated in similar way. The primitive atoms are $X = a$ and $X = s(Y)$. The interpretation universe can be taken to be $U = \{a, s(a), s(s(a)) \ldots\}$ with term equality as primitive. The transformation operator XB_P is then defined by :

$Pos_{i+1,even} = unl \quad (\{X : X = a\}$
$\qquad\qquad\qquad \oplus$
$\qquad\qquad\qquad \Pi_X(Neg^Y_{i,even} \bowtie \{\langle X, Y \rangle : X = s(Y)\}))$

$Neg_{i+1,even} = unl \quad (\{X : X \neq a\}$
$\qquad\qquad\qquad \bowtie$
$\qquad\qquad\qquad (Pos^Y_{i,even} \oplus \{\langle X, Y \rangle : X \neq s(Y)\}) \div U^Y)$

So, starting with $Pos_{0,even} = Neg_{0,even} = \{\}$ we get:

$$Pos_{1,even} = \{a\} \qquad Neg_{1,even} = \{\}$$
$$Pos_{2,even} = \{a\} \qquad Neg_{2,even} = \{s(a)\}$$
$$Pos_{3,even} = \{s(s(a))\} \quad Neg_{2,even} = \{s(a)\}$$
...

The fixpoint, obtained at the first ordinal ω, corresponds to a model in Fitting's semantics which is also obtained at ω.

Example 3 presents another interesting case to compare our semantics to Fitting's one. In fact, for this example the operator XB gives:

$$Pos_{1,p} = \{\} \qquad Neg_{1,p} = \{b\}$$
$$Pos_{1,q} = \{b\} \qquad Neg_{1,q} = \{a\}$$
$$Pos_{2,p} = \{\} \qquad Neg_{2,p} = \{b,s(b)\}$$
$$Pos_{2,q} = \{b,s(b)\} \qquad Neg_{2,q} = \{a,s(a)\}$$
$$\ldots \qquad\qquad \ldots$$
$$Pos_{\omega,p} = \{\} \qquad Neg_{\omega,p} = \{\ldots s^k(b)\ldots\}$$
$$Pos_{\omega,q} = \{\ldots s^k(b)\ldots\} \qquad Neg_{\omega,q} = \{\ldots s^k(a)\ldots\}$$
$$Pos_{\omega+1,p} = \{\} \qquad Neg_{\omega+1,p} = \{\ldots s^k(b)\ldots,a\}$$
$$Pos_{\omega+1,q} = \{\ldots s^k(b)\ldots\} \qquad Neg_{\omega+1,q} = \{\ldots s^k(a)\ldots\}$$
$$\ldots \qquad\qquad \ldots$$

And the fixpoint is reached at $\omega + \omega$ where Neg_p becomes $\{b, s(b), \ldots, a, s(a), \ldots\}$ exactly as in Fitting's approach. Actually, in this case deducing the fact that $P(a)$ is false necessitates a computation of p and q for an *infinite* set of $x's$ and this explains that the fixpoint is reached strictly after ω.

Applying our operator to other examples from (Kunen, 1987) gives the same results. In particular, examples like

$$p \Leftrightarrow q \vee \neg q$$
$$q \Leftrightarrow q \qquad\qquad\qquad\qquad\qquad\qquad \text{Example 4}$$

$$p \Leftrightarrow \neg p \qquad\qquad\qquad\qquad\qquad\qquad \text{Example 5}$$

(which are pathological from the operational point of view) gives the birelation $\langle\{\},\{\}\rangle$ as fixpoint for p and q. This models correctly the operational behavior of a general computation mechanism which fails to produce any answer. This reinforces our believe that the B-semantics represents a good support for a methodology based on the *intended model*.

3.5. ANSWER EXTRACTION IN THE B-SEMANTICS

The birelational semantics defined above can be adapted easily to take into account the *use of the theory P as a program*, viz., as a device receiving inputs and producing output. Inputs and outputs can be considered as *relations*; each procedure p can then be assigned *a specialization function* taking as input a relation *In* and returning as output a couple of relations, *Outpos* and *Outneg*, consisting of the tuples of *In* for which p is true and false, respectively (with $Outpos \subseteq In$, $Outneg \subseteq In$ and $Outpos \cap Outneg = \emptyset$).

Such specialization can be formulated straightforwardly by means of the B-semantics. In fact, if the B-semantics associates with the procedure p a birelation $B_p = \langle Pos_p, Neg_p \rangle$, then the specialization SP_p produces, for any relation In, a birelation $\langle Outpos, Outneg \rangle$ representing the tuples of In which are included in Pos_p and Neg_p, respectively. So, $Outpos_p = In \cap Pos_p$, and $Outneg_p = In \cap Neg_p$.

In terms of fixpoint, the operator XB can be adapted to become an operator, say XS (for eXtend Specialization), that transforms one specialization into another. We simply have to push the operation "$\cap In$" inside the relation $\hat{B}(D)$.

4. Implementation and Combination Issues

In this section we show how the B-semantics represents a platform to develop and combine modules in classical programming languages and to reason about them. The idea is to consider an operational semantics of a given programming languages as a specialization or *an implementation* of the B-semantics. Such operational semantics are generally incomplete and sometimes even unsound. But since we are only interested in *particular* programs, we can derive sufficient conditions for soundness and completeness which allow existing languages to be used and to be combined safely.

4.1. IMPLEMENTATION STRATEGY

Each particular specialization of the general language is made according to a given scheme. We call this specialization scheme *an implementation strategy*. More precisely, an implementation strategy consists of:

1. A possible restriction on the language expressive power. For example, the language can be restricted to functional (relational) atoms only, to one single type, ...
2. A representation of the universe objects. For example, ground terms can be used to represent objects, the usual symbols $\{1, 2, \ldots\}$ can be used to represent $\boldsymbol{N}$, ... In addition, in a given domain the representation of the objects is related to a representation of the relations, e.g, by general terms, by constraints, by substitutions,...
3. An implementation of the fixpoint semantics based on the chosen representation. A computable version of one of the operators XB and XS must be first provided, essentially by providing computable versions of the operators $\bowtie$, $\oplus$, $\div$,..., specialized to the chosen representation.

Moreover, a fixpoint algorithm must be given in order to implement the fixpoint operator of the B-semantics.

Each implementation strategy represents thus a particular way to make the B-semantics operational. It takes profit of some particular features (representation choices, language restrictions, safe approximations/implementations) or more generally of the combination of the features.

4.2. EXAMPLES OF IMPLEMENTATION STRATEGIES

Hereafter, we show how existing logic languages fit into our scheme, i.e., they indeed represent a particular implementation of our language. The B-semantics would then provide a good support for the integration of heterogeneous modules in these different languages.

- Prolog resolution represents a simple and straight example of implementation strategies. In Prolog, objects are represented by terms, relations are represented by tuples of terms and labeled relations are also represented by (sets of) variables substitutions.
  Operators $\bowtie$ are Π are respectively implemented by means of the unification, and by the restriction of substitutions to some variables. Operator $\oplus$ is represented by the union of substitutions and implemented through backtracking inside the resolution algorithm. Negation is handled by an approximation based on the Closed World Assumption and implemented through negation by failure.
  Note that the CWA, formulated by $Outneg = In - Outpos$, is the base of family of implementations which aim at avoiding operator $\div$. Such implementations are generally unsound because $Outpos \cup Outneg = In$ does not hold in general. But restrictive hypothesis on the program and on the input relation In may suffice to ensure soundness and completeness (see (Deville, 1990)).
- First-order functional languages represent another implementation strategy which is more *restrictive* than Prolog strategy. In this case the functional character of the atoms leads to a representation of the relation by substitutions in which variables are always associated with ground terms. This allows us to optimize the algorithm further, and to get an implementation strategy based on pattern-matching instead of unification.
- Constraint based languages represent implementation strategies where the relations are represented by constraints and where the implementation of the semantics is achieved by means of constraints solving. As mentioned above, some constraint languages over *finite domains* (e.g., (Rauzy, 1995)) provide sound and complete implementations of the B-

semantics above without approximation (but with some restriction on the use of negation).
- Relational databases represent also good candidates to be implemented in our language. As classical databases involve only finite relations, the implementation will be greatly facilitated since our operator *XB* becomes continuous. Continuity in this case will not be affected by the negation, so one can use and derive positive and negatives facts. Relational databases with a negation based on the CWA represent then another implementation strategy more restrictive than the full B-semantics.

4.3. CLASSIFICATION OF IMPLEMENTATION STRATEGIES

We have shown how some existing logic languages are particular implementations of our language. We believe that several other existing logic languages also fit this framework. One of the future goal of this work is to study the different possible implementations of our language more deeply, to classify them and to have a kind of *library of interesting specialization choices* to support our methodology. Such classification could also highlight some new combinations of specialization choices which can lead –we hope– to new interesting implementations strategies.

Hereafter we outline a first classification for the different implementation strategies. We consider classification according to three axes: language restriction, object representation and computing strategy. Each of these axes involves several choices which are not always independent.

1) Different constraints restricting the language expressive power can be used to support making the language operational. We just try to keep a catalogue of usual ones:
    1.i.   using functional atoms only, no predicate except =;
    1.ii. using relational atoms only;
    1.iii. allowing positive literals only;
    1.iv. allowing negative literals but not through recursion;
    1.v.   untyped language.
2) The choice of a representation involves a representation of the *objects*, together with, a related representation of the *relations*. For example, when objects are represented by ground terms, a relation becomes a collection of tuples of ground terms which can, in turn, be represented by a set of substitutions. Since the semantics of the language is based on relational operators, representation of relations is a determinant factor for making the semantics implementable. Classifying the different representation choices can be made on different bases.

      2.i. One can distinguish *complete representations*, where every relation in the universe can be represented (this is possible only when domains are finite), from *incomplete representations*.

      2.ii. In the general case of incomplete representations, one can distinguish the representations which are *conservative* with respect to the different relational operators ($\oplus, \bowtie, \ldots$), i.e., where the operators always produce representable relations as results.

      2.iii. Another way to classify the different representation, is to distinguish *symmetrical representations* where the positive and negative parts have the same representation; from *non symmetrical representations*.

      2.iv. One can also distinguish *non redundant representations* where the different parts of the relation representation denote separate subsets of tuples; from *redundant representations*. An example of the former case is the representation of a relation by a collection of substitutions or a collection of equations.

3) Within the representation choices and taking into account the possible language restriction, different implementations can be considered. There are different ways to compute the fixpoint operator and to implement the relational operators ($\bowtie, \oplus, \ldots$)

      3.i. One can distinguish *complete* computation strategy, that lead to complete implementations, from *incomplete* ones. Usually, the completion is more problematic for the negative part of the algorithm. So, one can distinguish the cases where the completion of the negative part is achieved in parallel with the positive part, from the cases where the negative part completion is achieved by means of a technique based on the positive part (e.g., using backtracking, negation by failure ...).

      3.ii. One can also distinguish sound and unsound implementations. A particular case of unsound implementations, is the case of approximated computing of the fixpoint operator. For example, abstract interpretation techniques (Cousot & Cousot, 1977) allow complete but approximated implementation even for infinite domains.

      3.iii. Fixpoint computation strategy can be either ascending (bottom-up) or descending (top-down). The former strategies are often more adequate to compute the XB-semantics while the latter are better for the XS-semantics. Such strategies have been extensively explored for deductive data bases (Ullman, 1989).

4.4. CORRECTNESS OF IMPLEMENTATION STRATEGIES

Now let us define conditions, like completeness and consistency, the operational semantics should fulfill to be an adequate implementation/approximation of the original B-semantics. Such conditions are related to a particular computation strategy, a particular representation choice, and a particular language restriction.

Notation 7 *We will use the subscript A to denote a given representation of objects; and also within that representation, to denote relations, birelations, operators on relations,...*

The question is to determine the conditions to be satisfied by a given effective operator to be considered as a good implementation/approximation of the function *XS*. In fact, *XS* as defined above represents an ideal situation when the specialization computes *all* the tuples for which the B-Semantics can assign a truth value (those belonging to one of the birelation components *Pos* or *Neg*).

   Weakening of these conditions can achieved in different ways according to the different languages used. This leads to *sufficient correctness* conditions. For example, a natural weakening is to require a specialization to guarantee results only for input tuples belonging to *Pos*∪*Neg* which plays thus the role of a *precondition*, outside of which result is not guaranteed. More formally, we can define a sufficient completeness and consistency properties of an approximate specialization semantics *XS* w.r.t. the B-semantics as follows:

Definition 8   Let $B_A = \langle Pos_A, Neg_A \rangle$ be the birelation associated to p; let Pre_A denote $Pos_A \cup Neg_A$; let SP_A be any candidate specialization operator; let In_A be any relation; let $SP_A(In_A) = \langle Outpos_A, Outneg_A \rangle$ then:

1) SP_A is consistent w.r.t. B_A iff
   1.a) $Outpos_A \cap Pre_A \subseteq In_A \cap Pos_A$;
   1.b) $Outneg_A \cap Pre_A \subseteq In_A \cap Neg_A$.
2) SP_A is complete w.r.t. B_A iff
   2.a) $Outpos_A \cap Pre_A \supseteq In_A \cap Pos_A$;
   2.b) $Outneg_A \cap Pre_A \supseteq In_A \cap Neg_A$.

4.5. COMBINATION ISSUES

The ultimate goal of this work is to allow combining different *modules* having each its own operational semantics into one single program and to allow reasoning about that program. As proposed above, the different operational semantics can be seen as *particular implementations* of the B-semantics.

Combining different modules having each a different operational semantics can be achieved via the concept of primitives. A module having one particular operational semantics makes use of primitives having other semantics which, in turn, use primitives of other semantics, and so on.

Of course, every *implementation strategy* should satisfy sufficient correctness conditions w.r.t. the original B-semantics. Furthermore, the *composition* itself should respect certain extra conditions to guarantee the correctness of the composed program as a whole. The definition of the B-semantics, together with, the related definition of implementation strategy provide a suitable framework to express such composition conditions and to deal with them. In fact, combining different implementation strategies involves the composition of different language restrictions, different representations, and different effective computation algorithms.

The composition of different computation strategies (generally based on the underlying different language restrictions) could necessitate some extra semantical *check*; while the composition of different representation necessitates a clean *interface*.

 - The different computation strategies are generally based on corresponding language restrictions. To combine different restrictions, the key idea is that a restrictive paradigm can use primitives in a less restrictive one with no need of extra check (the less restricted paradigm is simply under-used). For example, a paradigm allowing only relational atoms can use primitives of another paradigm having no such restriction. The reverse situation necessitates some extra check. For example, if a general paradigm uses primitives in a paradigm allowing only functional atoms, and if the used algorithm assumes that arguments are all ground, it is necessary to ensure that these conditions are fullfiled at call time.
 - The different paradigms can use different representations for objects and relations. An interface between the different representations will be needed in order to allow mixing these representations. This can be achieved in different ways. A straight and simple composition can be made when the representation used in the called primitives consists of a subset of the representation used in the calling program. For example, a logic program which represent object by terms can use a functional one using only ground terms; representation interfacing does not present a particular difficulties in this case.

The interface becomes more complex when the representations uses different abstraction levels. For example, in one representation, relations can be represented by equations (constraints) or by substitutions, and in the other representation, they can be represented by tuples. Clearly, in this case one equation (resp. one substitution) corresponds to a collection of tuples and a kind of *decomposition* (e.g. by solving the constraints) would be needed. If the calling paradigm is more general (resp. restrictive) than the called one, the decomposition should be applied on the call argument (resp. result).

So, ensuring the correctness of a composed program boils down to an incremental analysis of preconditions/postconditions satisfiability.

5. Conclusion and Future Work

The merits of the B-semantics as a framework to describe programs and to reason about them, lies in its independence of any object representation, its closeness to operational features by the use of a three-values logic, its clean formulation as a fixpoint of a monotonic operator and its symmetrical handling of positive and negative information.

Interesting works remain to be done in studying more deeply how different existing implementation strategies can fit in our framework and how to integrate those strategies adequately. New implementation strategies can also be investigated by starting with the general B-semantics and trying, by different ways, to determine sufficient conditions that make it implementable.

NAJI HABRA AND BAUDOUIN LE CHARLIER
Institut d'Informatique
University of Namur, Belgium

References

K. Apt. Logic programming. In J. Van Leeuwen, editor, *Handbook of Theoretical Computer Science :Formal Models and Semantics*, chapter 10, pages 493–574. Elsevier-The MIT Press, 1992.

P. Cousot and R. Cousot. Abstract interpretation: A unified lattice model for static analysis of programs by construction or approximation of fixpoints. In *Conference Record of Fourth ACM Symposium on Programming Languages (POPL'77)*, pages 238–252, Los Angeles, California, January 1977.

E.F. Codd. A Relational Model of Data for Large Shared Data Banks. *Communications of the ACM*, 13(6):377–387, 1970.

P. De Boeck and B. Le Charlier. Mechanical Transformation of Logic Definitions augmented with Type Information into Prolog Procedures: Some Experiments. In *Proceedings of (LOPSTR'93)*, Workshops in Computer Science. Springer Verlag, July 1993.

Y. Deville. *Logic Programming: Systematic Program Development*. International Series in Logic Programming. Addison-Wesley, Wokingham, United Kingdom, 1990.

M. Fitting. A Kripke-Kleene Semantics for Logic Programs. *Journal of Logic Programming*, 2(4):295–312, 1985.

J. Henrard and B. Le Charlier. FOLON: An environment for Declarative Construction of Logic Programs (extended abstract). In M. Bruynooghe and M. Wirsing, editors, *Proceedings of the Fourth International Workshop on Programming Language Implementation and Logic Programming (PLILP'92)*, Lecture Notes in Computer Science, Leuven, August 1992. Springer-Verlag.

K. Kunen. Negation in Logic Programming. *Journal of Logic Programming*, 4(4):289–308, 1987.

B. Le Charlier and S. Rossi. Extending the folon environment for automatically deriving totally correct prolog procedures from logic description. In *Proc. WLPE'95 Seventh Workshop on Logic Programming Environments, (in conjunction with ILPS'95)*, Portland, Oregon, USA, December 1995.

B. Le Charlier and P. Van Hentenryck. A general top-down fixpoint algorithm (revised version). Technical Report 93-22, Institute of Computer Science, University of Namur, Belgium, June 1993.

B. Le Charlier and P. Van Hentenryck. Experimental Evaluation of a Generic Abstract Interpretation Algorithm for Prolog. *ACM Transactions on Programming Languages and Systems (TOPLAS)*, January 1994.

J.W. Lloyd. *Foundations of Logic Programming*. Symbolic Computation Series. Springer-Verlag, Heidelberg, Germany, second edition, 1987.

A. Rauzy. Toupie : a Constraint Language for Model Checking. In Andreas Podelski, editor, *Constraint Programming : Basics and Trends, Proceedings of the 1994 Châtillon Spring School, Châtillon-sur-Seine, France, May 1994*, number 910 in Lecture Notes in Computer Science. Springer-Verlag, March 1995.

A. Tarski and F.B. Thompson. Some General Properties of Cylindric Algebras. *Bulletin of the Amer. Math. Soc.*, 58:65, 1952.

J.D. Ullman. *Principles of Database and Knowldge Base Systems*. Computer Science Press, 1989.

XIAO JUN CHEN

MODEL CHECKING ACTL CONSTRAINED PROCESSES

Abstract. Logical constrained processes are loose specifications for partially defined systems: they are a combination of open terms in process algebras and a constraint constructed on temporal logics. Using open terms for loose specifications provides a way to bring advantages from process algebras, yet we are not restricted to a single equivalence class due to the appearances of free variables. Furthermore, these variables are constrained by logical formulae expressing the properties they should give respect to in further refinements. In this paper, we discuss the specifications and verifications of logical constrained processes using (i) regular process (sequential and non-deterministic process) as open terms and (ii) ACTL, an action-based version of CTL, as the underlying logic for constraints. The problem of verifying a (global) property in such logical constrained process is shown to be reducible into the classical validity/satisfiability verification of ACTL formulae. The transformation follows the style of the algorithm for CTL model checking.

1. Motivation

The study of partially defined systems provides the theoretical foundation for a top-down design methodology. The specifications of partially defined systems (so-called *loose* specifications) have been previously discussed e.g. in (Boudol & Larsen, 1992; Jensen *et al.*, 1993; Larsen & XinXin, 1991). In this paper, we consider partially defined systems specified as logical constrained processes (Chen & De Nicola, 1995): a combination of open terms in process algebras (Bergstra & Klop, 1989; Milner, 1989) and a constraint constructed on temporal logics (Emerson & Halpern, 1986; Kozen, 1983; Manna & Pnueli, 1989).

Using open terms for loose specifications provides a way to bring advantages from process algebras, which are generally recognized as a convenient tool for describing concurrent systems. In process algebras, the correctness

F. Baader and K.U. Schulz (eds.), Frontiers of Combining Systems, 377–388.
© 1996 *Kluwer Academic Publishers.*

of an implementation with respect to a specification is often established
by relying on behavioural equivalences: this limits the possible implementations to a single equivalence class. Thus, we use *open* terms, where free
variables represent the underspecified subparts. In this context, an open
term represents a set of equivalence classes instead of a single one as a
closed term does.

However, the set of equivalence classes represented by an open term is
rather large: the free variables are completely undefined. At this point, one
would expect some local constraints on these free variables to express the
properties they should hold. We adopt logical constraints to describe these
local properties because process algebras require too much implementation
details with respect to logical languages: they are more suitable to describe
system behaviours rather than system properties.

In this paper, we discuss the specifications and verifications of logical constrained processes using (i) regular process (sequential and non-
deterministic process) as open terms and (ii) ACTL (De Nicola *et al.*,
1993), an action-based version of CTL (Browne *et al.*, 1988; Emerson &
Halpern, 1986; Emerson & Srinivasan, 1989), as the underlying logic for
constraints. CTL is a well-known propositional branching time temporal
logic interpreted on Kripke Structure. It has an expressivity good enough
for many application domains, while the model checking for its formulas
takes linear time with respect to the length of the formula and the number
of states and edges in the model. ACTL has the similar framework as CTL
but it is interpreted on labelled transition systems, which is the base of the
semantics of processes we consider.

Model checker provides a good way to reason about the truth of a logic
formula in a given process. Considering loose specifications, this concept has
been extended to the so-called *loose model checking* (Chen & De Nicola,
1995): to verify whether a property *must/may* be satisfied in a partially
defined system.

Remark that the introduction of loose model checking does not depend
on the models we use: open terms or any other loose specifications. Its logical explanation is the *validity checking in a given context*, which stands
between model checking problem and validity problem: The system being
partially defined, the model checker should answer, according to the incomplete information from the system, *yes* (must be satisfied, valid), *no* (may
not be satisfiedm not valid) or *I don't know* (may be satisfied, satisfiable)
as in validity/satisfiability checking, instead of the usual *yes* (satisfied) and
no (not satisfied) as in model checking.

The following example illustrates an application of *loose model checking*:
consider a series of refinements

$$S_1 \leq S_2 \leq \ldots \leq S_{i-1} \leq S_i \ldots$$

Suppose we require that property φ hold in the final implementation. If on step i, we discover that S_i *can not* satisfy φ, then we know that no matter what further refinements we make, φ will never be satisfied: one should go backwards and find another refinement of S_{i-1} (rather than S_i) which *may* satisfy φ.

In this paper, we consider the loose model checking problem of ACTL logical constrained processes. We show that the problem can be reduced into the validity problem of ACTL formula. This transformation follows the style of the algorithm for CTL model checking.

The rest of the paper is organized as follows: Section 2 introduces the logical constrained processes in ACTL as loose specifications for partially defined systems; Section 3 gives the transformation from ACTL loose model checking problem for logical constrained processes into ACTL validity one. The last section is dedicated to conclusion and future work.

2. Logical Constrained Processes in ACTL

In this section, we first introduce open terms in regular processes and their interpretation on labelled transition systems extended by open states. Then we give a short review of ACTL, and finally, we define logial constrained processes with logical constraints written in ACTL.

2.1. REGULAR PROCESSES

Let A be a set of *actions* ranged over by a, b, ... and let V be a set of variables ranged over by x, y, ... The set of regular processes over A and V, is generated by the following grammar:

$$t ::= Nil \mid a.t \mid t + t \mid x \mid x \Leftarrow t$$

Here, (i) *Nil* denotes the terminated process; (ii) $a.t$ denotes a process which can only perform action a and then behave like t; (iii) $t + t'$ denotes the alternative composition of t and t'; (iv) $x \Leftarrow t$ is used for recursive definitions.

We will use $\mathcal{T}$ to denote this set of terms, ranged over by $t(X_1, \ldots, X_n)$ where X_i are free variables, and when there is no confusion, we also use t to range over it. The set of closed terms is denoted by $\mathcal{P}$, ranged over by p, q, r, ...

The regular processes are interpreted on labelled transition systems (Milner, 1989; Plotkin, 1981), which provide an intuitive way to describe the activities and changes of programs and systems.

Definition 2.1 A *labelled transition system* is a quadruple $(S, A, \rightarrow, s_0)$ where S is a countable set of states, A is a countable set of elementary actions, $\rightarrow \subseteq S \times A \times S$ is a set of transitions, and $s_0 \in S$ is the initial state.

Typically, S is a set of program states, and for each $a \in A$, the relationship (s, a, t) indicates that s can evolve to t under the observation of a. In the following, a transition (s, a, t) will be denoted as $s \xrightarrow{a} t$. A labelled transition system will also be called transition system for short. The transition relation is given through a set of inference rules (Note that free variables are associated with no transitions, just like *Nil*):

$$
\begin{array}{l}
(Act)\ a.t \xrightarrow{a} t \\
(Sum)\ t \xrightarrow{a} t'\ \text{implies}\ (t + t'') \xrightarrow{a} t',\ (t'' + t) \xrightarrow{a} t' \\
(Con)\ t \xrightarrow{a} t',\ x \Leftarrow t\ \text{implies}\ x \xrightarrow{a} t'
\end{array}
$$

The transition system can be extended by divergency in order to deal with infinite internal action, undefinedness, etc. In (Steffen, 1989), a transition system is equipped with a mapping $\Uparrow$ which associates with each action a, a unary predicate $\Uparrow a$ to express divergence potential: $s \Uparrow a$ is intended to mean that s can be triggered by means of an offer of a action to evolve autonomously for ever without responding to the environment.

In our setting, the states may be marked with free variables to denote that the states are open. A state open to X, notation $\_ \uparrow \_$: $t(X_1, \ldots, X_n) \times \{X_1, \ldots, X_n\}$ is defined inductively by

$$
\begin{array}{l}
(D1)\ X \uparrow X \\
(D2)\ t \uparrow X\ \text{implies}\ (t + t') \uparrow X,\ (t' + t) \uparrow X \\
(D3)\ t \uparrow X\ \text{implies}\ (Y \Leftarrow t) \uparrow X
\end{array}
$$

2.2. A SHORT REVIEW OF ACTL

Many temporal and modal logics have been proposed as suitable for specifying system properties. These logics, usually interpreted on Kripke Structures or labelled transition systems, have been equipped with model checkers to prove the satisfiability of formulae and thus system properties: a (finite) system is considered as a potential model for the formula expressing the desired property. CTL is a well-known branching time temporal logic interpreted on Kripke Structure. ACTL is the action-based version of CTL: it has the similar framework as CTL but it is interpreted on labelled transition systems.

In order to define logic ACTL, an auxiliary logic of actions is introduced. The *action formulae* over A is defined by the following grammar where χ, χ' range over action formulae and $a \in A$:

$$\chi ::= a \mid \neg \chi \mid \chi \vee \chi'$$

The satisfaction of χ by an action a, notation $a \models \chi$, is defined inductively by

$$
\begin{array}{lll}
a \models b & \text{iff} & a = b \\
a \models \neg\chi & \text{iff} & a \not\models \chi \\
a \models \chi \vee \chi' & \text{iff} & a \models \chi \text{ or } a \models \chi'
\end{array}
$$

The syntax of ACTL is generated by the following grammar, where ϕ, ϕ', ... range over ACTL formulae, and χ, χ' are action formulae:

$$\phi ::= tt \mid \neg\phi \mid \phi \wedge \phi' \mid \exists\gamma \mid \forall\gamma$$

$$\gamma ::= X_\chi\phi \mid \phi_\chi U_{\chi'}\phi' \mid \phi_\chi U\phi'$$

Let $(S, A, \rightarrow, s_0)$ be an lts where s_0 is the initial state. A *run*

$$\pi = q_0 \xrightarrow{a_1} q_1 \ldots \xrightarrow{a_n} q_n$$

is a partial computation of S starting from a state q_0 and currently in state q_n. This computation can be extended (if q_0 is not a final state) and we write $\pi \xrightarrow{a} \pi'$ when π' is π with a transition $q_n \xrightarrow{a} q_{n+1}$ added. We use $\pi_{|i}$ (defined as $q_0.a_1.q_1 \ldots q_{i-1}$), and $act(\pi, i)$ (defined as a_i) to denote the i-th prefix of π, and the i-th action. We use $\Pi_S(q)$ for the set of all runs in S starting from q, and Π_S for the set of all runs in S. The *satisfaction* of a formula φ in run π at time n, written $\pi, n \models \varphi$ is given inductively by:

$$
\begin{array}{lll}
\pi, n \models tt & always \\
\pi, n \models \neg\phi & iff & \pi, n \not\models \phi \\
\pi, n \models \phi_1 \wedge \phi_2 & iff & \pi, n \models \phi_1 \text{ and } \pi, n \models \phi_2 \\
\pi, n \models \exists\gamma & iff & \exists\pi' \in \Pi_S \text{ s.t. } \pi'_{|n} = \pi_{|n}, \; \pi', n \models \gamma \\
\pi, n \models \forall\gamma & iff & \forall\pi' \in \Pi_S. \; \pi'_{|n} = \pi_{|n} \text{ implies } \pi', n \models \gamma \\
\pi, n \models \phi_\chi U_{\chi'}\phi' & iff & \exists k > n \text{ s.t. } \pi, k \models \phi', \; act(\pi, k) \models \chi', \\
& & \forall n \leq i \leq k - 1. \; \pi, i \models \phi, \; act(\pi, i) \models \chi \\
\pi, n \models \phi_\chi U\phi' & iff & \exists k \geq n \text{ s.t. } \pi, k \models \phi', \\
& & \forall n \leq i \leq k - 1. \; \pi, i \models \phi, \; act(\pi, i) \models \chi \\
\pi, n \models X_\chi\phi & iff & \pi, n + 1 \models \phi, \; act(\pi, n) \models \chi
\end{array}
$$

The indexed *next* modality $X_\chi\phi$ says that the next state of the run is reached by an action in χ, and in the next state, the formula ϕ holds. The indexed *until* modality $\phi_\chi U_{\chi'}\phi'$ says that along the run, all states will satisfy ϕ and reached by actions in χ, until a state that satisfies ϕ' and reached by an action in χ'.

Example 2.1 $\neg\exists tt_{\neg a}U_b tt$ expresses that there is no run beginning by actions all different from a until b, i.e. b can only happen when a has happened.

2.3. LOGICAL CONSTRAINTS

Given logic L, we introduce *logical constraint w.r.t.* L which will be used as constraints for the underspecified subparts (also called unknown processes in (Jensen *et al.*, 1993)).

Definition 2.2 An *n-ary logical constraint w.r.t.* L is constructed on n-ary vectors of formulas of L, $[\varphi_1, \ldots, \varphi_n]$ where $\varphi_i \in L$, and logical connective $\vee$.

The requirement that the final implementation of the n unknown processes (denoted by $[p_1, \ldots, p_n]$ or $\vec{P}$) satisfy a given n-ary logical constraint is expressed by means of satisfaction relation $\models_v$:

Definition 2.3 The satisfaction relation between an n-ary vector of processes and an n-ary local constraint is defined inductively below:

$$[p_1, \ldots, p_n] \models_v [\varphi_1, \ldots, \varphi_n] \quad \textit{iff } p_i \models \varphi_i \ \forall i. \ 1 \le i \le n$$
$$[p_1, \ldots, p_n] \models_v \Phi_1 \vee \Phi_2 \quad \textit{iff } [p_1, \ldots, p_n] \models_v \Phi_1 \ \textit{or } [p_1, \ldots, p_n] \models_v \Phi_2$$

Remark 2.1 We also use logical connectives $\wedge$, $\neg$, etc. in logical constraints. They are interpreted by

$$\vec{P} \models_v \neg\varphi \qquad \textit{iff } \vec{P} \not\models_v \varphi$$
$$\vec{P} \models_v \varphi_1 \wedge \varphi_2 \quad \textit{iff } \vec{P} \models_v \varphi_1 \ \textit{and } \vec{P} \models_v \varphi_2$$

It is easy to check that

(1) $\neg[\varphi_1, \ldots, \varphi_n] \leftrightarrow \bigvee_{1 \le i \le n}[tt, \ldots, \neg\varphi_i, \ldots, tt]$;
(2) $[\varphi_1, \ldots, \varphi_n] \wedge [\phi_1, \ldots, \phi_n] \leftrightarrow [\varphi_1 \wedge \phi_1, \ldots, \varphi_n \wedge \phi_n]$;
(3) $[\varphi_1, \ldots, \varphi_n] \vee [\phi_1, \ldots, \phi_n]$ implies $[\varphi_1 \vee \phi_1, \ldots, \varphi_n \vee \phi_n]$, but not vice versa.

Note that any logical constraints constructed by vectors of formulas and logical connectives $\wedge$, $\vee$, $\neg$ etc., can be rewritten into normal form $\bigvee \bigwedge \varphi_i$ where φ_i are the vectors of formulas or their negations. So according to (1), the $\neg$ connective in logical constraint is derivable. Furthermore, due to (2), neither $\wedge$ connective is necessary: any logical constraint in form $\bigvee \bigwedge \varphi_i$ can be rewritten into form $\bigvee \varphi_i'$ where φ_i' are vectors of formulas. Finally, by (3), logical constraint in form $\bigvee \varphi_i$ cannot be rewritten into pure vector of formulas.

The following fact is immediate:

Proposition 2.1 Given a logic L, the satisfiability/validity problem for logical constraints written on L is reducible to the satisfiability/validity problem of L.

In the following, we will simply use $\models$ for $\models_v$ when there is no confusion. Furthermore, we will use tt for $[tt, \ldots, tt]$, ff for $\neg tt$.

Definition 2.4 A *logical constrained process* is a pair (t, Φ) where t is an open term with n free variables and Φ is an *n-ary logical constraint*.

Example 2.2 Suppose we have a (probably unreliable) communication line (medium): it receives a message from one port and then sends it out at the other port. The logical constrained process

$$(t \Leftarrow receive.(send.t + X_1), \; [\neg \exists tt_{\neg send} U_{receive} tt])$$

describes a kind of such medium. Here open term

$$t \Leftarrow receive.(send.t + X_1)$$

expresses that the medium can do nothing but to receive a message. But after having received a message, it may send the message out and then repeat its communnication task t (which are the expected correct behaviour), or it may have some other behaviours expressed by X_1. The logical constraint on X_1 states that the process cannot receive the (next) message before sending out the (previous) one. This essentially says that no message will be lost. Note that however, neither duplicated messages nor modified ones, are excluded from this constraint.

3. Verifying Logical Constrained Process in ACTL

The truth of a global property in a logical constraint process is defined by means of *may* and *must* satisfaction relation:

1. (t, Φ) *must* satisfy ϕ, notation $(t, \Phi) \models_{must} \phi$, means that ϕ is satisfied by all ground instances of t obtained by substituting its free variables with a vector of closed terms constrained by Φ;
2. (t, Φ) *may* satisfy ϕ, notation $(t, \Phi) \models_{may} \phi$, means that ϕ is satisfied by some ground instances of t obtained by substituting its free variables with a proper vector of closed terms constrained by Φ;

Formally, the satisfaction relation between a logical constrained process (t, Φ) and a formula ϕ is defined as follows:

Definition 3.1 $(t, \Phi) \models_{must} \phi$ *iff* $\forall \vec{p}$ *s.t.* $\vec{p} \models \Phi : t[\vec{p}] \models \phi$;

$(t, \Phi) \models_{may} \phi$ *iff* $\exists \vec{p}$ *s.t.* $\vec{p} \models \Phi$ and $t[\vec{p}] \models \phi$.

Due to the fact that

$not\ (t, \Phi) \models_{must} \varphi$ is equivalent to $(t, \Phi) \models_{may} \neg \varphi$

for $i = 1$ to length(φ_0)

    for each subformula φ of φ_0 of length i

    case on the form of φ

        (1) $\varphi = tt$: for each $s \in S$,

           add (tt, tt) to $L(s)$;

        (2) $\varphi = \phi_1 \wedge \phi_2$: for each $s \in S$,

           if $(\Phi_i, \phi_i) \in L(s)$ for $i = 1, 2$,

           then add $(\Phi_1 \wedge \Phi_2, \phi_1 \wedge \phi_2)$ to $L(s)$;

        (3) $\varphi = \neg \phi$: for each $s \in S$,

           if $(\Phi, \phi) \in L(s)$,

           then add $(\neg \Phi, \varphi)$ to $L(s)$

           else add (tt, φ) to $L(s)$.

        (4) $\varphi = \exists X_\chi \phi$: for each $s \in S$,

           if $\exists a \models \chi$ s.t. $s \xrightarrow{a}$, or $\exists i$ s.t. $s {\uparrow} X_i$

           then add

           $(\bigvee \{\Phi \mid s \xrightarrow{a} s', a \models \chi, (\Phi, \phi) \in L(s')\} \vee \bigvee \{X_i \leftarrow \varphi \mid s {\uparrow} X_i\}, \varphi)$

           to $L(s)$

        (5) $\varphi = \forall X_\chi \phi$: for each $s \in S$,

           if $\exists a \models \chi$ s.t. $s \xrightarrow{a}$, or $\exists i$ s.t. $s {\uparrow} X_i$

           then add

           $(\bigwedge \{\Phi \mid s \xrightarrow{a} s', a \models \chi, (\Phi, \phi) \in L(s')\} \wedge \bigwedge \{X_i \leftarrow \varphi \mid s {\uparrow} X_i\}, \varphi)$

           to $L(s)$

           else add (tt, φ) to $L(s)$

        (6) $\varphi = \exists \phi_\chi U \phi'$: (see table 2)

        (7) $\varphi = \forall \phi_\chi U \phi'$: (see table 3)

        (8) $\varphi = \exists \phi_\chi U_{\chi'} \phi'$: (similar as (6))

        (9) $\varphi = \forall \phi_\chi U_{\chi'} \phi'$: (similar as (7))

    end of case

  end

end

Table 1.   Transformation of loose model checking in ACTL

(6) $\varphi = \exists\phi_\chi U\phi'$:

  for each $s \in S$

    if $(\Phi, \phi') \in L(s)$

    then add $(\Phi \vee \bigvee\{X_i \leftarrow \varphi \mid s\uparrow X_i\}, \varphi)$ to $L(s)$

  end

  for $j = 1$ to $card(S)$

    for each $s \in S$

      if $(\Phi_1, \phi') \in L(s)$, $\exists a \models \chi$ s.t. $s \xrightarrow{a}$, and $\forall a \models \chi, s'$ s.t. $s \xrightarrow{a} s'$: $\exists\Phi'. (\Phi', \varphi) \in L(s')$

      then add $(\Phi_1 \wedge (\bigvee\{\Phi \mid s \xrightarrow{a} s', a \models \chi, (\Phi, \varphi) \in L(s')\} \vee \bigvee\{X_i \leftarrow \varphi \mid s\uparrow X_i\}), \varphi)$

      end

    end

  for each $s \in S$

    if $(\Phi, \varphi) \notin L(s)$ for some Φ, and $s\uparrow X_i$ for some i

    then add $(X_i \leftarrow \varphi, \varphi)$ to $L(s)$

  end

Table 2.   Transformation of loose model checking in ACTL: subtable 1

we only need to discuss one of the satisfaction relations. In the rest of the paper, we concentrate on *must* satisfaction relation.

Given an extended transition system and a formula φ to be verified, we use pairs (Φ, φ) of logical constraint Φ and logical formula φ to label the states s. The intuitive meaning of $(\Phi, \varphi) \in L(s)$ is that φ is true in state s if the free variables satisfy Φ. The main part of the procedure of labelling is given in Table 3, with two portions in Table 3 and 3. Cases (8) and (9) are similar as cases (6) and (7) respectively, and are not included just for brevity.

Remark 3.1 We simply use $X_i \leftarrow \varphi$ to denote the vector of formula with φ as its i-th value and tt the others, and we use $s \xrightarrow{a}$ as an abbreviation of $\exists s'$ s.t. $s \xrightarrow{a} s'$.

Lemma 3.1 (Φ_0, φ) *is in* $L(t)$ *iff* $(\forall \vec{p} . \ \vec{p} \models \Phi_0$ *iff* $t[\vec{p}] \models \varphi)$

The detailed proof for the correctness of this lemma is not presented here, since it has the same structure of CTL model checking. We only mention some points here.

(7) $\varphi = \forall \phi_\chi U \phi'$:

    for each $s \in S$

        if $(\Phi, \phi') \in L(s)$

        then add $(\Phi \wedge \bigwedge \{X_i \leftarrow \varphi \mid s \uparrow X_i\}, \varphi)$ to $L(s)$

    end

    for $j = 1$ to $card(S)$

        for each $s \in S$

            if $(\Phi_1, \phi') \in L(s)$, $\exists a \models \chi$ s.t. $s \xrightarrow{a}$, and $\forall a \models \chi, s'$ s.t. $s \xrightarrow{a} s' : \exists \Phi'. (\Phi', \varphi) \in L(s')$

            then add $(\Phi_1 \wedge \bigwedge \{\Phi \mid s \xrightarrow{a} s', a \models \chi, (\Phi, \varphi) \in L(s')\} \wedge \bigwedge \{X_i \leftarrow \varphi \mid s \uparrow X_i\}, \varphi)$

        end

    end

    for each $s \in S$

        if $(\Phi, \varphi) \notin L(s)$ for some Φ

        then

            if $s \uparrow X_i$ for some i

            then add $(X_i \leftarrow \varphi, \varphi)$ to $L(s)$

            else add (tt, φ) to $L(s)$

    end

Table 3.   Transformation of loose model checking in ACTL: subtable 2

- In CTL model checking, the truth of a formula ϕ in s depends on the truth of its subformulae in s and/or on the truth of ϕ or its subformulae in the successive states. Similarly, here the constraint to make ϕ true in s depends on the constraints which make the subformulae true, the constraints which make ϕ true in successive states. For example, in case (2), we have

    if ϕ_i is true iff free variables satisfy Φ_i (for $i = 1, 2$),

    then $\phi_1 \wedge \phi_2$ is true iff free variables satisfy $\Phi_1 \wedge \Phi_2$.

- Due to the open states, we can not label a state for an existential formula just by considering one possibility: in ACTL, the truth of $\phi = \exists X_a \phi'$ in $a.p_1 + a.p_2$ depends on the truth of ϕ' only in one of the successive states, p_1 or p_2. But here, suppose ϕ' is true in p_i under constraint Φ_i, then we need the disjunction of Φ_i as the constraint for ϕ, because both p_1 and p_2 have the possibility to hold ϕ'.

– The constraint to make ϕ true in s depends also on whether s is an open state. Suppose $s = a.p + X$ (s is open to X). The truth of $\phi = \exists X_a \phi'$ in s does not depend only on whether p satisfies ϕ', but also depend on whether the constraint makes X always satisfy ϕ.

– For each formula ϕ, if a state is not labelled by (Φ, ϕ) for some Φ, it means that ϕ is *not valid* in s.

According to this lemma, the following result is immediate.

Theorem 3.1 If (Φ_0, φ) is in $L(t)$, then $(t, \Phi) \models_{\text{must}} \varphi$ *iff* $\Phi \to \Phi_0$ *is valid.*

4. Conclusion and Future Work

Process algebra provides a good way for describing system behaviours. For partially defined systems, however, they ask for too much implementation details. A successful extension of process algebras for *loose* specifications, so-called graphical specification, is presented in (Boudol & Larsen, 1992). A graphical specification is a labelled transition system equipped with a predicate on transitions. As a specification, such a system is intended to describe a whole class of possible realizations, rather than a particular process. Then the transition relation *allows* a model to perform the specified transitions, but the *existence* of these transitions is not necessarily required. On the other hand, if the specification does not mention a particular transition, then this transition is disallowed from any possible model. The predicate on transitions asserts that some transitions are required: these transitions *must exist* in any model of the specification.

Logical constrained process is another attempt to extend process algebras to specify partially defined systems. It describes the implemented part of the system in process algebra, leaving underspecified subparts constrained by logical formulae. The comparison between logical constrained process and graphical specification is presented in (Chen & De Nicola, 1995).

In this paper, we have discussed the verification of logical constrained processes when ACTL formulae are used for the logical constraints. This discussion is based on a transformation of loose model checking into validity verification. This transformation leads our discussion on verification in given context to the classical validity discussions in logic.

For further extension of this work, it seems interesting to investigate loose model checking in logical constrained processes where parallel processes are allowed and especially syncronization between processes are allowed.

Acknowledgements

The author gratefully acknowledge Rocco De Nicola for his encouragement and helpful discussions on this topic, and the anonymous referees for useful comments on the previous version of this paper.

XIAO JUN CHEN
Dipartimento di Scienze dell'Informazione
Università di Roma "La Sapienza", Italia

References

J. Bergstra and J. Klop. Process theory based on bisimulation semantics. In *LNCS 354*, pages 50–122. Springer-Verlag, 1989.

G. Boudol and K.G. Larsen. Graphical versus logical specifications. *Theoretical Computer Science*, 106:3–20, 1992.

M.C. Browne, E.M. Clarke, and O. Grumberg. Characterizing finite Kripke structure in propositional temporal logic. *Theoretical Computer Science*, 59:115–131, 1988.

X.J. Chen and R. De Nicola. Loose specifications and their verification. Technical report, Univ. di Roma "La Sapienza", 1995. Presented at II Express Workshop. submitted.

E.A. Emerson and J.Y. Halpern. "sometimes" and "not never" revisited: on branching time versus linear time temporal logic. *Journal of ACM*, 33(1):151–178, 1986.

E.A. Emerson and J. Srinivasan. Branching time temporal logic. In J. de Bakker, P. de Roever, and G. Rozenberg, editors, *Linear Time, Branching Time and Partial Order in Logics and Models for Concurrency, LNCS 354*, pages 123–172. Springer-Verlag, 1989.

O.H. Jensen, J.T. Lang, C. Jeppesen, and K.G. Larsen. Model construction for implicit specifications in modal logic. In *CONCUR'93*, pages 247–261, 1993.

D. Kozen. Results on the propositional μ-calculus. *Theoretical Computer Science*, 27(2):333–354, 1983.

K. G. Larsen and L. XinXin. Compositionality through an operational semantics of contexts. *Journal of Logic and Computation*, 1(6):761–795, 1991.

Z. Manna and A. Pnueli. The anchored version of the temporal framework. In J. de Bakker, P. de Roever, and G. Rozenberg, editors, *Linear Time, Branching Time and Partial Order in Logics and Models for Concurrency, LNCS 354*, pages 201–284. Springer-Verlag, 1989.

R. Milner. *Communication and Concurrency*. Prentice Hall, London, 1989.

R. De Nicola, A. Fantechi, S. Gnesi, and G. Ristori. An action-based framework for verifying logical and behavioural properties of concurrent systems. *Computer Networks and ISDN Systems*, 25:761–778, 1993.

G. Plotkin. A structural approach to operational semantics. Technical report, Computer Science Deot. Aarhus Univ. Denmark, 1981. DAIMI-FN-19.

B. Steffen. Characteristic formulae. *Proceedings ICALP*, LNCS 372:723–732, 1989.